PERGAMON INTERNATIONAL LIBRARY
of Science, Technology, Engineering and Social Studies
The 1000-volume original paperback library in aid of education, industrial training and the enjoyment of leisure

Publisher: Robert Maxwell, M.C.

Theory and Design of Broadband Matching Networks

THE PERGAMON TEXTBOOK INSPECTION COPY SERVICE

An inspection copy of any book published in the Pergamon International Library will gladly be sent to academic staff without obligation for their consideration for course adoption or recommendation. Copies may be retained for a period of 60 days from receipt and returned if not suitable. When a particular title is adopted or recommended for adoption for class use and the recommendation results in a sale of 12 or more copies, the inspection copy may be retained with our compliments. If after examination the lecturer decides that the book is not suitable for adoption but would like to retain it for his personal library, then a discount of 10% is allowed on the invoiced price. The Publishers will be pleased to receive suggestions for revised editions and new titles to be published in this important International Library.

APPLIED ELECTRICITY AND ELECTRONICS
General Editor: P. HAMMOND

Some other Titles of interest

ABRAHAMS, J. R. & COVERLEY, G. P.
Semiconductor Circuits: Theory, Design and Experiments

ABRAHAMS, J. R. & COVERLEY, G. P.
Semiconductor Circuits: Worked Examples

BADEN FULLER, A. J.
Microwaves

BADEN FULLER, A. J.
Engineering Field Theory

BADEN FULLER, A. J.
Worked Examples

BROOKES, A. M. P.
Basic Instrumentation for Engineers and Physicists

CRANE, P. W.
Electronics for Technicians

CRANE, P. W.
Worked Examples in Basic Electronics

GATLAND, H. B.
Electronic Engineering Applications of Two Port Networks

HAMMOND, P.
Applied Electromagnetism

HAMMOND, P.
Electromagnetism for Engineers

HANCOCK, N. N.
Matrix Analysis of Electrical Machinery 2nd Edition

HARRIS, D. J. & ROBSON, P. N.
The Physical Basis of Electronics

HINDMARSH, J.
Electrical Machines and their Applications 2nd Edition

HOWSON, D. P.
Mathematics for Electrical Circuit Analysis

PRIDHAM, G. J.
Solid State Circuits

SPARKES, J. J.
Junction Transistors

The terms of our inspection copy service apply to all the above books. A complete catalogue of all books in the Pergamon International Library is available on request.

Theory and Design of Broadband Matching Networks

by

WAI-KAI CHEN

*Professor of Electrical Engineering,
Department of Electrical Engineering,
Ohio University, Athens, Ohio*

PERGAMON PRESS
OXFORD · NEW YORK · TORONTO
SYDNEY · PARIS · FRANKFURT

U.K.	Pergamon Press Ltd., Headington Hill Hall, Oxford OX3 OBW, England
U.S.A.	Pergamon Press Inc., Maxwell House, Fairview Park, Elmsford, New York 10523, U.S.A.
CANADA	Pergamon of Canada Ltd., Box 9600, Don Mills M3C 2T9, Ontario, Canada
AUSTRALIA	Pergamon Press (Aust.) Pty. Ltd., 19a Boundary Street, Rushcutters Bay, N.S.W. 2011, Australia
FRANCE	Pergamon Press SARL, 24 rue de Ecoles, 75240 Paris, Cedex 05, France
WEST GERMANY	Pergamon Press GmbH, 6242 Kronberg/Taunus, Pferdstrasse 1, Frankfurt-am-Main, West Germany

Copyright © 1976 Pergamon Press Ltd

All Rights Reserved. No part of this publication may be reproduced, stored in a retrieval system or transmitted in any form or by any means: electronic, electrostatic, magnetic tape, mechanical, photocopying, recording or otherwise, without permission in writing from the publishers

First edition 1976

Library of Congress Cataloging in Publication Data

Chen, Wai-kai, 1936–
The theory and design of broadband matching networks.

(Applied electricity & electronics series)
(Pergamon international library)
Includes bibliographical references and indexes.
1. Electric networks. 2. Broadband amplifiers
I. Title.
TK454.2.C426   1976   621.319′2   75-40107
ISBN 0-08-019702-7
ISBN 0-08-019918-6 pbk.

Printed in Great Britain by A. Wheaton & Co. Exeter

To Shiao-Ling and Jerome and Melissa

Contents

Preface   xiii

Chapter 1. Foundations of Network Theory   1

1. Basic network postulates   2
    1.1. Real-time function postulate   3
    1.2. Time-invariance postulate   4
    1.3. Linearity postulate   5
    1.4. Passivity postulate   6
    1.5. Causality postulate   9
    1.6. Reciprocity postulate   10
2. Matrix characterizations of n-port networks   11
    2.1. The impedance matrix   12
    2.2. The admittance matrix   13
    2.3. The hybrid matrix   14
    2.4. The indefinite-admittance matrix   15
3. Power gains   21
4. Hermitian forms   23
5. The positive-real matrix   28
6. Frequency-domain conditions for passivity   39
7. Conclusions   43
   Problems   45
   References   46

Chapter 2. The Scattering Matrix   48

1. A brief review of the transmission-line theory   49
2. The scattering parameters of a one-port network   50
    2.1. Basis-dependent reflection coefficients   52
    2.2. Basis-independent reflection coefficient   54
    2.3. The factorization of the para-hermitian part of $z(s)$ 57

2.4. Alternative representation of the basis-independent reflection coefficient	62
2.5. The normalized reflection coefficient and passivity	64
3. The scattering matrix of an n-port network	66
3.1. Basis-dependent scattering matrices	70
3.2. Basis-independent scattering matrix	74
3.3. The scattering matrices and the augmented n-port networks	77
3.4. Alternative representation of the basis-independent scattering matrix	80
3.5. Physical interpretation of the normalized scattering parameters	82
3.6. The normalized scattering matrix and passivity	88
3.7. The normalized scattering parameters of a lossless two-port network	90
4. The bounded-real scattering matrix	91
5. Interconnection of multi-port networks	98
6. Conclusions	106
Problems	107
References	113

Chapter 3. Approximation and Ladder Realization — 115

1. The Butterworth response	116
1.1. Poles of the Butterworth function	118
1.2. Coefficients of the Butterworth polynomials	120
1.3. Butterworth networks	122
1.4. Butterworth LC ladder networks	125
2. The Chebyshev response	132
2.1. Chebyshev polynomials	132
2.2. Equiripple characteristic	134
2.3. Poles of the Chebyshev function	138
2.4. Coefficients of the polynomial $p(y)$	142
2.5. Chebyshev networks	144
2.6. Chebyshev LC ladder networks	146
3. Elliptic functions	152
3.1. Jacobian elliptic functions	152
3.2. Jacobi's imaginary transformations	154

3.3. Periods of elliptic functions	155
3.3.1. The real periods	157
3.3.2. The imaginary periods	158
3.4. Poles and zeros of the Jacobian elliptic functions	159
3.5. Addition theorems and complex arguments	162
4. The elliptic response	165
4.1. The characteristic function $F_n(\omega)$	167
4.2. Equiripple characteristic in passband and stopband	174
A. Maxima and minima in the passband	176
B. Maxima and minima in the stopband	177
C. Transitional band	178
4.3. Poles and zeros of elliptic response	183
4.4. Elliptic networks	190
5. Frequency transformations	196
5.1. Transformation to high-pass	197
5.2. Transformation to band-pass	200
5.3. Transformation to band-elimination	204
6. Conclusions	206
Problems	207
References	215

Chapter 4. Theory of Broadband Matching: The Passive Load   217

1. The Bode–Fano–Youla broadband matching problem	218
2. Youla's theory of broadband matching: preliminary considerations	219
3. Basic constraints on $\rho(s)$	222
4. Bode's parallel RC load	224
4.1. Butterworth transducer power-gain characteristic	225
4.2. Chebyshev transducer power-gain characteristic	235
4.3. Elliptic transducer power-gain characteristic	246
4.4. Equalizer back-end impedance	257
5. Proof of necessity of the basic constraints on $\rho(s)$	259
6. Proof of sufficiency of the basic constraints on $\rho(s)$	264
7. Design procedure for the equalizers	267
8. Darlington type-C load	274
8.1. Butterworth transducer power-gain characteristic	274
8.2. Chebyshev transducer power-gain characteristic	281

8.3. Elliptic transducer power-gain characteristic   287
8.4. Equalizer back-end impedance   290
9. Constant transducer power gain   292
10. Conclusions   306
Problems   307
References   311

Chapter 5. Theory of Broadband Matching: The Active Load   313

1. Special class of active impedances   314
2. General configuration of the negative-resistance amplifiers   316
3. Nonreciprocal amplifiers   319
   3.1. Design considerations for N_α 321
   3.2. Design considerations for N_β 323
   3.3. Design considerations for N_c 324
   3.4. Illustrative examples   326
      3.4.1. The tunnel diode amplifier: maximally-flat transducer power gain   338
      3.4.2. The tunnel diode amplifier: equiripple transducer power gain   346
   3.5. Extension and stability   355
4. Transmission-power amplifiers   357
   4.1. Tunnel diode in shunt with the load   358
      4.1.1. Transducer power gain: $R_2 > R$ 359
      4.1.2. Transducer power gain: $R_2 < R$ 368
   4.2. Tunnel diode in shunt with the generator   370
      4.2.1. Transducer power gain: $R_1 > R$ 372
      4.2.2. Transducer power gain: $R_1 < R$ 372
   4.3. Stability   373
   4.4. Sensitivity   375
      4.4.1. Tunnel diode in shunt with the load   375
      4.4.2. Tunnel diode in shunt with the generator   377
5. Reciprocal amplifiers   378
   5.1. General gain-bandwidth limitations   379
   5.2. Cascade connection   382
6. Amplifiers using more than one active impedance   388
   6.1. Nonreciprocal amplifiers   391

6.2. Reciprocal amplifiers	394
7. Conclusions	397
Problems	398
References	409

Appendices

Appendix A. The Butterworth Response — 411

Appendix B. The Chebyshev Response — 413

Appendix C. The Elliptic Response — 417

Symbol index — 423

Subject index — 427

Preface

OVER the past two decades, we have witnessed a rapid development of solid-state technology with its apparently unending proliferation of new devices. In order to cope with this situation, a steady stream of new theory, being general and independent of devices, has emerged. One of the most significant developments is the introduction of scattering techniques to network theory. The purpose of this book is to present a unified and detailed account of this theory and its applications to the design of broadband matching networks and amplifiers. It was written primarily as a late text in network theory as well as a reference for practicing engineers who wish to learn how the modern network theory can be applied to the design of many practical circuits. The background required is the usual undergraduate basic courses in networks as well as the ability to handle matrices and functions of a complex variable.

In the book, I have attempted to extract the essence of the theory and to present those topics that are of fundamental importance and that will transcend the advent of new devices and design tools. The guiding light throughout the book has been mathematical precision. Thus, all the assertions are rigorously proved; many of these proofs are believed to be new and novel. I have tried to give a balanced treatment between the mathematical aspects and the physical postulates which motivate the work, and to present the material in a concise manner, using discussions and examples to illustrate the concepts and principles involved. The book also contains some of the personal contributions of the author that are not available elsewhere in the literature.

The scope of this book should be quite clear from a glance at the table of contents. Chapter 1 introduces many fundamental concepts related to linear, time-invariant n-port networks, defines *passivity* in terms of the universally encountered physical quantities *time* and *energy*, and reviews briefly the general characterizations of an *n*-port network. Its time-domain passivity conditions are then translated into the equivalent frequency-domain passivity criteria, which are to be employed to obtain the fundamental limitations on its behavior and utility. Thus, this chapter, as the title implies, may be taken as the foundation for any subsequent network study as well as for the material treated in the remainder of the book.

Chapter 2 gives a fairly complete exposition of the scattering matrix associated with an *n*-port network, starting from a one-port network and using the concepts from transmission-line theory. Fundamental properties of the scattering matrix and its relation to the power transmission among the ports are then derived. The results are indispensable in developing the theory of broadband matching to be treated in the last two chapters.

In seeking fundamental limitations on network or device behavior, performance criteria are often overly idealistic and are not physically realizable. To avoid this difficulty, Chapter 3 considers the approximation problem along with a discussion of the approximating functions. It is shown that the ideal low-pass brick-wall type of gain response can be approximated by three popular rational function approximation schemes: the maximally-flat (Butterworth) response, the equiripple (Chebyshev) response, and the elliptic (Cauer-parameter) response. This is followed by presenting the corresponding ladder network realizations which are attractive from an engineering viewpoint in that they are unbalanced and contain no coupling coils. Explicit formulas for element values of these ladder networks with Butterworth or Chebyshev gain characteristic are given, which reduce the design of these networks to simple arithmetic. Confining attention to the low-pass gain characteristic is not to be deemed restrictive as it may appear. This is demonstrated by considering frequency transformations that permit low-pass characteristic to be converted to a high-pass, band-pass, or band-elimination characteristic.

Using the results developed in the first three chapters, Chapter 4 treats Youla's theory of broadband matching in detail, illustrating every phase of the theory with fully worked out examples. In particular, the fundamental gain-bandwidth limitations for Bode's parallel RC load and Darlington's type-C load are established in their full generality. The extension of Youla's theory to active load impedance is taken up in Chapter 5. It is demonstrated that with suitable manipulations of the scattering parameters, the theory can be applied to the design of negative-resistance amplifiers. This is especially significant in view of the continuing development of new one-port active devices such as the tunnel diode. Many readers will find the perusal of this chapter to be a gratifying and stimulating experience.

In selecting the level of presentation, considerable attention has been given to the fact that many readers may be encountering these topics for the first time. Thus basic introductory material has been included. For example, since many readers are not familiar with the subject of elliptic functions, in Chapter 3 on Approximation and Ladder Realization, an entire section is devoted to the discussion of elliptic functions and some of their fundamental properties that are needed in subsequent analysis. In fact, the section on elliptic response has never been so concisely and systematically treated elsewhere.

The text has grown out of a graduate course entitled "Linear Network Theory" organized at Ohio University. Over the period of years, the material has naturally evolved and up-dated into a shape quite different from the original. However, the basic objective of establishing the fundamentals in this area has remained unchanged throughout. There is little difficulty in fitting the book into a one-semester, or two-quarter course in linear network theory and design. It can be used equally well as a text in advanced network synthesis. For example, as an advanced text in modern network synthesis, Chapters 2, 4 and 5 plus some sections of Chapter 3 would serve for this purpose. Some of the later chapters are also suitable as topics for advanced seminars.

A special feature of the book is that results of direct practical value are included. They are design curves and tables for networks having Butterworth, Chebyshev or elliptic response. These results are extremely useful in that many of the design procedures may be reduced to simple arithmetic and that they find great use in the conduct of research. For example, it is often necessary to check one's hypothesis by specific examples; here they are ready at hand.

A variety of problems has been given at the end of each chapter, some of which are routine applications of results derived in the text. Others, however, require considerable extension of the text material. In all, there are 271 problems.

Much of the material in the book was developed from my research during the past few years. It is a pleasure to acknowledge publicly the research support of the Ohio University Baker Fund Awards Committee. Thanks are also due to many friends and colleagues who reviewed various portions of my manuscript and gave useful suggestions: among them are Professor M. E. Van Valkenburg of University of Illinois, Professor L. O. Chua of University of California at Berkeley, Professor S. P. Chan of University of Santa Clara, and Professor B. J. Leon of Purdue University. I am also indebted to many graduate students who have made valuable contributions to the improvement of this book. Special thanks are due to Mr. S. W. Leung who plotted some of the gain curves in Chapter 4, and to my doctoral students Dr. S. Chandra who gave the complete book a careful reading and Major T. Chaisrakeo who assisted me in computing the elliptic response as well as in many other ways. Finally, I wish to thank my wife and children for their patience and understanding to whom this book is dedicated.

Athens, Ohio WAI-KAI CHEN

CHAPTER 1

Foundations of Network Theory

An *electrical network* is a structure composed of a finite number of interconnected *elements* with a set of *ports* or *accessible terminal pairs* at which voltages and currents may be measured and the transfer of electromagnetic energy into or out of the structure can be made. The elements are idealizations of actual physical devices such as resistors, capacitors, inductors, transformers and generators; and obey the established laws of physics relating various physical quantities such as current, voltage and so forth. Fundamental to the concept of a *port* is the assumption that the instantaneous current entering one terminal of the port is always equal to the instantaneous current leaving the other terminal of the port. A network with n such accessible ports is called an *n-port network* or simply an *n-port*, as depicted in Fig. 1.1. In this chapter, we introduce many fundamental concepts related to linear, time-invariant n-port networks. We first define *passivity* in terms of the universally encountered physical quantities *time* and *energy* and review the general characterizations of an n-port network. We then translate the time-domain passivity conditions into the equivalent frequency-domain passivity criteria, which are to be employed to obtain the fundamental limitations on its behavior and utility.

Since in this book we deal exclusively with linear, lumped and time-invariant n-port networks, the adjectives "linear", "lumped" and "time-invariant" are to be omitted in the discussion unless they are used for emphasis. Much of the discussion and results obtained in the first two chapters are sufficiently general to be applicable to general linear systems.

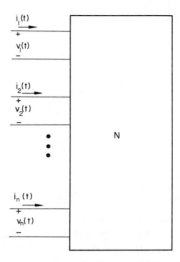

FIG. 1.1. The general symbolic representation of an n-port network N.

1. Basic network postulates

From the historical evolution of network theory, the physical nature of a network can best be described by a set of postulates, which make the theory as simple and as powerful as possible.

Referring to the general symbolic representation of an n-port network N of Fig. 1.1, in which the port voltages $v_k(t)$ and currents $i_k(t)$ can be conveniently represented by the *port-voltage* and *port-current vectors*,

$$v(t) = [v_1(t), v_2(t), \ldots, v_n(t)]', \qquad (1.1a)$$

$$i(t) = [i_1(t), i_2(t), \ldots, i_n(t)]', \qquad (1.1b)$$

respectively, where the prime denotes the matrix transpose. We say that the two n-vectors $v(t)$ and $i(t)$ constitute an *admissible signal pair*, written as $[v(t), i(t)]$, for the n-port network N. We shall generally be concerned with n-port networks that satisfy the following constraints on $v(t)$ and $i(t)$.

1.1. Real-time function postulate

It simply states that if the excitation signals of an n-port are real functions of time, the response signals must also be real functions of time.

Although there is certainly no such thing as a nonreal signal in the real, physical world, it is important to bear in mind that in network theory we often work with signals that are functions of a complex variable, since the use of these signals has become a convenient artifice in the study of networks. For example, in the steady-state analysis of a one-port whose impedance is $z(s)$, it is customary to employ a voltage excitation $V(j\omega)$. Then according to the postulate, if the voltage signal has the form

$$v(t) = \text{Re } V(j\omega)e^{j\omega t} = |V(j\omega)| \cos(\omega t + \theta), \qquad (1.2)$$

where $V(j\omega) = |V(j\omega)|e^{j\theta}$ and Re means the *real part of*, the response current signal must also be a real function of time. In fact, following the usual conventions, the steady-state current is given by

$$i(t) = \text{Re}\left[\frac{V(j\omega)}{z(j\omega)}e^{j\omega t}\right] = \left|\frac{V(j\omega)}{z(j\omega)}\right| \cos(\omega t + \theta - \phi), \qquad (1.3)$$

where $z(j\omega) = |z(j\omega)|e^{j\phi}$.

We remark that the complex variable $s = \sigma + j\omega$ is often referred to as the *complex frequency*. With this designation, if we refer simply to frequency, it is not clear whether we mean s or ω. To emphasize the distinction, people often say *real frequency* to mean ω, which is the *imaginary* part of s. The real part σ of s, misleading as it may be, is called the *imaginary frequency*, and was in general use before 1930. Another convention is to name ω *radian frequency* and σ *neper frequency*, thus avoiding the near metaphysical names. But, no matter what we call them, the two components of frequency add together to give complex frequency. For the present, we shall use the term real frequency for ω. When we speak of the real-frequency axis, we mean the $j\omega$-axis of the complex frequency plane.

1.2. Time-invariance postulate

Intuitively, an n-port network N is considered time-invariant if a given excitation produces the same response no matter when it is applied. Formally, we say that an n-port N is *time-invariant* if for every admissible signal pair $[v_1(t), i_1(t)]$ and for every real finite constant τ, there is an admissible signal pair $[v_2(t), i_2(t)]$ such that

$$v_1(t) = v_2(t + \tau), \qquad (1.4a)$$

$$i_1(t) = i_2(t + \tau). \qquad (1.4b)$$

An n-port that is not time-invariant is called *time-varying*. In other words, an n-port is time-invariant if its terminal behavior is invariant to a shift in the time origin. Thus, if the parameters of an n-port, which is devoid of any initial conditions, are constant then the n-port is time-invariant. The converse, however, is not necessarily true. It is quite easy to conceive of an n-port with time-varying physical elements which exhibits a port behavior that is time-invariant. Figure 1.2 shows a one-port composed of a series connection of

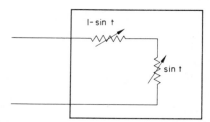

FIG. 1.2. A one-port network with time-varying physical elements which exhibits a port behavior that is time-invariant.

two time-varying resistors, whose input impedance is one ohm. According to the above definition, this one-port is considered to be time-invariant. Suppose, however, that another two-port is formed from this one-port, as shown in Fig. 1.3. This new two-port becomes time-varying. Also, in general, n-ports with initial stored energies that affect port behavior must be considered to be time-varying from the port behavior standpoint.

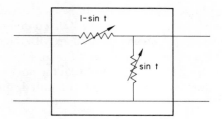

FIG. 1.3. A time-varying two-port network.

1.3. Linearity postulate

Generally speaking, a linear n-port is one in which the response is proportional to the excitation. More precisely, an n-port is said to be *linear* if for all admissible signal pairs

$$[v_1(t), i_1(t)] \quad \text{and} \quad [v_2(t), i_2(t)] \qquad (1.5a)$$

and for all real finite constants c_1 and c_2, then

$$[c_1 v_1(t) + c_2 v_2(t), c_1 i_1(t) + c_2 i_2(t)] \qquad (1.5b)$$

is an admissible signal pair. In other words, a linear n-port obeys the principle of superposition, and its admissible signal pairs comprise a linear space. Quite often, an n-port is called *nonlinear* if it is not linear. However, we must bear in mind that almost all nonlinear analysis techniques include linear case in their domain of applicability as well. Thus, care must be taken to assure the proper interpretation of the term "nonlinear".

Consider the one-port of Fig. 1.4, in which the capacitor is initially

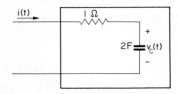

FIG. 1.4. A nonlinear one-port network in which the capacitor is initially charged to a voltage $V_0 \neq 0$.

charged to a voltage $v_C(0+) = V_0$. It is easy to confirm that the one-port N is nonlinear so long as $V_0 \neq 0$. To demonstrate that the superposition principle is intimately tied up with the idea of linearity, we apply a voltage source composed of a series connection of two identical batteries, each having V_0 volts, at the port of Fig. 1.4 at time $t = 0$. The port current can easily be computed, and is given by

$$i(t) = V_0 e^{-0.5t}. \tag{1.6}$$

Now suppose that we apply either one of the two batteries alone, the resulting current will be zero. Adding these two zero responses yields $i(t) = 0$. Thus, the superposition principle is not valid for nonlinear n-ports. However, if the initial capacitor voltage is considered as an excitation, a two-port can be formed from this one-port, as shown in Fig. 1.5. This new two-port becomes linear,

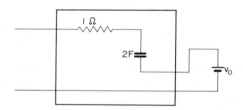

FIG. 1.5. A linear two-port network formed from the nonlinear one-port network of Fig. 1.4.

and we can apply the principle of superposition to obtain the correct result of (1.6). Thus, as with a time-varying n-port being rendered time-invariant, a nonlinear n-port can often be rendered linear by means of the extraction of internal sources at newly formed ports.

1.4. Passivity postulate

Passivity is a qualitative measure of n-ports that are incapable of delivering energy at any time. The concept is crucial to the study of network synthesis, and is not quite so well understood. For many people, the concept of passivity is closely tied up to the non-presence of internal sources, which is not quite what is desired. It is best defined in terms of energy at the ports.

An *n*-port network is said to be *passive* if for *all* admissible signal pairs $[v(t), i(t)]$

$$\mathscr{E}(t) = \mathscr{E}(t_0) + \int_{t_0}^{t} v'(x)i(x)dx \geq 0 \qquad (1.7)$$

for *all* initial time t_0 and *all* time $t \geq t_0$, where $\mathscr{E}(t_0)$ denotes the energy stored in the *n*-port at the initial time t_0.

Thus, when an *n*-port is completely quiescent with no stored energy at some very early time starting at $t_0 = -\infty$, an alternate definition of passivity becomes

$$\mathscr{E}(t) = \int_{-\infty}^{t} v'(x)i(x)dx \geq 0, \qquad (1.8)$$

meaning that a passive *n*-port is one for which the total input energy delivered never goes negative for any time after the excitation is applied. An *n*-port that is not passive is called *active*. We emphasize that we need only find one admissible signal pair such that the condition (1.7) or (1.8) is violated for at least one time t to demonstrate activity of an *n*-port.

The transformer, the gyrator and the linear time-invariant resistor, capacitor and inductor with nonnegative element values are examples of passive networks. A negative resistor is an example of an active one-port.

Figure 1.6 is a low-frequency small-signal model of a transistor,

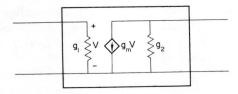

FIG. 1.6. A low-frequency small-signal model of a transistor.

whose port voltages and currents are governed by the equations

$$i_1 = g_1 v_1, \tag{1.9a}$$

$$i_2 = g_m v_1 + g_2 v_2, \tag{1.9b}$$

where $g_1 > 0$ and $g_2 > 0$. The input power to the transistor is given by

$$\begin{aligned}p &= v_1 i_1 + v_2 i_2 \\ &= g_1 v_1^2 + g_m v_1 v_2 + g_2 v_2^2.\end{aligned} \tag{1.10}$$

Choosing

$$v_2 = -\frac{g_m}{2g_2} v_1, \tag{1.11}$$

and substituting it in (1.10) yield

$$p = (4g_1 g_2 - g_m^2) v_1^2 / 4g_2. \tag{1.12}$$

Equation (1.12) shows that if

$$g_m^2 > 4g_1 g_2, \tag{1.13}$$

the input power p is always negative for arbitrary nonzero v_1. The input energy for these voltages is consequently negative, indicating that this transistor model is active.

We remark that, by direct application of the definition, it is often difficult to test passivity. However, equivalent conditions in the frequency domain for linear time-invariant n-ports will be developed in the later sections. They are much simpler to apply.

Finally, we mention two variations on the passivity of an n-port. An n-port is said to be *strictly passive* if the equality is not attained in (1.7) or (1.8) for all nonzero admissible signal pairs. Similarly, an n-port is *lossless* if the equality is attained in (1.7) or (1.8) for all finite admissible signal pairs.

A positive resistor is an example of a strictly passive one-port. The capacitor, the inductor and the transformer are examples of lossless one-ports.

1.5. Causality postulate

The term "causality" connotes the existence of a *cause-effect* relationship. It simply states that if an n-port is *causal*, it cannot yield any response until after the excitation is applied. More precisely, if the currents of an n-port are excitation, and voltages are response, then the n-port is *causal* if for all admissible signal pairs

$$[v_1(t), i_1(t)] \quad \text{and} \quad [v_2(t), i_2(t)], \tag{1.14}$$

$-\infty < t < \infty$, and for any t_0 such that

$$i_1(t) = i_2(t), \quad t \geq t_0, \tag{1.15}$$

then

$$v_1(t) = v_2(t), \quad t \geq t_0. \tag{1.16}$$

This is similarly valid for other excitation and response variables. Examples of the causal networks are the one-port resistor, the one-port capacitors, and the one-port inductors. An ideal transformer of turns ratio $k:1$ is an example of a two-port that is not causal under current (voltage) excitation-voltage (current) response. To see this, consider the signal pairs

$$\left\{ v_1(t) = \begin{bmatrix} kv_a(t) \\ v_a(t) \end{bmatrix}, \quad i_1(t) = \begin{bmatrix} i_a(t) \\ -ki_a(t) \end{bmatrix} \right\} \tag{1.17a}$$

and

$$\left\{ v_2(t) = \begin{bmatrix} kv_b(t) \\ v_b(t) \end{bmatrix}, \quad i_2(t) = \begin{bmatrix} i_b(t) \\ -ki_b(t) \end{bmatrix} \right\}, \tag{1.17b}$$

which are clearly admissible for the two-port. Choosing $i_a(t) = i_b(t)$ and $v_a(t) \neq v_b(t)$ for $t \geq t_0$ shows that (1.16) cannot be satisfied.

At this point, one naturally is led to ask whether the previous five postulates are completely independent. It turns out that they are not. It can be shown (Youla, 1960) that except for a few mostly trivial cases, causality is a consequence of linearity and passivity. In other words, a linear passive n-port must be causal except for a few extremely rare cases.

1.6. Reciprocity postulate

Unlike the other five postulates that are defined in terms of the time domain behavior, reciprocity postulate is primarily associated with network topology. For some n-ports, the response produced at one of the ports due to an excitation at another is invariant if the positions of excitation and response are interchanged. More formally, an n-port is said to be *reciprocal* if for all admissible signal pairs

$$[v_1(t), i_1(t)] \quad \text{and} \quad [v_2(t), i_2(t)], \quad (1.18)$$

then

$$\int_{-\infty}^{\infty} v_1'(t) i_2(t-x) dx = \int_{-\infty}^{\infty} v_2'(t) i_1(t-x) dx. \quad (1.19)$$

Otherwise, it is called *nonreciprocal*. This definition is evolved from the Lorenz reciprocity condition for linear time-invariant electromagnetic media. The integrals in (1.19) are the familiar convolution integrals. It can be shown that this definition coincides with the conventional one in the case where n-ports are linear and time-invariant.

The ideal transformer of turns ratio $k:1$ is a reciprocal two-port, since from (1.17) we have

$$\int_{-\infty}^{\infty} v_1'(t) i_2(t-x) dx = 0 = \int_{-\infty}^{\infty} v_2'(t) i_1(t-x) dx. \quad (1.20)$$

The transistor model of Fig. 1.6 is an example of a nonreciprocal two-port. We remark that reciprocity and passivity are independent postulates. Passive n-ports may be reciprocal or nonreciprocal, as are the active n-ports. For example, consider an ideal gyrator with gyration resistance r, as shown in Fig. 1.7. The port voltages and currents of the gyrator are governed by the equations

$$v_1 = ri_2, \quad (1.21a)$$
$$v_2 = -ri_1. \quad (1.21b)$$

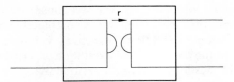

FIG. 1.7. A lossless nonreciprocal gyrator with gyration resistance r.

To show that the gyrator is nonreciprocal, let

$$v_m(t) = \begin{bmatrix} v_{m1}(t) \\ v_{m2}(t) \end{bmatrix} \quad \text{and} \quad i_m(t) = \begin{bmatrix} i_{m1}(t) \\ i_{m2}(t) \end{bmatrix} \quad (1.22)$$

for $m = 1, 2$. Then we have

$$\int_{-\infty}^{\infty} v_1'(t) i_2(t-x) dx = -\int_{-\infty}^{\infty} i_1'(t) v_2(t-x) dx$$
$$= -\int_{-\infty}^{\infty} v_2'(t) i_1(t-x) dx, \quad (1.23)$$

showing that (1.19) is not satisfied. Thus, the gyrator is nonreciprocal.

2. Matrix characterizations of *n*-port networks

In the preceding section, we have presented intuitively appealing axiomatic time-energy approach to the qualitative characterization of *n*-port networks. However, in many situations such as the topics to be treated in this book, these descriptions are not very useful since they are difficult to apply. In the present section, we discuss the general characterizations of the class of linear, time-invariant *n*-port networks. Since synthesis techniques are almost exclusively in the frequency domain, we consider only the frequency-domain representations of the *n*-ports.

Referring to the network of Fig. 1.1, an *n*-port network N is associated with 2*n*-port variables, *n*-port voltages and *n*-port currents. We can take any *n*-port variables as the excitation, and the remaining n-port variables as the response. For linear, time-invariant n-ports, their port behavior is completely characterized by giving the relationships among the port voltages and currents. Depending upon the choice of the excitation and response variables, various matrix descriptions are possible, each being presented briefly in a section. For our purposes, we shall deal with the Laplace-transformed variables and assume that all the initial conditions have been set to zero, since they are equivalent to sources.

2.1. The impedance matrix

Suppose that the port currents are the excitation and voltages are the response. Then the matrix $Z(s)$ relating the transform $V(s)$ of the port-voltage vector $v(t)$ of (1.1a) to the transform $I(s)$ of the port-current vector $i(t)$ of (1.1b) is called the *open-circuit impedance matrix* or simply the *impedance matrix* of the n-port N, that is,

$$V(s) = Z(s)I(s). \quad (1.24)$$

The elements of $Z(s)$ are known as the *open-circuit impedance parameters*. The reason for the term "open-circuit" follows directly from the observation that the jth row and kth column element z_{jk} of $Z(s)$ can be interpreted as

$$z_{jk} = \left.\frac{V_j}{I_k}\right|_{I_x = 0,\, x \neq k}, \quad (1.25)$$

where V_j and I_k denote the transforms of the port voltage $v_j(t)$ and current $i_k(t)$. Equation (1.25) represents the driving-point or transfer impedance when all of the ports except the kth one are open-circuited.

As an illustration, consider the transistor equivalent network of Fig. 1.8. Applying (1.25) yields

$$z_{11} = \left.\frac{V_1}{I_1}\right|_{I_2=0} = \frac{4s+2}{46s+1}, \quad (1.26a)$$

$$z_{21} = \left.\frac{V_2}{I_1}\right|_{I_2=0} = \frac{4s-20}{46s+1}. \quad (1.26b)$$

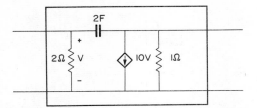

FIG. 1.8. A small-signal network model of a transistor.

In a similar manner, we can compute z_{22} and z_{12}. The open-circuit impedance matrix of the two-port is then given by

$$Z(s) = \frac{1}{46s+1} \begin{bmatrix} 4s+2 & 4s \\ 4s-20 & 4s+1 \end{bmatrix}. \qquad (1.27)$$

2.2. The admittance matrix

Instead of taking the currents as the excitation, we can choose the port voltages as the excitation. This leads to the relation

$$I(s) = Y(s)V(s). \qquad (1.28)$$

The coefficient matrix $Y(s)$ is called the *short-circuit admittance matrix* or simply the *admittance matrix* of the *n*-port N. The elements of $Y(s)$ are known as the *short-circuit admittance parameters*. As before, the term "short-circuit" follows from the observation that the jth row and kth column element y_{jk} of $Y(s)$ can be interpreted as

$$y_{jk} = \frac{I_j}{V_k}\bigg|_{V_x=0, x \neq k}, \qquad (1.29)$$

which corresponds to the driving-point or transfer admittance when all of the ports except the kth one are short-circuited. Clearly, if $Z(s)$ is nonsingular, then

$$Y(s) = Z^{-1}(s). \qquad (1.30)$$

As an example, in Fig. 1.8 we can use (1.29) to compute y_{11} and y_{21}, giving

$$y_{11} = \frac{I_1}{V_1}\bigg|_{V_2=0} = 0.5 + 2s, \qquad (1.31a)$$

$$y_{21} = \frac{I_2}{V_1}\bigg|_{V_2=0} = 10 - 2s. \qquad (1.31b)$$

In a similar fashion, we can compute y_{22} and y_{12}. The short-circuit admittance matrix is given by

$$Y(s) = \begin{bmatrix} 0.5 + 2s & -2s \\ 10 - 2s & 1 + 2s \end{bmatrix}, \qquad (1.32)$$

which is the inverse of $Z(s)$ given in (1.27).

2.3. The hybrid matrix

The hybrid matrix is the generalization of the impedance and admittance matrices discussed above. The excitation variables are either the port voltages or currents, and the response variables are their complements. Thus, there are many possibilities, depending upon the choice of the variables. For our purposes, it is sufficient to consider only the two-ports. If I_1 and V_2 are chosen as the excitation, we have the defining relation

$$\begin{bmatrix} V_1 \\ I_2 \end{bmatrix} = \begin{bmatrix} h_{11} & h_{12} \\ h_{21} & h_{22} \end{bmatrix} \begin{bmatrix} I_1 \\ V_2 \end{bmatrix}. \qquad (1.33)$$

The coefficient matrix is referred to as the *hybrid matrix*, and its elements are known as the *hybrid parameters*. As before, they can be determined by the relations

$$h_{11} = \frac{V_1}{I_1}\bigg|_{V_2=0}, \qquad h_{12} = \frac{V_1}{V_2}\bigg|_{I_1=0}, \qquad (1.34a)$$

$$h_{21} = \frac{I_2}{I_1}\bigg|_{V_2=0}, \qquad h_{22} = \frac{I_2}{V_2}\bigg|_{I_1=0}. \qquad (1.34b)$$

Again, consider the two-port network of Fig. 1.8. Applying (1.34) yields the hybrid matrix

$$H(s) = \frac{1}{4s+1} \begin{bmatrix} 2 & 4s \\ 20-4s & 46s+1 \end{bmatrix}. \qquad (1.35)$$

Another type of mixed variables is defined by the relation

$$\begin{bmatrix} V_1 \\ I_1 \end{bmatrix} = \begin{bmatrix} A & B \\ C & D \end{bmatrix} \begin{bmatrix} V_2 \\ -I_2 \end{bmatrix}, \qquad (1.36)$$

whose coefficient matrix is called the *transmission* or *chain matrix*, its elements being the *transmission* or *chain parameters*. They are also known as the *ABCD parameters*. The first two names come from the fact that they are the natural ones to use in a cascade, tandem, or chain connection of two-ports, since the transmission matrix of two two-ports connected in cascade is equal to the product of the transmission matrices of the individual two-ports (Problem 1.1). As in (1.25), (1.29) and (1.34), the transmission parameters can similarly be interpreted; the results being left as an exercise (Problem 1.19).

As an example, the transmission matrix of the two-port of Fig. 1.8 is given by

$$T(s) = \frac{1}{2s-10} \begin{bmatrix} 2s+1 & 1 \\ 23s+0.5 & 2s+0.5 \end{bmatrix}. \qquad (1.37)$$

By interchanging the roles of the excitation and the response vectors in (1.33) and (1.36), we obtain two more matrices called the *inverse hybrid matrix* and the *inverse transmission* or *chain matrix*, respectively. Since they are trivially different, the details are omitted.

2.4. The indefinite-admittance matrix

As mentioned at the beginning of this chapter, fundamental to the concept of a port is the assumption that the instantaneous current entering one terminal of the port is always equal to the instantaneous current leaving the other terminal of the port. However, we recognize that upon the interconnection of networks, this port constraint may be violated. Thus, it is sometimes desirable and more advantageous to consider $(n+1)$-terminal networks, as depicted in Fig. 1.9. In such a case, it would be useful to have a description of the terminal behavior rather than the port behavior.

Referring to Fig. 1.9, let $\hat{I}(s)$ and $\hat{V}(s)$ be the $(n+1)$-vectors

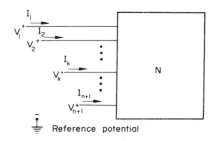

FIG. 1.9. The general symbolic representation of an $(n+1)$-terminal network.

denoting the terminal currents I_k and voltages V_k, being measured between terminals k and some arbitrary but unspecified reference point. Since the network is linear, terminal currents can be expressed in terms of terminal voltages by

$$\hat{I}(s) = \hat{Y}(s)\hat{V}(s). \tag{1.38}$$

The coefficient matrix $\hat{Y}(s)$ is called the *indefinite-admittance matrix*, since as in (1.28), its jth row and kth column element $\hat{y}_{jk}$ can be interpreted as

$$\hat{y}_{jk} = \left. \frac{I_j}{V_k} \right|_{V_x=0, x \neq k}, \tag{1.39}$$

which are nearly the same as the short-circuit admittance parameters given in (1.29). We now proceed to discuss several properties associated with the matrix $\hat{Y}(s)$.

First, we show that each column sum of $\hat{Y}(s)$ is equal to zero. For this we add all the $n+1$ equations of (1.38) to yield

$$\sum_{i=1}^{n+1}\sum_{j=1}^{n+1} \hat{y}_{ji}V_i = \sum_{m=1}^{n+1} I_m = 0. \quad (1.40)$$

The last equation is obtained by applying Kirchhoff's current law. Setting all the terminal voltages to zero except the kth one gives

$$V_k \sum_{j=1}^{n+1} \hat{y}_{jk} = 0. \quad (1.41)$$

Since $V_k \neq 0$, it follows that the sum of the elements of each column of $\hat{Y}(s)$ equals zero. Thus, the indefinite-admittance matrix is singular.

Next we show that each row sum of $\hat{Y}(s)$ is also zero. Without loss of generality, we assume that the network is connected. For, if not, we can apply the following procedure to each of its connected subnetworks, yielding the desired property. To justify the zero-row-sum property, we apply a voltage generator V_k at the kth terminal and set all other terminals open. Then, obviously, all the terminal currents are zero except the kth one, and all the terminal voltages are equal to V_k. This leads to

$$V_k \sum_{i=1}^{n+1} \hat{y}_{ji} = 0, \qquad j \neq k. \quad (1.42)$$

Since $V_k \neq 0$, the sum of the elements of the jth row of $\hat{Y}(s)$ is equal to zero. In a similar manner, we can show that the kth row sum is also equal to zero.

Now we show that the preceding two properties would also imply the equality of all the cofactors of the elements of the indefinite-admittance matrix. This gives rise to the following definition.

DEFINITION 1.1. *Equicofactor matrix.* A square matrix is said to be an *equicofactor matrix* if all of the cofactors of its elements are equal.

The equicofactor matrix is also referred to as the *zero-row-sum and zero-column-sum matrix*, the reason being given below.

LEMMA 1.1. *Let A be a square matrix of order n, $n > 1$, at least one of its cofactors being nonzero. Then A is an equicofactor matrix if and only if the sum of the elements of every row and every column equals zero.*

Proof. Necessity. Let

$$A = [a_{ij}], \tag{1.43}$$

and let c be the value of the cofactor of an element of A. Then by the row expansion of det A, we obtain

$$\det A = \sum_{j=1}^{n} a_{ij} c = c \sum_{j=1}^{n} a_{ij}, \qquad i = 1, 2, \ldots, n. \tag{1.44}$$

Adding all the columns of A to its last column and using (1.44) yield a matrix, each of its last-column elements being $(\det A)/c$. Let the resulting matrix be denoted by A''. Applying the column expansion to the last column of A'' gives

$$\det A'' = \det A = \sum_{j=1}^{n} c(\det A)/c = n(\det A). \tag{1.45}$$

This implies that $(n-1)(\det A) = 0$. Since $n > 1$, $\det A = 0$. From (1.44) we have

$$\sum_{j=1}^{n} a_{ij} = 0. \tag{1.46a}$$

Similarly, by considering the transpose of A, we arrive at

$$\sum_{j=1}^{n} a_{ji} = 0. \tag{1.46b}$$

Sufficiency. Let A_{ij} be the cofactor of the element a_{ij} of A. We first show that $A_{ix} = A_{iy}$ for all i, x and y. Without loss of generality, assume that $x > y$. Since each row sum is zero, we may replace the elements a_{jy} in the submatrix A_{ix} obtained from A by deleting the ith row and xth column by

$$-\sum_{\substack{k=1 \\ k \neq y}}^{n} a_{jk} \tag{1.47}$$

FOUNDATIONS OF NETWORK THEORY

without changing the value of A_{ix}. Let the submatrix thus obtained be denoted by A''_{ix}. Now adding all the columns of A''_{ix} to the column y, and then shifting column y to the right of column $x-1$ if $y \neq x-1$, result in

$$\det A''_{ix} = (-1)^{x-y-1}(-1)(\det A_{iy}), \qquad (1.48)$$

A_{iy} being the submatrix obtained from A by deleting the row i and column y. From (1.48), we have

$$A_{ix} = (-1)^{i+x}(-1)^{x-y-1}(-1)(\det A_{iy}) = A_{iy}. \qquad (1.49)$$

Likewise, by considering the transpose of A, we can show that $A_{xi} = A_{yi}$ for all i, x and y. Thus, we conclude that all the cofactors of the elements of A are equal. This completes the proof of the lemma.

We remark that if the value of the cofactors is zero, the zero-row-sum and zero-column-sum properties are not necessary. In other words, an equicofactor matrix having all its cofactors equal to zero does not necessarily mean that the sum of the elements of every row and every column equals zero. However, the converse is still true. As an example, consider the nth-order matrix A, all whose elements are 1. Evidently, for $n > 2$ A is an equicofactor matrix but each of its row sums or column sums is n.

The above results on the indefinite-admittance matrix are summarized below.

THEOREM 1.1. *The indefinite-admittance matrix of a linear, lumped and time-invariant n-port network is an equicofactor matrix.*

We illustrate the above results by the following example.

EXAMPLE 1.1. Figure 1.10 is the network model of a transistor. Using (1.39), the elements of the indefinite-admittance matrix $\hat{Y}(s)$ can easily be determined, and are given in

$$\hat{Y}(s) = \begin{bmatrix} sC_1 + sC_2 & -sC_2 & -sC_1 \\ -sC_2 + g_m & g + sC_2 & -g - g_m \\ -sC_1 - g_m & -g & g + sC_1 + g_m \end{bmatrix}. \qquad (1.50)$$

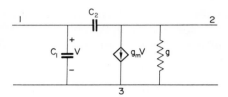

Fig. 1.10. A small-signal network model of a transistor.

It is easy to confirm that the sum of the elements of every row and every column equals zero, and that all of its nine cofactors are equal to

$$s^2 C_1 C_2 + s(C_2 g_m + C_1 g + C_2 g). \quad (1.51)$$

Observe that the admittance g_m enters the indefinite-admittance matrix in a rectangular pattern which is not necessarily centered upon the main diagonal. The admittance g_m is actuated by the voltage V from terminal 1 to terminal 3 that correspond to the first and third columns of (1.50) and affects the currents of the terminals 2 and 3 associated with the second and third rows of the matrix. Thus, the matrix (1.50) can easily be written down by inspection from the terminal labelings of the controlled sources.

To obtain the indefinite-admittance matrix of a network possessing inaccessible internal terminals from that with all terminals accessible, we must suppress these internal terminals to yield only the accessible ones. In terms of network equations, this procedure is equivalent to eliminating unwanted variables from the equations of (1.38). To this end, we partition the matrix equation (1.38) as follows:

$$\begin{bmatrix} I_a \\ I_b \end{bmatrix} = \begin{bmatrix} \hat{Y}_{11} & \hat{Y}_{12} \\ \hat{Y}_{21} & \hat{Y}_{22} \end{bmatrix} \begin{bmatrix} V_a \\ V_b \end{bmatrix}, \quad (1.52)$$

where the elements in I_b and V_b correspond to terminals to be suppressed. Since suppressing a terminal is equivalent to open-circuiting that terminal, this requires that we set $I_b = 0$. Using this and assuming that $\hat{Y}_{22}$ is nonsingular, we obtain the new indefinite-admittance matrix as

$$Y'' = \hat{Y}_{11} - \hat{Y}_{12}\hat{Y}_{22}^{-1}\hat{Y}_{21}. \tag{1.53}$$

In particular, if a single terminal is to be suppressed, the procedure is exceedingly simple. Under this situation, to suppress the kth terminal, the ith row and jth column element y''_{ij} of Y'' is given by

$$y''_{ij} = \hat{y}_{ij} - \hat{y}_{ik}\hat{y}_{kj}/\hat{y}_{kk}. \tag{1.54}$$

Finally, we mention that to obtain the admittance matrix of a common-terminal n-port network derived from an $(n+1)$-terminal network of Fig. 1.9 with one terminal taken as the common ground, we simply delete the row and the column of Y'' corresponding to the chosen ground terminal.

3. Power gains

Refer to the general representation of a two-port network N as shown in Fig. 1.11. The simplest measure of power flow in N is the

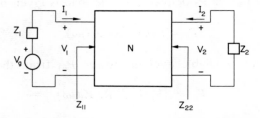

FIG. 1.11. The general representation of a two-port network for the measure of power flow.

power gain G_p, which is defined as the ratio of the average power delivered to the load to the average power entering the input port. Clearly, the power gain is a function of the two-port parameters and the load impedance, and it does not depend upon the source impedance. For a passive lossless two-port network, $G_p = 1$.

The second measure of power flow is called the *available power* gain G_A which is defined as the ratio of the maximum available average power at the output port of N to the maximum available average power at the source. Thus, it is a function of the two-port parameters and the source impedance, being independent of the load impedance.

The third and most useful measure of power flow is called the *transducer power gain* G which is defined as the ratio of average power delivered to the load to the maximum available average power at the source. It is a function of the two-port parameters, the load and the source impedances. The transducer power gain is the most meaningful description of the power transfer capabilities of a two-port network as it compares the power delivered to the load with the power which the source is capable of supplying under optimum conditions. To illustrate this definition, we shall derive an expression for G in terms of the impedance parameters z_{ij} of N.

Referring to Fig. 1.11, the input impedance Z_{11} of N when the output is terminated in Z_2 is related to the z_{ij} and Z_2 by

$$Z_{11} = z_{11} - z_{12}z_{21}/(z_{22} + Z_2). \quad (1.55)$$

The average power P_2 delivered to the load is given by

$$P_2 = |I_2(j\omega)|^2 \operatorname{Re} Z_2(j\omega). \quad (1.56)$$

The maximum available average power P_{a1} from the source is obtained as

$$P_{a1} = \tfrac{1}{4}|V_g(j\omega)|^2/\operatorname{Re} Z_1(j\omega), \quad (1.57)$$

which represents the average power delivered by the given source to a conjugately matched load $\bar{Z}_1(j\omega)$. Combining these yields

$$G = \frac{P_2}{P_{a1}} = \frac{4|z_{21}|^2 (\operatorname{Re} Z_1)(\operatorname{Re} Z_2)}{|(z_{11} + Z_1)(z_{22} + Z_2) - z_{12}z_{21}|^2}, \quad (1.58)$$

the variable $s = j\omega$ being dropped in the expression, for simplicity. In terms of the admittance parameters y_{ij} of N, G becomes

$$G = \frac{4|y_{21}|^2(\text{Re } Y_1)(\text{Re } Y_2)}{|(y_{11} + Y_1)(y_{22} + Y_2) - y_{12}y_{21}|^2}, \quad (1.59)$$

where $Y_1 = 1/Z_1$ and $Y_2 = 1/Z_2$.

In a similar manner, we can evaluate the other two power gains. For completeness, we list below the results for these gains:

$$G_p = \frac{|z_{21}|^2 \text{ Re } Z_2}{|z_{22} + Z_2|^2 \text{ Re } Z_{11}} = \frac{|y_{21}|^2 \text{ Re } Y_2}{|y_{22} + Y_2|^2 \text{ Re } Y_{11}}, \quad (1.60)$$

$$G_A = \frac{|z_{21}|^2 \text{ Re } Z_1}{|z_{11} + Z_1|^2 \text{ Re } Z_{22}} = \frac{|y_{21}|^2 \text{ Re } Y_1}{|y_{11} + Y_1|^2 \text{ Re } Y_{22}}, \quad (1.61)$$

where

$$Y_{11} = y_{11} - y_{12}y_{21}/(y_{22} + Y_2), \quad (1.62)$$

and by interchanging the roles of 1 and 2 in (1.55) and (1.62) we have Z_{22} and Y_{22}. The variable $s = j\omega$ is again dropped in the above expressions.

4. Hermitian forms

In this section, we shall digress slightly into a discussion of the hermitian forms, which are needed in the subsequent presentation. In order to indicate that the results are general, we use a general notation.

Let

$$A = [a_{ij}] \quad (1.63)$$

be a hermitian matrix of order n and let $X = [x_j]$ be a complex n-vector. Then the scalar expression

$$X^*AX = \sum_{i=1}^{n} \sum_{j=1}^{n} a_{ij}\bar{x}_i x_j, \quad (1.64)$$

where

$$X^* = \bar{X}', \quad (1.65)$$

$$\bar{X} = [\bar{x}_j], \quad (1.66)$$

and $\bar{x}_j$ denotes the complex conjugate of x_j, is called a *hermitian form*. The matrix A is referred to as the *matrix of the hermitian form* (1.64). Even though A and X are complex, the hermitian form X^*AX is real. To see this, we take the conjugate of X^*AX. Recall that since X^*AX is a scalar, $(X^*AX)' = X^*AX$. This shows that

$$\overline{X^*AX} = (X^*AX)^* = X^*A^*X = X^*AX, \tag{1.67}$$

meaning that a hermitian form is always real for any complex X. However, for a given A, the sign associated with such a form normally depends on the values of X. It may happen that, for some A, its hermitian form remains of one sign, independent of the values of X. Such forms are called *definite*. Since definiteness of a hermitian form must be an inherent property of its matrix A, it is natural then to refer to the matrix A as *definite*. We now consider two subclasses of the class of definite hermitian matrices.

DEFINITION 1.2. *Positive-definite matrix.* An $n \times n$ hermitian matrix A is called a *positive-definite matrix* if

$$X^*AX > 0 \tag{1.68}$$

for all complex n-vectors $X \neq 0$.

DEFINITION 1.3. *Nonnegative-definite matrix.* An $n \times n$ hermitian matrix A is called a *nonnegative-definite matrix* if

$$X^*AX \geq 0 \tag{1.69}$$

for all complex n-vectors X.

Evidently, the class of positive-definite matrices is part of the class of nonnegative-definite matrices. Very often, a nonnegative-definite matrix is also called *positive-semidefinite*, but we must bear in mind that some people define a positive-semidefinite matrix A as one that satisfies (1.69) for all X, provided there is at least one $X \neq 0$ for which the equality holds. In the latter case, it is evident that positive definiteness and positive semidefiniteness are mutually exclusive. Together, they form the class of nonnegative-definite matrices. Thus, care must be taken to assure the proper interpretation of the term "positive-semidefiniteness".

As an example, consider the hermitian matrix

$$A = \begin{bmatrix} 0 & 1+j2 \\ 1-j2 & 0 \end{bmatrix}, \quad (1.70)$$

whose associated hermitian form is given by

$$X^*AX = (1+j2)\bar{x}_1 x_2 + (1-j2)x_1 \bar{x}_2$$
$$= 2 \operatorname{Re}\left[(1+j2)\bar{x}_1 x_2\right]. \quad (1.71)$$

Choosing $x_1 = 1$ and $x_2 = 1$ gives $X^*AX = 2$, and choosing $x_1 = 1$ and $x_2 = j1$ yields $X^*AX = -4$. This shows that the matrix (1.70) is not definite, since the sign of its hermitian form depends upon the choice of X. On the other hand, the real symmetric matrix

$$A = \begin{bmatrix} 3 & -1 & -1 \\ -1 & 2 & 1 \\ -1 & 1 & 2 \end{bmatrix}, \quad (1.72)$$

which is also hermitian, is positive-definite; since according to (1.64) its hermitian form can be expressed as (Problem 1.3)

$$X^*AX = |3x_1 - x_2 - x_3|^2/3 + |5x_2 + 2x_3|^2/15 + 7|x_3|^2/5, \quad (1.73)$$

which is clearly nonnegative for all values of x's, and positive if the x's are not all zero. Thus, (1.72) is a nonnegative or positive-definite matrix.

The example illustrates a procedure for determining whether or not a given hermitian matrix is positive or nonnegative definite by converting a hermitian form into an expression involving squares of magnitudes. This process is long and tedious and really does not help very much. In the following, we present simple criteria for testing a given hermitian matrix for positive-definiteness and nonnegative-definiteness in terms of the elements of the matrix.

In a matrix A of order n, define a *principal minor* of order r to be the determinant of a submatrix consisting of the rows $i_1, i_2, \ldots, i_r$ and the columns $i_1, i_2, \ldots, i_r$. A *leading principal minor* of order r is the determinant of the submatrix consisting of the first r rows and first r columns. For $n = 3$, the principal minors of A are listed below:

Principal minors of order 1

$$a_{11}, \quad a_{22}, \quad a_{33};$$

Principal minors of order 2

$$\begin{vmatrix} a_{11} & a_{12} \\ a_{21} & a_{22} \end{vmatrix}, \quad \begin{vmatrix} a_{11} & a_{13} \\ a_{31} & a_{33} \end{vmatrix}, \quad \begin{vmatrix} a_{22} & a_{23} \\ a_{32} & a_{33} \end{vmatrix};$$

Principal minor of order 3

$$\begin{vmatrix} a_{11} & a_{12} & a_{13} \\ a_{21} & a_{22} & a_{23} \\ a_{31} & a_{32} & a_{33} \end{vmatrix}.$$

Of the seven principal minors, only the three leading ones in the groups are the leading principal minors.

We now present Sylvester's criteria for positive-definiteness and nonnegative-definiteness of a hermitian matrix.

THEOREM 1.2. *A hermitian matrix is positive-definite if and only if all of its leading principal minors are positive. A hermitian matrix is nonnegative-definite if and only if all of its principal minors are nonnegative.*

We remark that for a positive-definite matrix, not only its leading principal minors are positive, but all of its principal minors are also positive. In other words, for a positive-definite matrix, the positiveness of its leading principal minors also implies the positiveness of all of its principal minors. However, for a nonnegative-definite matrix, the nonnegativeness of its leading principal minors does not necessarily imply the nonnegativeness of all of its principal minors. As an example, consider a 2×2 matrix A, in which $a_{11} = a_{12} = a_{21} = 0$ and a_{22} is real and negative. Clearly, the two leading principal minors are nonnegative, but A is not nonnegative-definite, since $X^*AX = a_{22}|x_2|^2 < 0$. Another example is given in Problem 1.17.

To confirm that the hermitian matrix A of (1.72) is positive-definite, we compute its leading principal minors:

$$3>0, \qquad \begin{vmatrix} 3 & -1 \\ -1 & 2 \end{vmatrix} = 5>0, \qquad \det A = 7>0.$$

From the theorem we conclude that A is indeed a positive-definite matrix.

EXAMPLE 1.2. The hermitian matrix

$$Z_h(j\omega) = \begin{bmatrix} r_e + r_b & r_b + j\dfrac{a}{2\omega C} \\ r_b - j\dfrac{a}{2\omega C} & r_b \end{bmatrix} \tag{1.74}$$

is associated with a transistor amplifier. We wish to determine conditions under which the matrix $Z_h(j\omega)$ is nonnegative-definite. This requires that the following inequalities be satisfied:

$$r_e + r_b \geqq 0, \tag{1.75a}$$

$$r_b \geqq 0, \tag{1.75b}$$

$$r_e r_b - \frac{a^2}{4\omega^2 C^2} \geqq 0. \tag{1.75c}$$

For $r_e > 0$ and $r_b > 0$, the matrix remains nonnegative-definite for all the real frequencies

$$\omega \geqq \frac{a}{2Cr_e^{1/2}r_b^{1/2}}. \tag{1.76}$$

In the later sections, we show that passivity of an n-port is closely related to the nonnegative-definiteness of the hermitian part of its immittance (impedance or admittance) matrix or hybrid matrix.

5. The positive-real matrix

After a brief discussion of the hermitian forms, we are now led to the concept of a positive-real matrix, which is the matrix version of a positive-real function.

For an $n \times n$ matrix $A(s)$, the associated hermitian matrix

$$A_h(s) = \tfrac{1}{2}[A(s) + A^*(s)] \qquad (1.77)$$

is called the *hermitian part* of $A(s)$, and the symmetric matrix

$$A_s(s) = \tfrac{1}{2}[A(s) + A'(s)] \qquad (1.78)$$

is called the *symmetric part* of $A(s)$.

DEFINITION 1.4. *Positive-real matrix.* An $n \times n$ matrix function $A(s)$ of the complex variable s is said to be a *positive-real matrix* if it satisfies the following three conditions:

(i) (Each element of) $A(s)$ is analytic in the open RHS (right-half of the complex s-plane), i.e. Re $s > 0$.†
(ii) $\bar{A}(s) = A(\bar{s})$ for all s in the open RHS.
(iii) Its hermitian part $A_h(s)$ is nonnegative-definite for all s in the open RHS.

The second condition is equivalent to stating that each element of $A(s)$ is real when s is real, and for rational $A(s)$ it is always satisfied if all the coefficients of its elements are real. A natural question arises at this point as to whether or not all of the three conditions are independent. The answer is affirmative if general functions are considered. However, if $A(s)$ is a matrix of rational functions, then the analyticity requirement is redundant, since, as will be shown shortly, each element of $A(s)$ is devoid of poles in the open RHS.

†Likewise, LHS stands for the left-half of the complex s-plane. These abbreviations will be used throughout the remainder of the book.

DEFINITION 1.5. *Positive-real function.* A positive-real matrix of order 1 is called a *positive-real function.*

We illustrate the above definitions by the following examples.

EXAMPLE 1.3. Consider the impedance matrix

$$Z(s) = \begin{bmatrix} r_e + r_b & r_b \\ r_b & r_b + 1/sC \end{bmatrix} \quad (1.79)$$

of a passive two-port, whose hermitian part is given by

$$Z_h(s) = \tfrac{1}{2}[Z(s) + Z^*(s)]$$

$$= \begin{bmatrix} r_e + r_b & r_b \\ r_b & r_b + \dfrac{\sigma}{C(\sigma^2 + \omega^2)} \end{bmatrix}. \quad (1.80)$$

We now proceed to investigate the three conditions of Definition 1.4. Since the only singularity of $Z(s)$ is on the $j\omega$-axis at $s = 0$, $Z(s)$ is analytic in the open RHS. As there are real coefficients, $\bar{Z}(s) = Z(\bar{s})$. Finally, according to Theorem 1.2, $Z_h(s)$ is nonnegative-definite for all s in the open RHS if and only if the following three inequalities are satisfied:

$$r_e + r_b \geqq 0, \quad (1.81a)$$

$$r_b \sigma^2 + \sigma/C + r_b \omega^2 \geqq 0, \quad \text{Re } s > 0, \quad (1.81b)$$

$$\det Z_h(s) \geqq 0, \quad \text{Re } s > 0. \quad (1.81c)$$

The last inequality (1.81c) is equivalent to

$$r_e r_b \sigma^2 + (r_e + r_b)\sigma/C + r_e r_b \omega^2 \geqq 0 \quad (1.82)$$

for $\sigma > 0$. Thus, for positive r_e, r_b, and C, the hermitian matrix $Z_h(s)$ is nonnegative-definite for all s in the open RHS, and the impedance matrix $Z(s)$ is positive-real.

EXAMPLE 1.4. Consider the impedance matrix

$$Z(s) = \begin{bmatrix} r_e + r_b & r_b \\ r_b + a/sC & r_b + 1/sC \end{bmatrix} \quad (1.83)$$

associated with a transistor amplifier, where r_e, r_b and C are positive and $a \leq 1$ is real. This matrix is essentially the same as that given in (1.79) except that a term a/sC is inserted in the (2, 1)-position. The hermitian part of $Z(s)$ is given by

$$Z_h(s) = \begin{bmatrix} r_e + r_b & r_b + \dfrac{a(\sigma + j\omega)}{2C(\sigma^2 + \omega^2)} \\ r_b + \dfrac{a(\sigma - j\omega)}{2C(\sigma^2 + \omega^2)} & r_b + \dfrac{\sigma}{C(\sigma^2 + \omega^2)} \end{bmatrix}. \quad (1.84)$$

Clearly, the first two constraints of Definition 1.4 are satisfied. The third constraint is equivalent to the following three inequalities:

$$r_e + r_b \geq 0, \quad (1.85a)$$

$$Cr_b\sigma^2 + \sigma + Cr_b\omega^2 \geq 0, \quad (1.85b)$$

$$4C^2 r_e r_b \sigma^2 + 4C[r_e + (1-a)r_b]\sigma + 4C^2 r_e r_b \omega^2 - a^2 \geq 0 \quad (1.85c)$$

for $\sigma > 0$. Equations (1.85a) and (1.85b) are always satisfied, but (1.85c) may not. For, if we choose sufficiently small σ and ω, the inequality (1.85c) will be violated for $a \neq 0$. Thus, we conclude that the impedance matrix $Z(s)$ is not positive-real.

From the above two examples, we recognize that, without introducing additional theory, it is sometimes difficult to apply the definition to test the positive-realness of a matrix. The first two constraints of the definition are relatively easy to check. However, the third one is somewhat difficult to verify, since the nonnegative-definiteness of its hermitian part must be investigated for all s in the open RHS. In the following, we show that if $A(s)$ is rational, only the $j\omega$-axis points need be tested for the nonnegative-definiteness of $A_h(s)$.

FOUNDATIONS OF NETWORK THEORY 31

THEOREM 1.3. *An $n \times n$ rational matrix $A(s)$ is positive-real if and only if the following four conditions are satisfied:*

(i) *$\bar{A}(s) = A(\bar{s})$.*
(ii) *(Each element of) $A(s)$ has no poles in the open RHS.*
(iii) *Poles of (any element of) $A(s)$ on the $j\omega$-axis, if they exist, are simple, and the associated residue matrix K evaluated at each of these poles is hermitian and nonnegative-definite.*
(iv) *$A_h(j\omega)$ is nonnegative-definite whenever it is defined.*

Proof. Necessity. Assume that $A(s)$ is positive-real. Then conditions (i) and (ii) are certainly satisfied. To establish condition (iii), let $j\omega_0$ be an mth-order finite pole of $A(s)$, and expand each element of $A(s)$ in a Laurent series about $j\omega_0$. As s approaches to $j\omega_0$, the terms containing $(s - j\omega_0)^{-m}$ become dominant, and we can write

$$X^*A(s)X = \frac{X^*KX}{(s - j\omega_0)^m}, \quad (1.86)$$

where K is the coefficient matrix associated with the term $(s - j\omega_0)^{-m}$ in the Laurent series expansion of $A(s)$ about the pole $j\omega_0$. Since $A(s)$ is positive-real, $A_h(s)$ is nonnegative-definite for all s in the open RHS, meaning that

$$X^*A_h(s)X = \tfrac{1}{2}X^*A(s)X + \tfrac{1}{2}X^*A^*(s)X = \text{Re } X^*A(s)X$$
$$\geq 0, \quad \text{Re } s > 0. \quad (1.87)$$

Write

$$s - j\omega_0 = re^{j\theta}, \quad (1.88a)$$

$$X^*KX = ke^{j\phi}. \quad (1.88b)$$

Substituting (1.86) in (1.87) in conjunction with (1.88) yields the inequality

$$\text{Re } X^*A(s)X = kr^{-m} \cos(m\theta - \phi) \geq 0 \quad (1.89)$$

for $-\pi/2 < \theta < \pi/2$. Clearly, this inequality can be satisfied only if $m = 1$ and $\phi = 0$ (Problem 1.5). This means that the $j\omega$-axis poles of

$A(s)$ are simple, and that K denotes the matrix of the residues of the elements of $A(s)$ evaluated at the pole $j\omega_0$. Since from (1.88b)

$$X^*KX = k \geqq 0 \tag{1.90}$$

for any complex constant X, it follows that K is hermitian and nonnegative-definite (Problem 1.6). The above argument must also hold for a pole at the infinity, since such a pole can be brought to the origin by considering the matrix $A(1/s)$ (Problem 1.8). Thus, condition (iii) is satisfied. To show that condition (iv) cannot be violated, we observe that since $A(s)$ is positive-real, $X^*A_h(s)X$ is nonnegative, and is a continuous function of σ and ω for all s in the open RHS. As a limit of a continuous nonnegative function, $X^*A_h(j\omega)X$ must be nonnegative whenever it exists. Consequently, the four conditions are necessary for $A(s)$ to be positive-real.

Sufficiency. Assume that $A(s)$ is a rational matrix satisfying the four conditions. We show that $A(s)$ is positive-real. For this, it is sufficient to prove that $A_h(s)$ is nonnegative-definite for all s in the open RHS. From (1.87), this is equivalent to showing that

$$\text{Re } X^*A(s)X \geqq 0 \quad \text{for} \quad \text{Re } s > 0. \tag{1.91}$$

Now consider a contour C_x enclosing the RHS, with its boundary being the $j\omega$-axis except for arbitrarily small indentations into the open RHS around the $j\omega$-axis poles of $X^*A(s)X$, as depicted in Fig. 1.12. For s on one of the indentations of C_x near a finite $j\omega$-axis pole at $j\omega_0$, (1.86) applies. By condition (iii), we have $m = 1$ and $X^*KX \geqq 0$ for all constant X. This shows that

$$\text{Re } X^*A(s)X = kr^{-1}\cos\theta \geqq 0 \tag{1.92}$$

for $-\pi/2 \leqq \theta \leqq \pi/2$. In other words, (1.91) is satisfied on the indentations of C_x. Evidently, this is also valid for the pole at the infinity by replacing $s - j\omega_0$ by $1/s$. Using this in conjunction with condition (iv) shows that (1.91) holds for all the points on the contour C_x. Now consider the function

$$f(s) = e^{-X^*A(s)X}. \tag{1.93}$$

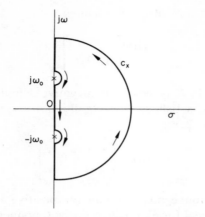

FIG. 1.12. A contour enclosing the RHS with its boundary being the $j\omega$-axis except for arbitrarily small indentations into the open RHS around the $j\omega$-axis poles of $X^*A(s)X$.

Since by condition (ii), $X^*A(s)X$ is analytic within and on the boundary of the region formed by the contour C_x, $f(s)$ is also analytic within and on the boundary of the region. Appealing to the maximum modulus theorem of the theory of a complex variable, the maximum value of

$$|f(s)| = e^{-\operatorname{Re} X^*A(s)X}, \qquad (1.94)$$

which corresponds to the minimum value attained by $\operatorname{Re} X^*A(s)X$, for all s within and on the boundary of the region occurs on C_x. Thus, we conclude that

$$\operatorname{Re} X^*A(s)X = X^*A_h(s)X \geqq 0 \quad \text{for} \quad \operatorname{Re} s > 0. \qquad (1.95)$$

This completes the proof of the theorem.

We illustrate this theorem by the following examples.

EXAMPLE 1.5. We wish to test the positive-realness of the matrix (1.79) by the conditions of Theorem 1.3. Clearly, conditions (i) and (ii) are satisfied. To test condition (iii), we first compute the residue matrix K for the pole at $s = 0$, yielding

$$K = \begin{bmatrix} 0 & 0 \\ 0 & 1/C \end{bmatrix}. \qquad (1.96)$$

The residue matrix K is clearly nonnegative-definite for $C > 0$, and condition (iii) is satisfied. For condition (iv), we compute the hermitian part of $Z(s)$:

$$Z_h(j\omega) = \begin{bmatrix} r_e + r_b & r_b \\ r_b & r_b \end{bmatrix}, \qquad (1.97)$$

which is again nonnegative-definite for positive r_e and r_b. Thus, $Z(s)$ is positive-real for all positive (in fact, nonnegative) r_e, r_b and C.

EXAMPLE 1.6. Using Theorem 1.3, test the positive-realness of the matrix $Z(s)$ given in Example 1.4. Again, conditions (i) and (ii) are trivially satisfied. For condition (iii), we compute the residue matrix K at the pole $s = 0$, which gives

$$K = \begin{bmatrix} 0 & 0 \\ a/C & 1/C \end{bmatrix}. \qquad (1.98)$$

Since this matrix is not hermitian, condition (iii) is not satisfied.

On the $j\omega$-axis, the hermitian part $Z_h(j\omega)$ of $Z(s)$ is given by (1.74). $Z_h(j\omega)$ cannot be nonnegative-definite for all ω since for sufficiently small ω, $\det A_h(j\omega) < 0$. This shows that condition (iv) is violated, and the matrix $Z(s)$ is not positive-real, a fact that was pointed out in Example 1.4.

As mentioned at the beginning of this section, if $A(s)$ is a rational matrix, then not all of the three conditions of Definition 1.4 are independent. In fact, the analyticity requirement is redundant. To justify this assertion, we first assume that s_0 is an mth-order finite pole of $A(s)$ in the open RHS. Then following the argument used in (1.86)–(1.89) would show that there exists a point s in a neighborhood of s_0 such that

$$X^*A_h(s)X = \operatorname{Re} X^*A(s)X < 0, \qquad (1.99)$$

violating the condition that $A_h(s)$ is nonnegative-definite for all s in the open RHS. Thus, no such poles s_0 can exist, and $A(s)$ is analytic in the open RHS.

THEOREM 1.4. *If $A(s)$ is a nonsingular positive-real matrix of rational functions, then so is its inverse.*

Proof. Let $F(s) = A^{-1}(s)$. Since $A(s)$ is real for real s, it is evident that $F(s)$ is also real for real s. To show that $F(s)$ is analytic† in the open RHS, it is sufficient to prove that $F(s)$ is devoid of poles in the open RHS. Assume that s_0 is an open RHS pole of $F(s)$. Then $A(s_0)$ must be singular, and the associated homogeneous system of equations

$$A(s_0)X = 0 \qquad (1.100)$$

possesses a non-trivial solution $X_0 \neq 0$. This shows that $X_0^*A(s_0)X_0 = 0$ or

$$X_0^*A_h(s_0)X_0 = 0. \qquad (1.101)$$

As in (1.93), consider the exponential function

$$f(s) = e^{-X_0^*A(s)X_0}, \qquad (1.102)$$

which is analytic in the open RHS. Appealing to the maximum modulus theorem and using (1.94) show that $\operatorname{Re} X_0^*A(s)X_0$ cannot have a minimum in the open RHS unless it is identically constant. But clearly, $\operatorname{Re} X_0^*A(s_0)X_0 = 0$ is a minimum. Thus, under the stipulated hypothesis, $X_0^*A_h(s)X_0 = 0$ for all s in the open RHS. It follows that either $A(s)$ is identically singular or it is skew hermitian, $A(s) = -A^*(s)$. In the latter case, as $A(s)$ is independent of $\bar{s}$, being analytic, it is also independent of s, and thus it is identically singular, being singular at s_0 (Problem 1.16). This contradicts to the assumption that $A(s)$ is nonsingular. Consequently, $F(s)$ is analytic in the open RHS.

---

†Since the analyticity requirement is redundant, we need only show that $F_h(s)$ is nonnegative-definite for all s in the open RHS. It is included here for completeness.

To demonstrate that $F_h(s)$ is nonnegative-definite for all s in the open RHS, we observe that for each complex n-vector X_1, there is a unique X_2 such that

$$X_2 = F(s)X_1, \qquad (1.103)$$

whenever $F(s)$ exists. Then we have

$$X_1^* F_h(s) X_1 = X_2^* A^*(s) F_h(s) A(s) X_2 = X_2^* A_h(s) X_2 \geqq 0, \qquad (1.104)$$

which completes the proof of the theorem.

Before we turn our attention to link passivity with positive-realness, we consider a special case of Theorem 1.3, which is sufficiently important to be discussed separately.

COROLLARY 1.1. *A rational function $f(s)$ is positive-real if and only if the following four conditions are satisfied:*

(i) *$f(s)$ is real when s is real.*
(ii) *$f(s)$ has no poles in the open RHS.*
(iii) *Poles of $f(s)$ on the $j\omega$-axis, if they exist, are simple, and residues evaluated at these poles are real and positive.*
(iv) *Re $f(j\omega) \geqq 0$, $0 \leqq \omega \leqq \infty$.*

An alternative testing procedure that avoids the necessity of computing residues can be deduced from the above corollary by the well-known bilinear transformation,

$$w(s) = \frac{f(s)-1}{f(s)+1} \quad \text{or} \quad f(s) = \frac{1+w(s)}{1-w(s)}, \qquad (1.105)$$

which has the property that

$$\operatorname{Re} f(s) \gtreqless 0 \qquad (1.106a)$$

if and only if

$$|w(s)| \lesseqgtr 1, \qquad (1.106b)$$

respectively. We see that an alternative set of necessary and sufficient conditions for $f(s)$ to be positive-real is that

(i) $w(s)$ is real when s is real,
(ii) $w(s)$ has no poles in the closed RHS, i.e., Re $s \geq 0$,
(iii) $|w(j\omega)| \leq 1$, $0 \leq \omega \leq \infty$.

For computational purposes, these conditions are restated in terms of equivalent conditions better suited for use in testing.

COROLLARY 1.2. *A rational function represented in the form*

$$f(s) = \frac{p(s)}{q(s)} = \frac{m_1(s) + n_1(s)}{m_2(s) + n_2(s)}, \quad (1.107)$$

where $m_1(s)$, $m_2(s)$ and $n_1(s)$, $n_2(s)$ are the even and odd parts of the polynomials $p(s)$ and $q(s)$, respectively, is positive-real if and only if the following three conditions are satisfied:

(i) $f(s)$ *is real when s is real.*
(ii) $p(s) + q(s)$ *is strictly Hurwitz.*
(iii)

$$m_1(j\omega)m_2(j\omega) - n_1(j\omega)n_2(j\omega) \geq 0 \quad (1.108)$$

for all ω.

The testing of the second condition can easily be accomplished by means of the Hurwitz test, which states that a real polynomial is strictly Hurwitz if and only if the continued fraction expansion of the ratio of the even part to the odd part or the odd part to the even part of the polynomial gives only real and positive coefficients, and does not terminate prematurely. The third condition is satisfied if and only if the left-hand side of (1.108) does not have real positive roots of odd multiplicity. This may be determined by factoring it or by the use of Sturm's theorem, which can be found in most texts on elementary theory of equations.

COROLLARY 1.3. *A rational function $f(s)$ is positive-real if and only if*

$$|\arg f(s)| \leq |\arg s| \quad \text{for} \quad |\arg s| \leq \pi/2. \quad (1.109)$$

We illustrate the above results by the following example.

EXAMPLE 1.7. Test the following function to see if it is positive-real.

$$f(s) = \frac{2s^4 + 4s^3 + 5s^2 + 5s + 2}{s^3 + s^2 + s + 1}. \tag{1.110}$$

For illustrative purposes, we follow the three steps outlined in Corollary 1.2.

$$p(s) = (2s^4 + 5s^2 + 2) + (4s^3 + 5s) = m_1(s) + n_1(s), \tag{1.111a}$$

$$q(s) = (s^2 + 1) + (s^3 + s) = m_2(s) + n_2(s), \tag{1.111b}$$

$$p(s) + q(s) = (2s^4 + 6s^2 + 3) + (5s^3 + 6s) = m(s) + n(s). \tag{1.111c}$$

Condition (i) is clearly satisfied. To test condition (ii), we perform the Hurwitz test, which gives

$$\frac{m(s)}{n(s)} = \frac{2s^4 + 6s^2 + 3}{5s^3 + 6s}$$

$$= 2s/5 + \cfrac{1}{25s/18 + \cfrac{1}{108s/55 + \cfrac{1}{11s/18}}} \tag{1.112a}$$

Since the continued fraction expansion does not terminate prematurely and since all of its coefficients are real and positive, we conclude that the polynomial $p(s) + q(s)$ is strictly Hurwitz. Thus, condition (ii) is satisfied. Finally, to test condition (iii) we compute

$$m_1(j\omega)m_2(j\omega) - n_1(j\omega)n_2(j\omega) = 2\omega^6 - 2\omega^4 - 2\omega^2 + 2$$
$$= 2(\omega^2 + 1)(\omega^2 - 1)^2. \tag{1.112b}$$

The third condition is also satisfied, since (1.112b) does not possess any real and positive roots of odd multiplicity. Thus, the function $f(s)$ is positive-real.

6. Frequency-domain conditions for passivity

After this digression into a discussion of the hermitian forms and the positive-real matrices, we now turn our attention to the problem of translating the time-domain passivity criterion of §1.4 into the equivalent frequency-domain passivity conditions. The significance of passivity in the study of active networks is that activity is the formal negation of passivity. These conditions will be employed to obtain fundamental limitations on the behavior and utility of active devices.

Let N be a linear, time-invariant n-port network that can be characterized by an $n \times n$ admittance matrix $Y(s)$, which is not identically singular. This assumption is made solely for the convenience of subsequent manipulations. Other characterizations such as the impedance matrix and the hybrid matrix are equally valid. We recognize that the results obtained for $Y(s)$ are similarly valid for other matrices. In the case where the admittance matrix is identically singular, the original n-port network can then be replaced by an m-port subnetwork, $m < n$, for which the $m \times m$ admittance matrix would not be identically singular.

Referring to Fig. 1.1, let

$$v(t) = \text{Re}\,[V_0 e^{s_0 t}], \qquad (1.113a)$$

$$i(t) = \text{Re}\,[I_0 e^{s_0 t}], \qquad (1.113b)$$

where the real part of a matrix means the matrix of the real parts, and

$$s_0 = \sigma_0 + j\omega_0, \qquad (1.114a)$$

σ_0 and ω_0 being real, and

$$\sigma_0 \geqq 0. \qquad (1.114b)$$

Then we can write

$$v(t) = \tfrac{1}{2}[V_0 e^{s_0 t} + \bar{V}_0 e^{\bar{s}_0 t}], \qquad (1.115a)$$

$$i(t) = \tfrac{1}{2}[I_0 e^{s_0 t} + \bar{I}_0 e^{\bar{s}_0 t}]. \qquad (1.115b)$$

Using the relation

$$I_0 = Y(s_0)V_0, \qquad (1.116)$$

the instantaneous power into the n-port network is given by

$$\begin{aligned}v'(t)i(t) &= \tfrac{1}{4}[V_0'Y(s_0)V_0 e^{2s_0 t} + V_0^* \bar{Y}(s_0)\bar{V}_0 e^{2\bar{s}_0 t} \\ &\quad + V_0^* Y(s_0)V_0 e^{2\sigma_0 t} + V_0' \bar{Y}(s_0)\bar{V}_0 e^{2\sigma_0 t}] \\ &= \tfrac{1}{2} e^{2\sigma_0 t}\,\mathrm{Re}\,[V_0^* Y(s_0)V_0 + V_0' Y(s_0)V_0 e^{j2\omega_0 t}]. \qquad (1.117)\end{aligned}$$

To facilitate our discussion, two cases are distinguished.

Case 1. $\sigma_0 = 0$. Depending upon ω_0, two subcases arise. *Subcase 1.* $\omega_0 = 0$. Then the passivity condition (1.7) becomes

$$\begin{aligned}\mathscr{E}(t) &= \tfrac{1}{2}\mathrm{Re}\,[V_0^* Y(s_0)V_0 + V_0' Y(s_0)V_0](t - t_0) + \mathscr{E}(t_0) \\ &= V_0' Y(s_0)V_0(t - t_0) + \mathscr{E}(t_0) \geqq 0. \qquad (1.118)\end{aligned}$$

The second line follows from the fact that for $s_0 = 0$, V_0 and $Y(s_0)$ are real. For the above inequality to hold for all $t \geqq t_0$, it is necessary that

$$V_0' Y(s_0) V_0 \geqq 0 \qquad (1.119)$$

for all V_0. *Subcase 2.* $\omega_0 \neq 0$. Then the passivity condition (1.7) becomes

$$\mathscr{E}(t) = \mathscr{E}(t_0) + \tfrac{1}{2}\mathrm{Re}\,V_0^* Y(j\omega_0)V_0(t - t_0) + \frac{|V_0' Y(j\omega_0)V_0|}{4|\omega_0|}$$

$$\times \cos(2\omega_0 t + \underline{/V_0' Y(j\omega_0)V_0} \pm \pi/2) + C_1 \geqq 0, \qquad (1.120)$$

the constant C_1 is introduced so that, for $t = t_0$, $\mathscr{E}(t) = \mathscr{E}(t_0)$. Since the second term increases linearly without bound, ultimately dominating all other terms, for the inequality to hold for all $t \geqq t_0$, it is necessary that

$$\mathrm{Re}\,V_0^* Y(j\omega_0) V_0 \geqq 0 \qquad (1.121)$$

for all complex V_0.

Case 2. $\sigma_0 \neq 0$. Substituting (1.117) in (1.7) yields the passivity condition

$$\mathscr{E}(t) = \mathscr{E}(t_0) + \frac{1}{4}e^{2\sigma_0 t}\left[\frac{1}{\sigma_0}\text{Re } V_0^* Y(s_0) V_0 + \frac{1}{|s_0|}|V_0' Y(s_0) V_0|\right.$$
$$\left. \times \cos(2\omega_0 t + \underline{/V_0' Y(s_0) V_0} - \underline{/s_0})\right] + C_2 \geqq 0, \quad (1.122)$$

where as before the constant C_2 is introduced so that for $t = t_0$, $\mathscr{E}(t) = \mathscr{E}(t_0)$. Since the terms containing $\exp(2\sigma_0 t)$ grow exponentially without bound as t is increased, ultimately dominating the other terms, for (1.122) to hold for all $t \geqq t_0$, it is necessary that

$$\text{Re } V_0^* Y(s_0) V_0 + \frac{\sigma_0}{|s_0|}|V_0' Y(s_0) V_0| \cos(2\omega_0 t + \underline{/V_0' Y(s_0) V_0} - \underline{/s_0}) \geqq 0$$
$$(1.123)$$

for sufficiently large t. Even so, if $\omega_0 \neq 0$ the cosine function can assume the value -1 for some t, necessitating that

$$\text{Re } V_0^* Y(s_0) V_0 - \frac{\sigma_0}{|s_0|}|V_0' Y(s_0) V_0| \geqq 0. \quad (1.124)$$

For $\omega_0 = 0$, (1.123) becomes

$$\text{Re } V_0^* Y(s_0) V_0 \geqq 0, \quad (1.125)$$

since in this case V_0 is real. In fact, (1.125) is contained in (1.124), because for $\omega_0 = 0$, (1.124) can be written as

$$\text{Re } V_0^* Y(s_0) V_0 = V_0' Y(\sigma_0) V_0 \geqq |V_0' Y(\sigma_0) V_0|, \quad (1.126)$$

which is satisfied if and only if (1.125) holds.

Using the symbols defined in (1.77) and (1.78), the passivity conditions (1.119), (1.121), (1.124) and (1.125) can be summarized and rewritten by the single relation

$$V_0^* Y_h(s_0) V_0 - \frac{\sigma_0}{|s_0|}|V_0' Y_s(s_0) V_0| \geqq 0, \quad \text{Re } s_0 \geqq 0 \quad (1.127)$$

for all complex n-vectors V_0, the second term being zero for $s_0 = 0$. Thus, the hermitian part of the admittance matrix of a linear, passive and time-invariant n-port network is nonnegative-definite for all the complex frequencies in the closed RHS. This together with the facts that we deal exclusively with n-ports of real elements and that the admittance matrix is analytic in the open RHS, being devoid of poles in the open RHS from stability considerations, shows that the admittance matrix of such an n-port network must be positive-real. Proceeding as in the foregoing, we can show that the impedance matrix and the hybrid matrix of such an n-port are positive-real.

To be explicit, we state the following characterization for passivity.

THEOREM 1.5. *A necessary and sufficient condition for a linear, time-invariant n-port network possessing an admittance, impedance, or hybrid matrix to be passive is that the matrix be positive-real.*

If, in addition, we stipulate that the n-port network be lumped, then the positive-real condition is also sufficient for its realization.

THEOREM 1.6. *An $n \times n$ rational matrix is the admittance, impedance or hybrid matrix of a linear, time-invariant, passive and lumped n-port network if and only if it is positive-real.*

Sufficiency of this theorem is stated here for completeness; its proof amounts to constructing an n-port network having the prescribed positive-real matrix. For a detailed account of these techniques and all of their variations and ramifications, the reader is referred to Newcomb (1966), Belevitch (1968) and Anderson and Vongpanitlerd (1973). In general, for nonsymmetric positive-real matrix, the corresponding network realization is nonreciprocal, which employs gyrators. For a symmetric positive-real matrix, it can be shown that there exists a reciprocal and gyratorless n-port. We shall not discuss this aspect of the subject any further, since it would take us far afield into network synthesis.

EXAMPLE 1.8. Consider the bridged-Tee network of Fig. 1.13, whose impedance matrix is given by

FOUNDATIONS OF NETWORK THEORY

$$Z(s) = \frac{1}{4s(s+2)} \begin{bmatrix} s^2 + 6s + 4 & s^2 + 2s + 4 \\ s^2 + 2s + 4 & s^2 + 6s + 4 \end{bmatrix}. \quad (1.128)$$

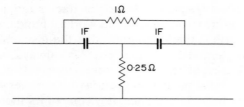

FIG. 1.13. A bridged-T network.

Expanding $Z(s)$ by partial fraction expansion yields

$$Z(s) = \begin{bmatrix} \frac{1}{4} & \frac{1}{4} \\ \frac{1}{4} & \frac{1}{4} \end{bmatrix} + \frac{1}{s} \begin{bmatrix} \frac{1}{2} & \frac{1}{2} \\ \frac{1}{2} & \frac{1}{2} \end{bmatrix} + \frac{1}{s+2} \begin{bmatrix} \frac{1}{2} & -\frac{1}{2} \\ -\frac{1}{2} & \frac{1}{2} \end{bmatrix}$$

$$= K_\infty + \frac{1}{s} K_0 + \frac{1}{s+2} K_{-2}, \quad (1.129)$$

where K_0 and K_{-2} are the residue matrices evaluated at the poles at $s = 0$ and $s = -2$, respectively. Clearly, K_0 and K_{-2} are symmetric, hence hermitian, and nonnegative-definite. On the real-frequency axis, the hermitian part of $Z(s)$ becomes

$$Z_h(j\omega) = \frac{1}{4(\omega^2 + 4)} \begin{bmatrix} \omega^2 + 8 & \omega^2 \\ \omega^2 & \omega^2 + 8 \end{bmatrix}. \quad (1.130)$$

Appealing to Theorem 1.2, it is easy to confirm that this matrix is nonnegative-definite for all real ω. Thus, according to Theorems 1.3 and 1.5, the two-port network of Fig. 1.13 must be passive.

7. Conclusions

We began the chapter by introducing six basic postulates describing the physical nature of a network: reality, time-invariance, linearity, passivity, causality and reciprocity. In this way, we can make the theory as simple and as powerful as possible. The passivity is defined in terms of the universally encountered physical quantities *time* and *energy*, which is then translated into the equivalent frequency-domain passivity criteria. To this end, we reviewed briefly the general frequency-domain characterizations of the class of linear time-invariant n-port networks. Fundamental to the concepts of ports is the assumption that the instantaneous current entering one terminal of the port is always equal to the instantaneous current leaving the other terminal of the port. We recognize that upon the interconnection of n-port networks, the port constraint may be violated. Thus, it is sometimes more desirable to consider n-terminal networks. For this we introduced the concepts of the indefinite-admittance matrix and its relations to the admittance matrix.

For the two-port networks, we defined several measures of power flow: power gain, available power gain and transducer power gain. We found that the transducer power gain is the most meaningful description of power transfer capabilities of a two-port network as it compares the power delivered to the load with the power which the source is capable of supplying under the optimum power-matching conditions. For computational purposes, we derived formulas for these gains in terms of the open-circuit impedance parameters, the short-circuit admittance parameters and the load and source impedances.

We then proceeded to show that a linear, time-invariant n-port network possessing an admittance matrix, impedance matrix, or hybrid matrix is passive if and only if these matrices are positive-real. The positive-real matrix, which is based on the concept of definiteness of the hermitian form of a matrix, plays a very important role in characterizing the passivity criterion of an n-port network. For this reason, we digressed into a discussion of the hermitian forms and the positive-real matrices, and presented some of their fundamental properties, which are needed in the subsequent analysis. The significance of passivity in our study of active networks is that activity is the formal negation of passivity.

After these preliminaries, we now proceed to discuss the scattering matrix and its use in the design of matching networks in the following chapters.

Problems

1.1. Show that the transmission matrix of two two-port networks connected in cascade is equal to the product of the transmission matrices of the individual two-port networks.

1.2. Show that if a two-port network is reciprocal, then the determinant of its transmission matrix has value 1.

1.3. Show that the hermitian form of the matrix (1.72) can be expressed as in (1.73).

1.4. Without invoking Theorem 1.2, show that the value of the determinant of a positive-definite matrix is always positive.

1.5. Justify the assertion that the inequality (1.89) holds for $|\theta| < \pi/2$ only if $m = 1$ and $\phi = 0$.

1.6. Show that the coefficient matrix K defined in (1.86) is hermitian and nonnegative-definite if $A(s)$ is positive-real.

1.7. For a square real matrix A, show that there exists a real symmetric matrix A_s such that

$$X'AX = X'A_s X. \quad (1.131)$$

1.8. Show that if the matrix $A(s)$ is positive-real, so is $A(1/s)$.

1.9. An $n \times n$ hermitian matrix A is called a *negative-definite matrix* if

$$X^*AX < 0 \quad (1.132)$$

for all complex n-vectors $X \neq 0$. Using this definition, show that A is negative-definite if and only if all the leading principal minors of odd orders are negative and all the leading principal minors of even orders are positive. (*Hint.* Apply Theorem 1.2.)

1.10. An $n \times n$ hermitian matrix A is called a *non-positive definite matrix* if

$$X^*AX \leq 0 \quad (1.133)$$

for all complex n-vectors X. Show that A is non-positive definite if and only if all the principal minors of odd orders are non-positive and all the principal minors of even orders are nonnegative. Very often, a non-positive definite matrix is also referred to as a *negative-semidefinite matrix*. Like positive semidefiniteness, care must be taken to assure the proper interpretation of the terms since many people define a negative-semidefinite matrix as one that satisfies (1.133) for all X provided there is at least one $X \neq 0$ for which the equality holds. [*Hint.* Apply Theorem 1.2.]

1.11. Test the following function to see if it is positive-real:

$$f(s) = \frac{2s^4 + 7s^3 + 11s^2 + 12s + 4}{s^4 + 5s^3 + 9s^2 + 11s + 6}. \quad (1.134)$$

1.12. A nonlinear time-invariant resistor is characterized by the relation

$$v(t) = i^2(t). \quad (1.135)$$

Show that this resistor is active.

1.13. The matrix

$$Y(s) = \frac{1}{q(s)} \begin{bmatrix} 4s^2 + 12s + 1 & -1 \\ -1 & 12s^2 + 8s + 1 \end{bmatrix}, \quad (1.136a)$$

where

$$q(s) = 12s^3 + 44s^2 + 28s + 5, \quad (1.136b)$$

is known as the admittance matrix of a passive two-port network. Show that it is a positive-real matrix.

1.14. Show that a two-port network composed of a pair of wires is not causal under either current excitation–voltage response or voltage excitation–current response.

1.15. Test the following matrices to see if they are positive-real:

(i) $$A(s) = \frac{1}{46s + 1} \begin{bmatrix} 4s + 2 & 4s \\ 4s - 20 & 4s + 1 \end{bmatrix}. \quad (1.137)$$

(ii) $$A(s) = \frac{1}{2s(s^2 + 1)} \begin{bmatrix} 2s^2 + 1 & 1 \\ 1 & 2s^2 + 1 \end{bmatrix}. \quad (1.138)$$

(iii) $$A(s) = \frac{1}{q(s)} \begin{bmatrix} 2s^3 + 2s^2 + 2s + 1 & -1 \\ -1 & 8s^3 + 2s^2 + 8s + 1 \end{bmatrix}, \quad (1.139a)$$

where

$$q(s) = 8s^3 + 10s^2 + 10s + 5. \quad (1.139b)$$

1.16. Assume that $X_0^* A_h(s) X_0 = 0$ for all s in the open RHS and for some fixed X_0. Show that $A(s)$ is identically singular if it is singular at s_0, Re $s_0 > 0$.

1.17. Using Definition 1.3, show that the matrix

$$A = \begin{bmatrix} 1 & 1 & 1 \\ 1 & 1 & 1 \\ 1 & 1 & 0 \end{bmatrix}, \quad (1.140)$$

whose leading principal minors are all nonnegative, is neither positive-definite nor nonnegative-definite.

1.18. Show that a hermitian matrix A is positive-definite if and only if any one of the following conditions is satisfied:
   (i) B^*AB is positive-definite for arbitrary nonsingular matrix B.
   (ii) A^n is positive-definite for every integer n.
   (iii) There exists a nonsingular matrix B such that $A = B^*B$.

1.19. Give a physical interpretation for the transmission parameters.

References

1. Anderson, B. D. O. and Vongpanitlerd, S. (1973) *Network Analysis and Synthesis: A Modern Systems Theory Approach*, Englewood Cliffs, N.J.: Prentice-Hall.

2. Belevitch, V. (1968) *Classical Network Theory*, San Francisco, Calif.: Holden-Day.
3. Bolinder, E. F. (1957) Survey of some properties of linear networks. *IRE Trans. Circuit Theory*, vol. CT-4, no. 3, pp. 70–78.
4. Chen, W. K. (1976) *Applied Graph Theory: Graphs and Electrical Networks*, Amsterdam, The Netherlands: North-Holland, 2nd edn.
5. Chen, W. K. (1972) On equicofactor and indefinite-admittance matrices. *Matrix Tensor Quart.*, vol. 23, no. 1, pp. 26–28.
6. Kuh, E. S. and Rohrer, R. A. (1967) *Theory of Linear Active Networks*, San Francisco, Calif.: Holden-Day.
7. Kuo, Y. L. (1968) A note on the n-port passivity criterion. *IEEE Trans. Circuit Theory*, vol. CT-15, no. 1, p. 74.
8. Newcomb, R. W. (1962) On causality, passivity and single-valuedness. *IRE Trans. Circuit Theory*, vol. CT-9, no. 1, pp. 87–89.
9. Newcomb, R. W. (1966) *Linear Multiport Synthesis*, New York: McGraw-Hill.
10. Raisbeck, G. (1954) A definition of passive linear networks in terms of time and energy. *J. Appl. Phys.*, vol. 25, no. 12, pp. 1510–1514.
11. Resh, J. A. (1966) A note concerning the n-port passivity condition. *IEEE Trans. Circuit Theory*, vol. CT-13, no. 2, pp. 238–239.
12. Rohrer, R. A. (1968) Lumped network passivity criteria. *IEEE Trans. Circuit Theory*, vol. CT-15, no. 1, pp. 24–30.
13. Youla, D. C. (1960) Physical realizability criteria. *IRE Trans. Circuit Theory*, vol. CT-7, Special Supplement, pp. 50–68.

CHAPTER 2

The Scattering Matrix

IN THE preceding chapter, we have indicated that the terminal behavior of a linear n-port network can be characterized by any one of the various sets of parameters such as the short-circuit admittance parameters, the open-circuit impedance parameters, the hybrid parameters, the transmission parameters, etc. However, not all of these parameters will exist. For example, a two-port network consisting only of two wires with a finite impedance connecting across them has no short-circuit admittance matrix but has the open-circuit impedance matrix. If the finite impedance is removed from the two-port network, the resulting two-port possesses neither the impedance matrix, the admittance matrix, the hybrid matrix, nor the transmission matrix. An ideal transformer possesses the hybrid matrix but neither the impedance nor the admittance matrix. The reason for this is that these parameters are defined in terms of the quantities that are obtained when one of the ports is short-circuited or open-circuited. In other words, they are defined with respect to the zero or infinite loading at the ports. The scattering parameters, on the other hand, are defined in terms of some finite stable loadings at the ports. Thus, they always exist for all nonpathological linear passive time-invariant networks.

The scattering parameters originated in the theory of transmission lines. They are defined in such a way that the various quantities of interest in power transmission have very simple expressions in terms of them. Thus, they are indispensable in the design of microwave networks. In fact, in microwave networks the concept of power is much more important than the concepts of voltage and current. This fact also makes the scattering parameters very useful in the design of power transmission networks.

THE SCATTERING MATRIX

In the present chapter, we shall discuss some of the fundamental properties of the scattering parameters associated with an n-port network, and show how they are closely related to the power transmissions among its ports. To motivate the discussion, we shall start from one-port network and use the concepts from transmission-line theory.

1. A brief review of the transmission-line theory

Consider a uniform lossless transmission line with a characteristic impedance z_0, as shown in Fig. 2.1. The line is connected between a

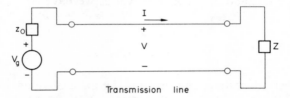

FIG. 2.1. A uniform lossless transmission line with a characteristic impedance z_0.

load impedance Z and a sinusoidal voltage source V_g having internal impedance z_0. The total voltage V or current I along the line may be regarded as the sum of voltages or currents in an incident or positively traveling wave and in a reflected or negatively traveling wave. The incident voltage V_i and current I_i are defined as those that would appear at the transmission line when it is terminated in its characteristic impedance z_0. Then

$$z_0 = V_i/I_i, \qquad (2.1a)$$

$$V_i = \tfrac{1}{2} V_g. \qquad (2.1b)$$

The reflected voltage V_r and current I_r are defined by the relations

$$V_r = V - V_i, \qquad (2.2a)$$

$$-I_r = I - I_i. \qquad (2.2b)$$

The negative sign associated with the reflected current I_r indicates that the positive direction for I_r is opposite to that for I_i. The ratio of the reflected voltage to the incident voltage is called the *voltage reflection coefficient* ρ^V. Likewise, the ratio of the reflected current to the incident current is the *current reflection coefficient* ρ^I. By eliminating unwanted variables in (2.1) and (2.2), these coefficients can easily be found, and are given by

$$\rho^I = (Z + z_0)^{-1}(Z - z_0), \qquad (2.3a)$$

$$\rho^V = -(Y + y_0)^{-1}(Y - y_0), \qquad (2.3b)$$

where $y_0 = 1/z_0$ and $Y = 1/Z$. For real z_0 we have

$$\rho^I = \rho^V, \qquad (2.4)$$

$$z_0 = V_r/I_r. \qquad (2.5)$$

The most interesting, and perhaps the most obvious, conclusion from the above relations is that there is no reflected voltage or current wave if the terminating impedance is exactly equal to the characteristic impedance of the line. All energy of the incident wave is then transferred to the load impedance, which cannot be distinguished from a line of infinite length and of characteristic impedance $z_0 = Z$.

2. The scattering parameters of a one-port network

Consider a one-port network N of Fig. 2.2 which is characterized by its driving-point impedance $Z(s)$. The one-port is loaded by a

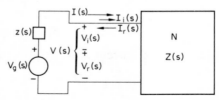

Fig. 2.2. A one-port network characterized by its impedance $Z(s)$.

voltage source $V_g(s)$ in series with an impedance $z(s)$, which may be considered as the Thévenin equivalent of another network. The one-port of Fig. 2.2 can also be viewed as a transmission line of infinitesimal length and of characteristic impedance $z(s)$ as depicted in Fig. 2.1. The amount of power that can be transferred from the source $V_g(s)$ to the one-port N depends on the impedance $Z(s)$. It is well known (Problem 2.1) that, on the real-frequency axis, the maximum power that a one-port can absorb is obtained when $Z(j\omega) = \bar{z}(j\omega)$, the complex conjugate of $z(j\omega)$. Under this situation, we say that the one-port N is *conjugately matched* to the load. We recognize that, with the exception that $z(s)$ is a frequency-independent constant, the conjugate impedance $\bar{z}(j\omega)$ cannot be realized by a passive network for all frequencies. Thus, for a complex $z(s)$ this maximum power transfer is attained only at a single sinusoidal frequency, which may be at any point on the $j\omega$-axis. To extend this optimal power-matching condition to the entire complex frequency plane, which assures the maximum energy absorption by the one-port under arbitrary excitation (transient or steady-state), we appeal to the theory of analytic continuation. Since the analytic continuation of the function $\bar{z}(j\omega) = z(-j\omega)$ is $z(-s)$, in the notation

$$z_*(s) = z(-s), \qquad (2.6)$$

the optimal matching one-port network is characterized by the complex impedance $z_*(s)$ for all s. In other words, the maximum energy absorption by the one-port under arbitrary excitation is attained when $Z(s) = z_*(s)$.

Define the function

$$r(s) = \tfrac{1}{2}[z(s) + z_*(s)] \qquad (2.7)$$

as the *para-hermitian part* of $z(s)$, and assume that it is not identically zero. This is true as long as $z(s)$ does not represent a lossless network. The assumption is necessary since, as will be shown shortly, the scattering parameter provides a quantitative measure of the deviation of the actual power-matching behavior of the network from the optimal power-matched condition, which cannot be meaningfully defined if $r(s) = 0$. On the real-frequency axis, the para-hermitian part $r(j\omega)$ of $z(j\omega)$ becomes its real part.

Using the concepts of transmission-line theory discussed in the foregoing, we now proceed to define the scattering parameters of a one-port, which are also known as the reflection coefficients of the one-port because of the attaching intuitive wave-propagation interpretation of various quantities.

2.1. Basis-dependent reflection coefficients

As in the transmission-line theory, the actual terminal voltage $V(s)$ of the one-port network N of Fig. 2.2 is regarded as the sum of an *incident voltage* $V_i(s)$ and a *reflected voltage* $V_r(s)$, and the actual terminal current $I(s)$ as the sum of an *incident current* $I_i(s)$ and a *reflected current* $I_r(s)$. The incident voltage $V_i(s)$ and the incident current $I_i(s)$ are defined as those that would appear at the terminal of N under the optimal power-matching condition, as depicted in Fig. 2.3. Thus, they are completely independent of the one-port

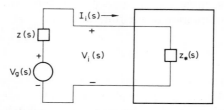

FIG. 2.3. The optimal power-matching situation for a one-port network.

itself. From Fig. 2.3, we obtain

$$V_i(s) = \tfrac{1}{2} r^{-1}(s) z_*(s) V_g(s), \qquad (2.8a)$$

$$I_i(s) = \tfrac{1}{2} r^{-1}(s) V_g(s), \qquad (2.8b)$$

$$V_i(s) = z_*(s) I_i(s). \qquad (2.8c)$$

Observe that the incident voltage and current cannot be meaningfully defined if $r(s)$ is identically zero. As in (2.2), the reflected voltage $V_r(s)$ and current $I_r(s)$ are defined by the relations

$$V_r(s) = V(s) - V_i(s), \qquad (2.9a)$$

$$-I_r(s) = I(s) - I_i(s). \qquad (2.9b)$$

Again, the negative sign associated with $I_r(s)$ indicates that the positive direction for $I_r(s)$ is chosen to be opposite to that for $I_i(s)$, for convenience. The incident and reflected quantities are depicted symbolically in Fig. 2.2. Finally, the *voltage* and *current reflection coefficients* $S^V(s)$ and $S^I(s)$ are defined according to the relations

$$V_r(s) = S^V(s) V_i(s), \qquad (2.10a)$$

$$I_r(s) = S^I(s) I_i(s). \qquad (2.10b)$$

The reflection coefficients $S^V(s)$ and $S^I(s)$ are also referred to as the *voltage-based* and *current-based scattering parameters*, respectively, of the one-port network N. The impedance $z(s)$ from which the incident voltage and current are defined is called the *reference impedance* of N.

From (2.8b) and (2.9b), we can easily derive the expression for $S^I(s)$ in terms of $Z(s)$ and $z(s)$, yielding

$$S^I(s) = I_r(s)/I_i(s) = 1 - I(s)/I_i(s) = 1 - 2r(s)/[Z(s) + z(s)]$$
$$= [Z(s) + z(s)]^{-1}[Z(s) - z_*(s)]. \qquad (2.11)$$

In a similar manner, we can show that

$$S^V(s) = -[Y(s) + y(s)]^{-1}[Y(s) - y_*(s)], \qquad (2.12)$$

where $y(s) = z^{-1}(s)$ and $Y(s) = Z^{-1}(s)$. Combining (2.11) and (2.12) gives

$$S^V(s) z_*(s) = z(s) S^I(s), \qquad (2.13)$$

showing that the voltage and current reflection coefficients are in general different. However, on the real-frequency axis, they differ only by a phase, which is equal to twice the angle of the reference impedance $z(s)$. We remark that under the optimal power-matching condition, $Z(s) = z_*(s)$, both reflection coefficients are zero, indicating that the reflected voltage or current is a measure of the deviation of the one-port terminal voltage or current, when under actual operation, from its value when optimally matched.

It is interesting to note that while the incident voltage and current are related by $z_*(s)$, the reflected voltage and current are related by the reference impedance $z(s)$ itself:

$$V_r(s) = S^V(s)V_i(s) = S^V(s)z_*(s)I_i(s) = z(s)S^I(s)I_i(s) = z(s)I_r(s). \tag{2.14}$$

Intuitively, this means that the incident waves see the impedance $z_*(s)$, as required by the optimal power match, and that the reflected waves see the reference impedance $z(s)$ itself. We emphasize that the transmission-line concepts that were used in the preceding discussion are artificial; all of the above results can be derived without attaching any interpretive significance to the quantities that reflect their intuitive origin by simply regarding (2.8a), (2.8b) and (2.9) as formal definitions of the variables $V_i(s)$, $I_i(s)$, $V_r(s)$ and $I_r(s)$.

2.2. Basis-independent reflection coefficient

In the previous section, we have defined two reflection coefficients. One is based on voltage and the other on current. The dual representation can be eliminated by introducing a normalization for which the normalized reflection coefficient becomes basis-independent.

Consider the para-hermitian part $r(s)$ of a real rational impedance $z(s)$. It is easy to see that $r(s)$ is an even function, being the ratio of two even polynomials. This means that the poles and zeros of $r(s)$ will appear in quadrantal symmetry with respect to both the real and imaginary axes. Thus, $r(s)$ can be expressed in factored form as

$$r(s) = h(s)h_*(s), \tag{2.15}$$

whose factorization will be discussed in detail in the following section.

THE SCATTERING MATRIX

The basis-independent *normalized incident wave* and *reflected wave* are defined by the relations

$$a(s) = h_*(s)I_i(s), \qquad (2.16a)$$

$$b(s) = h(s)I_r(s), \qquad (2.16b)$$

respectively. As in (2.10), the basis-independent *normalized reflection coefficient* $S(s)$, which is also called the *normalized scattering parameter*, is introduced by the equation

$$b(s) = S(s)a(s). \qquad (2.16c)$$

From (2.8c) and (2.14), we obtain

$$a(s) = h_*(s)I_i(s) = h_*(s)z_*^{-1}(s)V_i(s), \qquad (2.17a)$$

$$b(s) = h(s)I_r(s) = h(s)z^{-1}(s)V_r(s). \qquad (2.17b)$$

Combining (2.16) with (2.10b) yields

$$S(s) = h(s)S^I(s)h_*^{-1}(s). \qquad (2.18)$$

Since the scattering parameter $S(s)$ completely characterizes the one-port network N, the actual port voltage, current and the driving-point impedance can easily be expressed in terms of the normalized waves and the reflection coefficient by the equations

$$V(s) = V_i(s) + V_r(s) = z_*(s)h_*^{-1}(s)a(s) + z(s)h^{-1}(s)b(s), \qquad (2.19a)$$

$$I(s) = I_i(s) - I_r(s) = h_*^{-1}(s)a(s) - h^{-1}(s)b(s), \qquad (2.19b)$$

$$Z(s) = V(s)/I(s)$$
$$= [h(s) - h_*(s)S(s)]^{-1}[z_*(s)h(s) + z(s)h_*(s)S(s)]. \qquad (2.19c)$$

Solving for the normalized waves and the reflection coefficient in terms of the port voltage, current and the impedances gives

$$a(s) = \tfrac{1}{2}h^{-1}(s)[V(s) + z(s)I(s)], \qquad (2.20a)$$

$$b(s) = \tfrac{1}{2}h_*^{-1}(s)[V(s) - z_*(s)I(s)], \qquad (2.20b)$$

$$S(s) = h(s)h_*^{-1}(s) - 2h(s)[Z(s) + z(s)]^{-1}h(s). \qquad (2.20c)$$

As mentioned at the end of §2.1, much of the foregoing manipulation could have been avoided by simply regarding (2.20a) and (2.20b) as formal definitions of the variables $a(s)$ and $b(s)$. This approach is entirely consistent with the above exposition, yet it does not depend on the existence of the impedance characterization of the given one-port network N.

We illustrate the above results by the following example.

EXAMPLE 2.1. Consider the one-port network N of Fig. 2.4, which is loaded by another one-port network represented by its Thévenin equivalent. Choose $z = 4 \, \Omega$ as the reference impedance. Then from (2.8a) and (2.8b), the incident voltage and current are computed as

$$V_i(s) = \tfrac{1}{2} r^{-1}(s) z_*(s) V_g(s) = \tfrac{1}{2} \times 4^{-1} \times 4 V_g(s) = \tfrac{1}{2} V_g(s), \quad (2.21a)$$

$$I_i(s) = \tfrac{1}{2} r^{-1}(s) V_g(s) = \tfrac{1}{2} \times 4^{-1} V_g(s) = V_g(s)/8. \quad (2.21b)$$

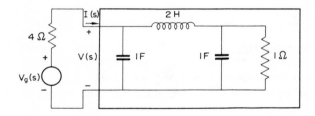

FIG. 2.4. A one-port network loaded by a resistive generator.

The normalized reflection coefficient is obtained through (2.18) in conjunction with (2.11), and is given by

$$S(s) = \frac{h(s)}{h(-s)} \cdot \frac{Z(s) - z(-s)}{Z(s) + z(s)} = \frac{8s^3 + 6s^2 + 6s + 3}{8s^3 + 10s^2 + 10s + 5}, \quad (2.22a)$$

where $h(s) = 2$ and

$$Z(s) = \frac{2s^2 + 2s + 1}{2s^3 + 2s^2 + 2s + 1}. \quad (2.22b)$$

THE SCATTERING MATRIX

Observe that $S(s)$ is analytic in the closed RHS. To compute the port voltage and current, we use (2.19a) and (2.19b) which can be simplified to

$$V(s) = 2[1 + S(s)]a(s) = \frac{(2s^2 + 2s + 1)V_g(s)}{8s^3 + 10s^2 + 10s + 5}, \qquad (2.23a)$$

$$I(s) = \tfrac{1}{2}[1 - S(s)]a(s) = \frac{(2s^3 + 2s^2 + 2s + 1)V_g(s)}{8s^3 + 10s^2 + 10s + 5}, \qquad (2.23b)$$

where $a(s) = h_*(s)I_i(s) = \tfrac{1}{4}V_g(s)$.

2.3. The factorization of the para-hermitian part of $z(s)$

In (2.15) we have indicated that the para-hermitian part of the reference impedance $z(s)$ can be expressed as the product of two functions $h(s)$ and $h_*(s)$. Clearly, this factorization is not unique. For example, consider the reference impedance

$$z(s) = \frac{s+4}{2(s+1)}, \qquad (2.24)$$

whose para-hermitian part can be factored in the following four different ways:

$$r(s) = \tfrac{1}{2}[z(s) + z_*(s)] = \frac{(s+2)(s-2)}{2(s+1)(s-1)}$$

$$= \left(\frac{s+2}{\sqrt{2}(s+1)}\right)\left(\frac{s-2}{\sqrt{2}(s-1)}\right) = \left(\frac{s-2}{\sqrt{2}(s+1)}\right)\left(\frac{s+2}{\sqrt{2}(s-1)}\right)$$

$$= \left(\frac{s+2}{\sqrt{2}(s-1)}\right)\left(\frac{s-2}{\sqrt{2}(s+1)}\right) = \left(\frac{s-2}{\sqrt{2}(s-1)}\right)\left(\frac{s+2}{\sqrt{2}(s+1)}\right). \qquad (2.25)$$

At this point, it is natural to ask "Which factorization shall we select for the normalization of the reflection coefficient?" To answer this question, we recall that the passivity and activity of a one-port network are defined in terms of the behavior of its immittance functions in the closed RHS, since no useful information can be obtained in this manner regarding the behavior of its immittance functions in the open LHS. Thus, our criterion in selecting the factorization is to make the basis-independent normalized reflection coefficient analytic in the open RHS for passive one-port networks, which would enable us to extend the passivity conditions on the real-frequency axis to the entire RHS.

To this end, we write explicitly the rational impedance $z(s)$ as the ratio of two polynomials:

$$z(s) = \frac{p(s)}{q(s)}. \tag{2.26}$$

Because the impedance $z(s)$, by assumption, is strictly passive, both $p(s)$ and $q(s)$ are Hurwitz polynomials. The para-hermitian part $r(s)$ becomes

$$r(s) = \frac{p(s)q_*(s) + p_*(s)q(s)}{2q(s)q_*(s)} = h(s)h_*(s), \tag{2.27}$$

showing that $r(s)$ is the ratio of two even polynomials. Thus, the zeros and poles of $r(s)$ must appear in quadrantal symmetry, i.e. they are symmetric with respect to both the real and the imaginary axes. Since $z(s)$ is positive real, $r(s)$ has no poles on the real-frequency axis. For, if there were such poles, the partial fraction expansion of $r(s)$ would contain an odd term

$$\frac{K}{s + j\omega_0} + \frac{K}{s - j\omega_0} = \frac{2Ks}{s^2 + \omega_0^2} \tag{2.28}$$

and $r(s)$ would be odd, which is contrary to the fact that $r(s)$ is even. On the real-frequency axis, for $z(s)$ to be positive real it is necessary that Re $z(j\omega) = r(j\omega) \geqq 0$ for all ω. As stated in Corollary 1.2 and the subsequent remark, this is satisfied if and only if $r(j\omega)$ has no real positive zeros of odd multiplicity in ω, or equivalently the $j\omega$-axis zeros of $r(s)$ must be of even multiplicity.

Observe from (2.20c) that, for a passive one-port network, its basis-independent normalized reflection coefficient $S(s)$ will be analytic in the open RHS if $h(s)$ and $h_*^{-1}(s)$ are both made analytic in the open RHS, since $Z(s) + z(s)$ together with its reciprocal is a positive-real function, thus being analytic in the open RHS. For $h(s)$

and $h_*^{-1}(s)$ to be analytic in the open RHS, it is necessary and sufficient that the zeros of $h(s)$ be restricted to the closed RHS and the poles of $h(s)$ to the open LHS, since $r(s)$ is devoid of any $j\omega$-axis poles. Thus, the poles and zeros of $r(s)$ can be uniquely distributed between $h(s)$ and $h_*(s)$ as follows: the open LHS poles of $r(s)$ belong to $h(s)$ whereas those in the open RHS belong to $h_*(s)$; the open RHS zeros of $r(s)$ belong to $h(s)$ whereas those in the open LHS belong to $h_*(s)$. The $j\omega$-axis zeros of $r(s)$, being of even multiplicity, are divided equally between $h(s)$ and $h_*(s)$. For example, in (2.25) the only permissible factorization is given by

$$r(s) = \left(\frac{s-2}{\sqrt{2}(s+1)}\right)\left(\frac{s+2}{\sqrt{2}(s-1)}\right) = h(s)h_*(s) \qquad (2.29)$$

with $h(s)$ being identified as $(s-2)/[\sqrt{2}(s+1)]$.

We remark that, although the distribution of the poles and zeros of $r(s)$ is unique, the decomposition of the para-hermitian part $r(s)$ of $z(s)$ into the factors $h(s)$ and $h_*(s)$ is not. In fact, if "surplus factors" are permitted we can have infinitively many decompositions with the property that both $h(s)$ and $h_*^{-1}(s)$ are analytic in the open RHS. It is unique only if no additional surplus factors are allowed. To see this, we consider a real all-pass function defined by the equation

$$\eta(s) = \pm \prod_{x=1}^{m} \left\{\frac{s-s_x}{s+s_x}\right\}, \qquad \operatorname{Re} s_x > 0, \qquad (2.30)$$

where the s_x ($x = 1, 2, \ldots, m$) occur in complex conjugate pairs if they are complex. The all-pass function $\eta(s)$ is analytic in the closed RHS and such that

$$\eta(s)\eta_*(s) = 1. \qquad (2.31)$$

Thus, if $\hat{h}(s)$ is a solution of (2.27) obtained by the unique distribution of the poles and zeros of $r(s)$ without using any additional surplus factors, then the product

$$h(s) = \eta(s)\hat{h}(s) \qquad (2.32)$$

is also a solution. In other words, for a given reference impedance $z(s)$, the decomposition of its para-hermitian part $r(s)$ into the factors $h(s)$ and $h_*(s)$ is uniquely determined once and for all, in accordance with the following three requirements:

(i) $h(s)$ is analytic in the open RHS;
(ii) $h_*(s)$ is analytic in the open LHS;
(iii) $h(s)$ is the ratio of two polynomials of minimal degree;

The choice that $h(s)$ and $h_*^{-1}(s)$ be analytic in the open RHS is not the only one available that will make the normalized reflection coefficient $S(s)$ analytic in the open RHS. In the following, we demonstrate that a different approach can lead essentially to the same result.

From (2.20c), observe that if $h(s)$ and

$$d(s) = h(s)h_*^{-1}(s) \qquad (2.33)$$

are analytic in the open RHS, $S(s)$ will be analytic in the open RHS. To this end, let $\hat{h}(s)$ be a solution of (2.27) obtained by assigning all the open LHS poles of $r(s)$ to $h(s)$, and all the open LHS zeros of $r(s)$ to $h(s)$ plus half of the $j\omega$-axis zeros of $r(s)$. Thus, $h(s)$ is the ratio of two Hurwitz polynomials. Then, as before, the most general solution of (2.27) is given by

$$h(s) = \eta(s)\hat{h}(s), \qquad (2.34)$$

which, with the proper choice of $\eta(s)$, results essentially in the same decomposition as in the previous case. To see this, we substitute (2.34) in (2.33), yielding

$$d(s) = \eta^2(s)\hat{h}(s)\hat{h}_*^{-1}(s). \qquad (2.35)$$

We recognize that while $\hat{h}_*^{-1}(s)$ may not be analytic in the open RHS, we can choose zeros of the function $\eta^2(s)$ to cancel the open RHS zeros of $\hat{h}_*(s)$. Specifying that the numerator or denominator polynomial of $\eta(s)$ be of minimal degree and such that its zeros cancel all the open RHS zeros of $\hat{h}_*(s)$, we determine uniquely

THE SCATTERING MATRIX

$h(s) = \eta(s)\hat{h}(s)$, which unlike the previous decomposition makes use of the surplus factors. We note that, with the above choice of $\eta(s)$, $d(s)$ becomes an all-pass function, as in the previous case.

EXAMPLE 2.2. Consider the positive-real impedance

$$z(s) = \frac{s+3}{s+2}, \qquad (2.36)$$

whose para-hermitian part $r(s)$ can be decomposed as

$$r(s) = h(s)h_*(s) = \frac{(s+\sqrt{6})(s-\sqrt{6})}{(s+2)(s-2)}, \qquad (2.37a)$$

where

$$\hat{h}(s) = \frac{s+\sqrt{6}}{s+2}. \qquad (2.37b)$$

Choose an all-pass function $\eta(s)$, whose numerator or denominator polynomial is of minimal degree, to cancel the open RHS zero of $\hat{h}_*(s)$, which is at $\sqrt{6}$. This gives

$$\eta(s) = \frac{s-\sqrt{6}}{s+\sqrt{6}}. \qquad (2.38)$$

The desired decomposition is then obtained as

$$h(s) = \eta(s)\hat{h}(s) = \frac{s-\sqrt{6}}{s+2}. \qquad (2.39)$$

From (2.33) or (2.35), it is easy to confirm that

$$d(s) = h(s)h_*^{-1}(s) = \frac{(s-2)(s-\sqrt{6})}{(s+2)(s+\sqrt{6})}, \qquad (2.40)$$

being an all-pass function, is analytic in the open RHS.

Suppose that we wish to compute the reflection coefficient of the one-port network N of Fig. 2.4 normalizing to the reference impedance $z(s)$ of (2.36). Then from (2.18) in conjunction with (2.11) and (2.22b), we have

$$S(s) = \frac{h(s)}{h_*(s)} S^I(s) = \frac{h(s)}{h_*(s)} \cdot \frac{Z(s) - z_*(s)}{Z(s) + z(s)}$$

$$= \frac{(s - \sqrt{6})(-2s^4 + 6s^3 + 2s^2 + 2s + 1)}{(s + \sqrt{6})(2s^4 + 10s^3 + 14s^2 + 12s + 5)}, \quad (2.41)$$

which is analytic in the closed RHS, as expected, while $S^I(s)$ is not.

2.4. Alternative representation of the basis-independent reflection coefficient

As indicated in §2.2, the basis-independent normalized incident and reflected waves are defined in terms of the incident current and reflected current. In this section, we demonstrate that the normalization can start from the voltage basis, and arrive at the same conclusion.

Referring to (2.26), the para-hermitian part $g(s)$ of the reference admittance $y(s) = 1/z(s)$ can be expressed explicitly as

$$g(s) = \frac{p(s)q_*(s) + p_*(s)q(s)}{2p(s)p_*(s)} = k(s)k_*(s), \quad (2.42)$$

where the factorization $k(s)k_*(s)$ is obtained by the procedure outlined in the preceding section with

$$k(s) = \frac{w(s)}{\sqrt{2}p(s)}, \quad (2.43)$$

$w_*(s)$ and $p(s)$ being real Hurwitz polynomials. The corresponding factorization $h(s)h_*(s)$ of $r(s)$, as given in (2.27), can be written explicitly as

$$h(s) = \frac{w(s)}{\sqrt{2}q(s)}, \quad (2.44)$$

$q(s)$ being a real Hurwitz polynomial. Substituting these in (2.17) in conjunction with (2.26) yields

$$a(s) = k_*(s)V_i(s), \tag{2.45a}$$
$$b(s) = k(s)V_r(s), \tag{2.45b}$$

which after combining with (2.10a) give

$$S(s) = k(s)S^V(s)k_*^{-1}(s). \tag{2.46}$$

Following (2.19), the actual port voltage, current and impedance are given by

$$V(s) = V_i(s) + V_r(s) = k_*^{-1}(s)a(s) + k^{-1}(s)b(s), \tag{2.47a}$$
$$I(s) = I_i(s) - I_r(s) = y_*(s)k_*^{-1}(s)a(s) - y(s)k^{-1}(s)b(s), \tag{2.47b}$$
$$Z(s) = [k(s) + k_*(s)S(s)][y_*(s)k(s) - y(s)k_*(s)S(s)]^{-1}. \tag{2.47c}$$

Solving for $a(s)$, $b(s)$ and $S(s)$ in terms of $V(s)$, $I(s)$, $Y(s)$, $y(s)$ and $k(s)$ yields

$$a(s) = \tfrac{1}{2}k^{-1}(s)[y(s)V(s) + I(s)], \tag{2.48a}$$
$$b(s) = \tfrac{1}{2}k_*^{-1}(s)[y_*(s)V(s) - I(s)], \tag{2.48b}$$
$$S(s) = -k(s)k_*^{-1}(s) + 2k(s)[Y(s) + y(s)]^{-1}k(s), \tag{2.48c}$$

where as before $Y(s) = 1/Z(s)$.

As an illustration, suppose that we wish to compute the reflection coefficient of the one-port network N of Fig. 2.4 normalizing to the reference admittance $y(s) = (s+2)/(s+3)$. Using the procedure outlined in the preceding section, we have

$$k(s) = \frac{s - \sqrt{6}}{s + 3}. \tag{2.49}$$

Then from (2.46) in conjunction with (2.12) and (2.22b), we obtain

$$\begin{aligned} S(s) &= -\frac{k(s)}{k_*(s)} \cdot \frac{Y(s) - y_*(s)}{Y(s) + y(s)} \\ &= \frac{(s - \sqrt{6})(-2s^4 + 6s^3 + 2s^2 + 2s + 1)}{(s + \sqrt{6})(2s^4 + 10s^3 + 14s^2 + 12s + 5)}, \end{aligned} \tag{2.50}$$

confirming (2.41).

2.5. The normalized reflection coefficient and passivity

As mentioned at the beginning of this chapter, the scattering parameters are found to be particularly useful for handling problems of power transfer in networks designed to be terminated with prescribed loads, since the various quantities of interest in power transmission have very simple expressions in terms of them. In this section, we derive a simple expression of power flow for a one-port network in terms of its normalized reflection coefficient.

Consider the one-port network N of Fig. 2.2. The average power entering N at a sinusoidal frequency can easily be expressed in terms of the normalized reflection coefficient by means of (2.19):†

$$\begin{aligned}
P_{av} &= \tfrac{1}{2} \operatorname{Re}\left[\bar{V}(j\omega)I(j\omega)\right] \\
&= \tfrac{1}{2} \operatorname{Re}\{[z(j\omega)h^{-1}(j\omega)\bar{a}(j\omega) + \bar{z}(j\omega)\bar{h}^{-1}(j\omega)\bar{b}(j\omega)] \\
&\quad [\bar{h}^{-1}(j\omega)a(j\omega) - h^{-1}(j\omega)b(j\omega)]\} \\
&= \tfrac{1}{2}[|a(j\omega)|^2 - |b(j\omega)|^2] + \tfrac{1}{2}\operatorname{Re}\left[\bar{b}(j\omega)\bar{h}^{-1}(j\omega)\bar{z}(j\omega)\bar{h}^{-1}(j\omega)a(j\omega)\right. \\
&\quad \left. - \bar{a}(j\omega)h^{-1}(j\omega)z(j\omega)h^{-1}(j\omega)b(j\omega)\right] \\
&= \tfrac{1}{2}[|a(j\omega)|^2 - |b(j\omega)|^2] \\
&= \tfrac{1}{2}|a(j\omega)|^2[1 - |S(j\omega)|^2].
\end{aligned} \qquad (2.51)$$

For a passive one-port network, the average power is always nonnegative for all sinusoidal frequencies. Thus, we have

$$|S(j\omega)|^2 \leq 1 \qquad (2.52)$$

for all ω. Since by the choice of the factor $h(s)$ used in the normalization, as described in §2.3, the reflection coefficient $S(s)$ is analytic in the open RHS, by appealing to the maximum modulus theorem in the theory of a complex variable, we conclude that the maximum magnitude of $S(s)$, $\operatorname{Re} s \geq 0$, occurs on the boundary, which is the $j\omega$-axis. This gives the most penetrating result on the reflection coefficient of a passive one-port network normalizing to a

---

†In contrast to the quantities used in § 3 of Chapter 1, which are rms values, the sinusoidal voltage and current used here are their maximum values. However, the choice does not affect our conclusions.

strictly passive reference impedance:

$$|S(s)|^2 \leq 1 \qquad (2.53)$$

for all s in the closed RHS. A complete characterization of the normalized reflection coefficient $S(s)$ will be presented in §4.

We now proceed to demonstrate that, on the real-frequency axis, the term $1 - |S(j\omega)|^2$ corresponds to the ratio of average power delivered to the one-port network to the maximum available average power at the source. The maximum available average power from the given source combination of $V_g(s)$ and $z(s)$ of Fig. 2.2 is given by

$$P_m = |V_g(j\omega)|^2/8r(j\omega), \qquad (2.54)$$

which corresponds to the average power delivered by the given source to a conjugately matched load. The average power delivered to the one-port network was computed in (2.51), and from (2.8b) and (2.16a) we obtain

$$\begin{aligned}P_{av} &= \tfrac{1}{2}|a(j\omega)|^2[1 - |S(j\omega)|^2] \\ &= [1 - |S(j\omega)|^2]P_m,\end{aligned} \qquad (2.55)$$

indicating that the power delivered to the one-port network may be regarded as being made up of the power in the incident wave P_i, less the power returned to the source by the reflected wave P_r. With these, (2.55) can be rewritten as

$$P_{av} = P_i - P_r, \qquad (2.56)$$

where $P_i = P_m$ and $P_r = |S(j\omega)|^2 P_m$. Thus, if there is no reflection ($S = 0$), all the power is transferred to the one-port, meaning that $|S(j\omega)|^2$ represents the fraction of the maximum available average power that is returned to the source, and thus provides a quantitative measure of the deviation of the actual power-matching behavior of the network from the optimal power-matched condition.

A reflection coefficient that is analytic in the closed RHS and possesses the property (2.53) is termed as a *bounded-real reflection coefficient*. A formal definition together with other properties will be presented in §4.

As indicated in (2.51), on the real-frequency axis, the expression

$|a(j\omega)|^2[1-|S(j\omega)|^2]$ represents twice the average input power in the sinusoidal steady state. Then the expression $|a(s)|^2[1-|S(s)|^2]$, as demonstrated in (2.53), can be thought of as an extension of this steady-state concept into the entire right-half of the complex-frequency plane.

As an illustration, consider the one-port network of Fig. 2.4. The reflection coefficient $S(s)$ of the one-port normalizing to the 4-Ω resistance was computed earlier in (2.22a). Its value on the $j\omega$-axis is obtained as

$$S(j\omega) = \frac{-3 + 6\omega^2 - j(6\omega - 8\omega^3)}{5 - 10\omega^2 + j(10\omega - 8\omega^3)}, \quad (2.57)$$

whose squared magnitude is given by

$$|S(j\omega)|^2 = 1 - \frac{16}{64\omega^6 - 60\omega^4 + 25}. \quad (2.58)$$

Since the inductor and capacitors of the one-port network are lossless, from (2.55) the fraction of the maximum available average power that is delivered to the one-port corresponds to the second term on the right-hand side of (2.58), and all of this power must be dissipated in the 1-Ω resistor inside the one-port.

3. The scattering matrix of an n-port network

The scattering parameters or reflection coefficients of a one-port network were discussed in detail in the foregoing. The extension of these concepts to an n-port network will be considered in the present section. The scattering matrix of an n-port network is merely the matrix version of the scalar reflection coefficient of a one-port network. As a matter of fact, in anticipation of this extension, most of the formulas derived in the preceding sections were written in the forms that could easily be extended to the n-port case, reducing considerably the effort required for making the extension.

We begin by considering the general representation of an n-port network N as depicted in Fig. 2.5. Assume that the n-port is characterized by its open-circuit impedance matrix $Z(s)$. This

THE SCATTERING MATRIX

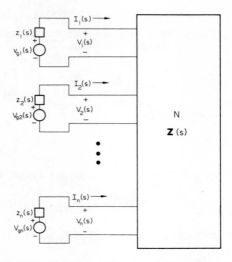

FIG. 2.5. The general representation of an n-port network N.

assumption, as will be seen shortly, is not essential for the existence of the scattering matrix; it is made here solely for the purpose of manipulation and will be abandoned later. In fact, the development henceforth is equally valid if the short-circuit admittance matrix or the hybrid matrix is used.

As in the one-port case, each of the n ports of N is assumed to be loaded by a strictly passive impedance $z_k(s)$ in series with a voltage source $V_{gk}(s)$ as depicted in Fig. 2.5. Since the $z_k(s)$'s are strictly passive, none of their para-hermitian parts $r_k(s)$ can be identically zero. The diagonal matrix

$$z(s) = \begin{bmatrix} z_1(s) & 0 & \cdots & 0 \\ 0 & z_2(s) & \cdots & 0 \\ \vdots & \vdots & \vdots & \vdots \\ 0 & 0 & \cdots & z_n(s) \end{bmatrix}, \quad (2.59)$$

whose kkth element is the reference impedance $z_k(s)$ of the kth port, is called the *reference impedance matrix* of N. Referring to Fig. 2.5, the port voltages, currents and sources are represented by the vectors

$$V(s) = \begin{bmatrix} V_1(s) \\ V_2(s) \\ \vdots \\ V_n(s) \end{bmatrix}, \quad I(s) = \begin{bmatrix} I_1(s) \\ I_2(s) \\ \vdots \\ I_n(s) \end{bmatrix}, \quad V_g(s) = \begin{bmatrix} V_{g1}(s) \\ V_{g2}(s) \\ \vdots \\ V_{gn}(s) \end{bmatrix}, \quad (2.60)$$

respectively. They are related by the matrix equation

$$V_g(s) = V(s) + z(s)I(s) = [Z(s) + z(s)]I(s). \quad (2.61)$$

We next consider the extensions of (2.6) and (2.7). For a square matrix $A(s)$, write

$$A_*(s) = A'(-s), \quad (2.62)$$

the prime as before denoting the matrix transpose. A matrix $A(s)$ is said to be *para-hermitian* if $A_*(s) = A(s)$. The matrix

$$\tfrac{1}{2}[A(s) + A_*(s)] \quad (2.63)$$

is called the *para-hermitian part* of $A(s)$ since it is para-hermitian. Observe that, on the real-frequency axis, $A_*(j\omega) = \bar{A}'(j\omega) = A^*(j\omega)$ and (2.63) becomes the hermitian part of $A(j\omega)$. Then the para-hermitian part of $z(s)$, which is

$$r(s) = \tfrac{1}{2}[z(s) + z_*(s)], \quad (2.64)$$

represents the extension of (2.7). Factorizing each diagonal element of $r(s)$ according to the procedure outlined in §2.3 yields

$$r(s) = h(s)h_*(s), \quad (2.65)$$

where $h(s)$ is a diagonal matrix, whose kkth element $h_k(s)$ together with $h_{k*}^{-1}(s)$ is analytic in the open RHS. As in (2.32), the most general solution of (2.65) having the desired analyticity is given by

$$h(s) = \eta(s)\hat{h}(s), \qquad (2.66)$$

where $\eta(s)$ and $\hat{h}(s)$ are diagonal matrices whose kkth entries are $\eta_k(s)$ and $\hat{h}_k(s)$, respectively. $\eta_k(s)$ is an all-pass function of the form (2.30) and $\hat{h}_k(s)$ is a solution of the kkth element $r_k(s)$ of $r(s)$ obtained by the unique distribution of the poles and zeros of $r_k(s)$ without introducing any surplus factors. Thus, (the elements of) $h(s)$ and $h_*^{-1}(s)$ are analytic in the open RHS. As in the one-port case, each of the elements of $\hat{h}(s)\hat{h}_*^{-1}(s)$ is an all-pass function and

$$\eta(s)\eta_*(s) = U_n, \qquad (2.67)$$

U_n denoting the identity matrix of order n. As an example, consider the reference impedance matrix

$$z(s) = \begin{bmatrix} s/(s+1) & 0 \\ 0 & 2/(2s+1) \end{bmatrix}, \qquad (2.68)$$

whose para-hermitian part can be decomposed as

$$r(s) = \tfrac{1}{2}[z(s) + z_*(s)] = \begin{bmatrix} s^2/(s^2-1) & 0 \\ 0 & 2/(1-4s^2) \end{bmatrix}$$

$$= \begin{bmatrix} s/(s+1) & 0 \\ 0 & \sqrt{2}/(2s+1) \end{bmatrix} \begin{bmatrix} s/(s-1) & 0 \\ 0 & \sqrt{2}/(1-2s) \end{bmatrix} \qquad (2.69a)$$

$$= h(s)h_*(s), \qquad (2.69b)$$

in which the first matrix in (2.69a) is identified as $\hat{h}(s)$ and the second as $\hat{h}_*(s)$. The most general solution of (2.69b) is given by

$$h(s) = \begin{bmatrix} \eta_1(s)s/(s+1) & 0 \\ 0 & \eta_2(s)\sqrt{2}/(2s+1) \end{bmatrix}, \qquad (2.70)$$

$\eta_1(s)$ and $\eta_2(s)$ being the all-pass functions of the form (2.30). It is easy to confirm that

$$\hat{h}(s)\hat{h}_*^{-1}(s) = \begin{bmatrix} (s-1)/(s+1) & 0 \\ 0 & -(s-\tfrac{1}{2})/(s+\tfrac{1}{2}) \end{bmatrix} \quad (2.71)$$

is a diagonal matrix, whose diagonal elements are all-pass functions.

3.1. Basis-dependent scattering matrices

By analogy with the one-port case, we begin by introducing the *incident-voltage vector* $V_i(s)$ and *incident-current vector* $I_i(s)$. They represent voltages $V_{i1}(s)$, $V_{i2}(s), \ldots, V_{in}(s)$ and currents $I_{i1}(s)$, $I_{i2}(s), \ldots, I_{in}(s)$ that would appear at the terminals of the n-ports under optimal power-matching condition, as depicted in Fig. 2.6. This

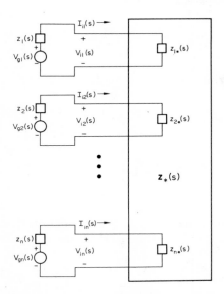

FIG. 2.6. The optimal power-matching situation for an n-port network.

gives

$$V_i(s) = z_*(s)I_i(s), \qquad (2.72a)$$

$$I_i(s) = \tfrac{1}{2}r^{-1}(s)V_g(s). \qquad (2.72b)$$

We remark that $r(s)$ is not identically singular since its diagonal elements $r_k(s)$, being the para-hermitian parts of $z_k(s)$, are not identically zero. The *reflected-voltage vector* $V_r(s)$ and the *reflected-current vector* $I_r(s)$ are defined by the relations

$$V_r(s) = V(s) - V_i(s), \qquad (2.73a)$$

$$-I_r(s) = I(s) - I_i(s). \qquad (2.73b)$$

The matrix relating the reflected-voltage vector $V_r(s)$ to the incident-voltage vector $V_i(s)$,

$$V_r(s) = S^V(s)V_i(s), \qquad (2.74)$$

is called the *voltage-based scattering matrix*. Likewise, the matrix relating the reflected-current vector $I_r(s)$ to the incident-current vector $I_i(s)$,

$$I_r(s) = S^I(s)I_i(s), \qquad (2.75)$$

is called the *current-based scattering matrix*. The elements of $S^V(s)$ and $S^I(s)$ are referred to as the *scattering parameters* of the n-port network. From (2.72) to (2.75) in conjunction with (2.61), we can easily deduce expressions for $S^I(s)$ and $S^V(s)$ in terms of the matrices $Z(s)$ and $z(s)$:

$$\begin{aligned}I_r(s) &= -I(s) + I_i(s) = -[Z(s)+z(s)]^{-1}V_g(s) + I_i(s) \\ &= \{U_n - 2[Z(s)+z(s)]^{-1}r(s)\}I_i(s),\end{aligned} \qquad (2.76)$$

giving

$$\begin{aligned}S^I(s) &= U_n - 2[Z(s)+z(s)]^{-1}r(s) \\ &= [Z(s)+z(s)]^{-1}[Z(s)-z_*(s)].\end{aligned} \qquad (2.77)$$

In a similar way, we can show that

$$S^V(s) = -U_n + 2[Y(s) + y(s)]^{-1}g(s)$$
$$= -[Y(s) + y(s)]^{-1}[Y(s) - y_*(s)], \quad (2.78)$$

where

$$Y(s) = Z^{-1}(s), \qquad y(s) = z^{-1}(s), \quad (2.79)$$

and $g(s)$ is the para-hermitian part of $y(s)$.

Observe the similarity between the formulas (2.77) and (2.78) and those in (2.11) and (2.12), and also the similarity between (2.77) and (2.88). Like (2.13), the current- and voltage-based scattering matrices are related by the equation (Problem 2.2)

$$S^V(s)z_*(s) = z(s)S^I(s). \quad (2.80)$$

Under the optimal power-matching condition, $Z(s) = z_*(s)$ and both scattering matrices become zero matrix, indicating that the reflected voltage and current vectors are a measure of the deviations of the port voltages and currents, when under actual operation, from their values when optimally power-matched.

As in the one-port case, we have

$$V_r(s) = S^V(s)V_i(s) = S^V(s)z_*(s)I_i(s)$$
$$= z(s)S^I(s)I_i(s) = z(s)I_r(s). \quad (2.81)$$

This shows that the reflected voltage and current vectors see the reference impedance matrix $z(s)$, while the incident voltage and current vectors see $z_*(s)$, as required by the ideal power match.

EXAMPLE 2.3. Consider an ideal transformer N with turns ratio m to 1 as shown in Fig. 2.7. We wish to compute the incident and reflected voltage and current vectors and the scattering matrices with respect to the reference impedance matrix

$$z(s) = \begin{bmatrix} R_1 & 0 \\ 0 & R_2 \end{bmatrix}. \quad (2.82)$$

THE SCATTERING MATRIX

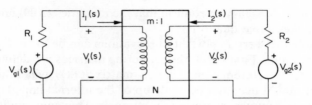

FIG. 2.7. An ideal transformer N with turns ratio $m:1$.

From Fig. 2.7, the port voltages and currents can easily be computed, and are given by

$$V(s) = \begin{bmatrix} V_1(s) \\ V_2(s) \end{bmatrix} = \frac{1}{R_1 + m^2 R_2} \begin{bmatrix} m^2 R_2 & m R_1 \\ m R_2 & R_1 \end{bmatrix} \begin{bmatrix} V_{g1}(s) \\ V_{g2}(s) \end{bmatrix}, \quad (2.83a)$$

$$I(s) = \begin{bmatrix} I_1(s) \\ I_2(s) \end{bmatrix} = \frac{1}{R_1 + m^2 R_2} \begin{bmatrix} 1 & -m \\ -m & m^2 \end{bmatrix} \begin{bmatrix} V_{g1}(s) \\ V_{g2}(s) \end{bmatrix}. \quad (2.83b)$$

From (2.72), the incident current and voltage vectors are obtained as

$$I_i(s) = \begin{bmatrix} I_{i1}(s) \\ I_{i2}(s) \end{bmatrix} = \tfrac{1}{2} r^{-1}(s) V_g(s) = \begin{bmatrix} \dfrac{1}{2R_1} & 0 \\ 0 & \dfrac{1}{2R_2} \end{bmatrix} \begin{bmatrix} V_{g1}(s) \\ V_{g2}(s) \end{bmatrix}, \quad (2.84a)$$

$$V_i(s) = \begin{bmatrix} V_{i1}(s) \\ V_{i2}(s) \end{bmatrix} = z_*(s) I_i(s) = \begin{bmatrix} 0.5 & 0 \\ 0 & 0.5 \end{bmatrix} \begin{bmatrix} V_{g1}(s) \\ V_{g2}(s) \end{bmatrix}. \quad (2.84b)$$

Substituting these in (2.73) yields the reflected voltage and current vectors

$$V_r(s) = \frac{1}{R_1 + m^2 R_2} \begin{bmatrix} m^2 R_2 - R_1 & 2 m R_1 \\ 2 m R_2 & R_1 - m^2 R_2 \end{bmatrix} \begin{bmatrix} V_{i1}(s) \\ V_{i2}(s) \end{bmatrix}, \quad (2.85a)$$

$$I_r(s) = \frac{1}{R_1 + m^2 R_2} \begin{bmatrix} m^2 R_2 - R_1 & 2 m R_2 \\ 2 m R_1 & R_1 - m^2 R_2 \end{bmatrix} \begin{bmatrix} I_{i1}(s) \\ I_{i2}(s) \end{bmatrix}, \quad (2.85b)$$

the coefficient matrices being the voltage-based and current-based scattering matrices $S^V(s)$ and $S^I(s)$, respectively, of the ideal

74   BROADBAND MATCHING NETWORKS

transformer. It is easy to confirm that the identities (2.80) and (2.81) hold for the above quantities.

A number of very important observations can be made from this simple example. First of all, the scattering matrices are defined with respect to a reference impedance matrix $z(s)$, which is quite arbitrary, and it need not be made up of the internal impedances of the voltage sources represented by their Thévenin equivalents. Secondly, the voltage- and current-based scattering matrices are not necessarily symmetric even for a reciprocal network, as demonstrated in the above example. This difficulty can easily be removed by considering the complex normalization to be presented in the following section. Thirdly, the computation of these scattering matrices from their definitions is usually cumbersome. Nevertheless, other techniques are available; they will be presented in later sections. Finally, we notice that the ideal transformer possesses neither the impedance matrix nor the admittance matrix, but it possesses the scattering matrices. In other words, even though the matrices $Z(s)$ and $Y(s)$ in (2.77) and (2.78) do not exist, the matrices $[Z(s) + z(s)]^{-1}$ and $[Y(s) + y(s)]^{-1}$ do; they represent the admittance and impedance matrices of some n-port networks derived from the original n-port network by augmentation. These results will be elaborated in § 3.3.

3.2. Basis-independent scattering matrix

In the preceding section, we have defined two scattering matrices. One is based on the voltage and the other on the current. In the present section, we introduce a normalization for which the normalized scattering matrix becomes basis-independent. To this end, we define the *normalized incident-wave vector* and the *normalized reflected-wave vector* as

$$a(s) = h_*(s)I_i(s), \qquad (2.86a)$$

$$b(s) = h(s)I_r(s), \qquad (2.86b)$$

respectively, where $h(s)$ is given in (2.65). Like (2.16c), the basis-independent *normalized scattering matrix* $S(s)$ is defined by

THE SCATTERING MATRIX

the relation
$$b(s) = S(s)a(s). \qquad (2.87)$$

The elements of $S(s)$ are called the *normalized scattering parameters* of the n-port network. Like the basis-dependent scattering matrices, $S(s)$ can also be expressed in terms of the impedance matrix $Z(s)$ and the reference impedance matrix $z(s)$:

$$b(s) = h(s)I_r(s) = h(s)S^I(s)I_i(s) = h(s)S^I(s)h_*^{-1}(s)a(s), \qquad (2.88)$$

giving

$$S(s) = h(s)S^I(s)h_*^{-1}(s) \qquad (2.89a)$$

$$= h(s)h_*^{-1}(s) - 2h(s)Y_{a1}(s)h(s) \qquad (2.89b)$$

$$= h(s)Y_{a1}(s)[Z(s) - z_*(s)]h_*^{-1}(s), \qquad (2.89c)$$

where

$$Y_{a1}(s) = [Z(s) + z(s)]^{-1}. \qquad (2.90)$$

Equations (2.89b) and (2.89c) follow directly from (2.77).

Like (2.19), we can express the port voltage and current vectors and the impedance matrix in terms of the normalized wave vectors and the normalized scattering matrix:

$$V(s) = V_i(s) + V_r(s) = z_*(s)h_*^{-1}(s)a(s) + z(s)h^{-1}(s)b(s), \qquad (2.91a)$$

$$I(s) = I_i(s) - I_r(s) = h_*^{-1}(s)a(s) - h^{-1}(s)b(s), \qquad (2.91b)$$

$$Z(s) = [z_*(s) + z(s)h^{-1}(s)S(s)h_*(s)][U_n - h^{-1}(s)S(s)h_*(s)]^{-1}. \qquad (2.91c)$$

The inverse relations of expressing the normalized waves in terms of the port voltage and current vectors $V(s)$ and $I(s)$ and the impedance matrices $Z(s)$ and $z(s)$ can easily be obtained from (2.91a) and (2.91b), and are given by

$$a(s) = \tfrac{1}{2}h^{-1}(s)[V(s) + z(s)I(s)], \qquad (2.92a)$$

$$b(s) = \tfrac{1}{2}h_*^{-1}(s)[V(s) - z_*(s)I(s)]. \qquad (2.92b)$$

Again, much of the foregoing manipulation could have been avoided by simply regarding (2.92) as the formal definitions of the wave vectors $a(s)$ and $b(s)$. This approach is entirely consistent with the above exposition, yet it does not depend on the existence of the impedance matrix $Z(s)$.

EXAMPLE 2.4. Consider the lossless two-port network N of Fig. 2.8. We wish to compute its scattering matrix normalizing to the load

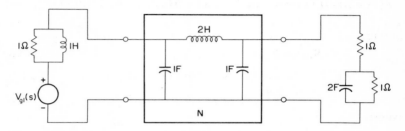

FIG. 2.8. A lossless reciprocal two-port network together with its loading.

impedances as shown in the figure. To this end, we first compute the reference impedance matrix $z(s)$ and the impedance matrix $Z(s)$ of N, and they are given by

$$z(s) = \begin{bmatrix} s/(s+1) & 0 \\ 0 & 2(s+1)/(2s+1) \end{bmatrix}, \qquad (2.93\text{a})$$

$$Z(s) = \frac{1}{2s(s^2+1)} \begin{bmatrix} 2s^2+1 & 1 \\ 1 & 2s^2+1 \end{bmatrix}. \qquad (2.93\text{b})$$

From (2.77), the current-based scattering matrix is obtained as

$$\begin{aligned} S^I(s) &= [Z(s) + z(s)]^{-1}[Z(s) - z_*(s)] \\ &= \frac{1}{\Delta(s)} \begin{bmatrix} \dfrac{-(s+1)q(s)}{s-1} & \dfrac{4(2s^2-1)(s+1)}{2s-1} \\ \dfrac{2s^2(2s+1)}{s-1} & \dfrac{(2s+1)q(-s)}{2s-1} \end{bmatrix}, \end{aligned} \qquad (2.94\text{a})$$

where
$$\Delta(s) = 4s^5 + 12s^4 + 18s^3 + 18s^2 + 7s + 2. \quad (2.94b)$$
$$q(s) = 4s^5 + 4s^4 + 2s^3 + 10s^2 + 3s + 2. \quad (2.94c)$$

The para-hermitian part of $z(s)$ is given by

$$r(s) = \tfrac{1}{2}[z(s) + z_*(s)] = \begin{bmatrix} s^2/(s^2-1) & 0 \\ 0 & 2(2s^2-1)/(4s^2-1) \end{bmatrix}$$
$$= h(s)h_*(s), \quad (2.95)$$

from which by applying the procedure outlined in §2.3 we obtain a desired factorization of $r(s)$ as

$$h(s) = \begin{bmatrix} s/(s+1) & 0 \\ 0 & (2s-\sqrt{2})/(2s+1) \end{bmatrix}. \quad (2.96)$$

Finally, substituting (2.94) and (2.96) in (2.89a) yields the basis-independent normalized scattering matrix

$$S(s) = h(s)S^I(s)h_*^{-1}(s)$$
$$= \frac{1}{\Delta(s)} \begin{bmatrix} -q(s) & 2s(2s-\sqrt{2}) \\ 2s(2s-\sqrt{2}) & (2s-\sqrt{2})q(-s)/(2s+\sqrt{2}) \end{bmatrix}. \quad (2.97)$$

It is significant to point out that, unlike the current-based scattering matrix (2.94a), the normalized scattering matrix (2.97) of the two-port network N is symmetric, reflecting the property that the two-port network itself is reciprocal.

3.3. The scattering matrices and the augmented n-port networks

As mentioned in the preceding section, the existence of the scattering matrices $S^I(s)$, $S^V(s)$ and $S(s)$ of an n-port network does not depend upon the existence of its impedance matrix $Z(s)$ or admittance matrix $Y(s)$. In other words, in (2.77), (2.78) and (2.89b), if $Z(s)$ or $Y(s)$ does not exist, the matrices $[Z(s)+z(s)]^{-1}$ and $[Y(s)+y(s)]^{-1}$, as will be seen shortly, always exist for all nonpathological passive networks. For this we show that these matrices can be interpreted as the short-circuit admittance matrix or the open-circuit impedance matrix of certain augmented n-port networks.

For a given n-port network N as shown in Fig. 2.5, let N_{a1} be the n-port network derived from N by augmentation, as depicted in Fig. 2.9. Denote by $Y_{a1}(s)$ the admittance matrix of N_{a1}. Under our

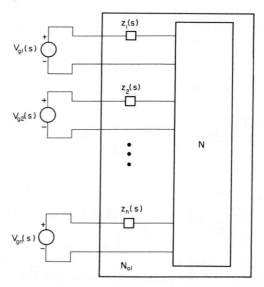

FIG. 2.9. The augmented n-port network N_{a1} of the n-port N of Fig. 2.5.

assumptions, $Y_{a1}(s)$ clearly exists for all passive, and most active, n-port networks. But $[Z(s)+z(s)]^{-1}$ also denotes the admittance matrix of N_{a1}. Thus, we conclude that

$$Y_{a1}(s) = [Z(s)+z(s)]^{-1}. \tag{2.98}$$

In a similar way, if we let N_{a2} be the augmented n-port network N, as depicted in Fig. 2.10, whose impedance matrix is $Z_{a2}(s)$, then

$$Z_{a2}(s) = [Y(s) + y(s)]^{-1}. \qquad (2.99)$$

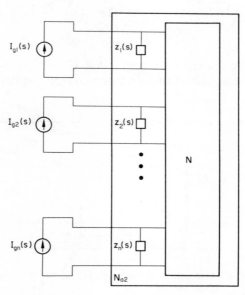

FIG. 2.10. The augmented n-port network N_{a2} of the n-port N of Fig. 2.5.

A significant consequence of the above interpretations is that if N is reciprocal, $Y_{a1}(s)$ and $Z_{a2}(s)$ are symmetric, and thus from (2.89b) the normalized scattering matrix $S(s)$ is also symmetric, a fact that was pointed out in Example 2.4. However, the current-based and the voltage-based scattering matrices are, in general, not symmetric even for reciprocal networks. This is one of the reasons for introducing a normalization for the scattering parameters; the others will be presented in the later sections.

EXAMPLE 2.5. To illustrate (2.98), we shall use the same problem considered in Example 2.4. From (2.93) we first compute the inverse of the matrix $[Z(s) + z(s)]$, which is given by

$$[Z(s) + z(s)]^{-1} = \frac{1}{\Delta(s)}$$
$$\cdot \begin{bmatrix} (s+1)(4s^4+8s^3+6s^2+6s+1) & -(s+1)(2s+1) \\ -(s+1)(2s+1) & (2s+1)(2s^4+2s^3+4s^2+s+1) \end{bmatrix}.$$
(2.100)

It is straightforward to confirm that (2.100) is the admittance matrix of the augmented two-port network N_{a1} as depicted in Fig. 2.11.

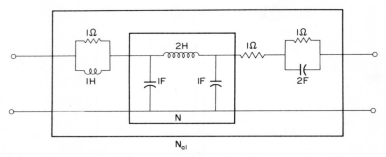

FIG. 2.11. The augmented two-port network N_{a1} of the two-port of Fig. 2.8.

3.4. Alternative representation of the basis-independent scattering matrix

As in §2.4, we demonstrate that the normalization can start from the voltage basis, and arrive at the same conclusion.

Let the para-hermitian part $g(s)$ of the reference admittance matrix $y(s)$ be factored according to the procedure outlined in §2.3, giving

$$g(s) = \tfrac{1}{2}[y(s) + y_*(s)] = k(s)k_*(s), \quad (2.101)$$

where $k(s)$ and $k_*^{-1}(s)$ are analytic in the open RHS. Combining

(2.72a) and (2.81) with (2.86) yields

$$a(s) = h_*(s)I_i(s) = h_*(s)z_*^{-1}(s)V_i(s) = k_*(s)V_i(s), \quad (2.102a)$$

$$b(s) = h(s)I_r(s) = h(s)z^{-1}(s)V_r(s) = k(s)V_r(s), \quad (2.102b)$$

from which we obtain

$$S(s) = k(s)S^V(s)k_*^{-1}(s). \quad (2.102c)$$

Following (2.91) we have

$$V(s) = V_i(s) + V_r(s) = k_*^{-1}(s)a(s) + k^{-1}(s)b(s), \quad (2.103a)$$

$$I(s) = I_i(s) - I_r(s) = y_*(s)k_*^{-1}(s)a(s) - y(s)k^{-1}(s)b(s), \quad (2.103b)$$

$$Y(s) = [y_*(s) - y(s)k^{-1}(s)S(s)k_*(s)][U_n + k^{-1}(s)S(s)k_*(s)]^{-1}. \quad (2.103c)$$

Solving for $a(s)$, $b(s)$ and $S(s)$ gives

$$a(s) = \tfrac{1}{2}k^{-1}(s)[y(s)V(s) + I(s)], \quad (2.104a)$$

$$b(s) = \tfrac{1}{2}k_*^{-1}(s)[y_*(s)V(s) - I(s)], \quad (2.104b)$$

$$S(s) = -k(s)k_*^{-1}(s) + 2k(s)[Y(s) + y(s)]^{-1}k(s). \quad (2.104c)$$

Alternatively, $S(s)$ can also be expressed as

$$S(s) = k(s)S^V(s)k_*^{-1}(s) \quad (2.105a)$$

$$= -k(s)Z_{a2}(s)[Y(s) - y_*(s)]k_*^{-1}(s). \quad (2.105b)$$

We remark that (2.105b) is valid only if $Y(s)$ exists, and that (2.104c) holds as long as $[Y(s) + y(s)]^{-1}$ exists. Recall that $[Y(s) + y(s)]^{-1}$ also denotes the impedance matrix $Z_{a2}(s)$ of the augmented n-port network N_{a2} as illustrated in Fig. 2.10. Like $Y_{a1}(s)$, $Z_{a2}(s)$ certainly exists if the given n-port network is passive, but it may be singular, indicating that $Y(s) + y(s)$ may not exist. From (2.104c), we can also deduce that if the given n-port network is reciprocal, its normalized scattering matrix is symmetric.

3.5. Physical interpretation of the normalized scattering parameters

Like the impedance parameters and the admittance parameters, which can be computed as the driving-point or transfer immittances of certain ports when the other ports are open-circuited or short-circuited, the normalized scattering parameters can similarly be interpreted as the reflection or transmission coefficients of certain ports when the other ports are terminated in their reference impedances. The details of these interpretations will be presented in this section.

Using the symbols defined at the beginning of §3, the mth equations of (2.92) can be written out explicitly as

$$a_m(s) = \tfrac{1}{2} h_m^{-1}(s)[V_m(s) + z_m(s)I_m(s)], \qquad (2.106a)$$

$$b_m(s) = \tfrac{1}{2} h_{m*}^{-1}(s)[V_m(s) - z_{m*}(s)I_m(s)]. \qquad (2.106b)$$

These two equations permit us to determine the conditions under which there will be no incident wave $a_m(s)$ or reflected wave $b_m(s)$ at port m. For $a_m(s)$ to vanish, it is necessary and sufficient that

$$\frac{V_m(s)}{-I_m(s)} = z_m(s). \qquad (2.107)$$

Referring to Fig. 2.5, this is equivalent to saying that if the mth port is terminated in its reference impedance $z_m(s)$, there will be no incident wave $a_m(s)$ at port m. In a similar manner, for $b_m(s)$ to vanish identically, we require that

$$\frac{V_m(s)}{-I_m(s)} = -z_{m*}(s), \qquad (2.108)$$

showing that if the mth port is terminated in $-z_{m*}(s)$, there will be no reflected wave $b_m(s)$ at the mth port. Write

$$S(s) = [S_{ij}(s)], \qquad S^I(s) = [S^I_{ij}(s)]. \qquad (2.109)$$

Then from (2.89a) we obtain

$$S_{ij}(s) = h_i(s)h_{j*}^{-1}(s)S_{ij}^I(s). \tag{2.110}$$

Using the defining equation (2.87) for the normalized scattering matrix, we see immediately that

$$S_{jj}(s) = \frac{b_j(s)}{a_j(s)}\bigg|_{a_x(s)=0 \text{ for } x \neq j}, \tag{2.111a}$$

and for $m \neq j$

$$S_{mj}(s) = \frac{b_m(s)}{a_j(s)}\bigg|_{a_x(s)=0 \text{ for } x \neq j}. \tag{2.111b}$$

Substituting (2.106) in (2.111a) yields

$$S_{jj}(s) = \frac{h_j(s)[V_j(s) - z_{j*}(s)I_j(s)]}{h_{j*}(s)[V_j(s) + z_j(s)I_j(s)]}\bigg|_{a_x(s)=0 \text{ for } x \neq j}$$
$$= \frac{h_j(s)}{h_{j*}(s)} \cdot \frac{Z_{jj}(s) - z_{j*}(s)}{Z_{jj}(s) + z_j(s)}, \tag{2.112}$$

$Z_{jj}(s)$ being the driving-point impedance looking into the jth port when all other ports are terminated in their reference impedances, as depicted in Fig. 2.12. From (2.11) and (2.18), $S_{jj}(s)$ is recognized as the basis-independent normalized reflection coefficient of the one-port network obtained from the original n-port network by terminating all of its ports except the jth one in their reference impedances, as illustrated in Fig. 2.12. The jth port is loaded by a voltage generator in series with the impedance $z_j(s)$.

In a similar manner, we examine the physical meaning of $S_{mj}(s)$, $m \neq j$. Substituting (2.106) in (2.111b) yields

$$S_{mj}(s) = \frac{h_j(s)[V_m(s) - z_{m*}(s)I_m(s)]}{h_{m*}(s)[V_j(s) + z_j(s)I_j(s)]}\bigg|_{a_x(s)=0 \text{ for } x \neq j}. \tag{2.113}$$

In the following, we show that, on the real-frequency axis, the squared magnitude of $S_{mj}(j\omega)$ represents the transducer power gain from port j to port m with all other $n-2$ ports being terminated in their reference impedances, as depicted in Fig. 2.13.

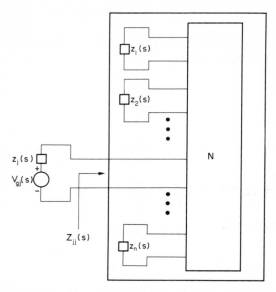

FIG. 2.12. A physical interpretation of the jth port reflection coefficient $S_{jj}(s)$ of the n-port network N of Fig. 2.5.

Consider the squared magnitude of $S_{mj}(j\omega)$ which can be written as

$$|S_{mj}(j\omega)|^2 = \frac{|I_m(j\omega)|^2 r_m(\omega)}{|V_{gj}(j\omega)|^2/4r_j(\omega)}\bigg|_{a_x(j\omega)=0 \text{ for } x \neq j} \equiv G_{jm}(\omega^2), \quad (2.114)$$

where $r_x(\omega) = \text{Re } z_x(j\omega)$. This is recognized to be the transducer power gain from port j to port m of the two-port network derived from the original n-port by terminating all of its ports except the mth and jth ones in their respective reference impedances. The jth port is loaded by a voltage generator in series with the impedance $z_j(s)$, while the mth port is terminated again in its reference impedance $z_m(s)$. The situation is depicted in Fig. 2.13.

Because of the above interpretation, the diagonal element $S_{jj}(s)$ of $S(s)$ is referred to as the *normalized reflection coefficient* of the port j, and the off-diagonal element $S_{mj}(s)$ as the *normalized transmission coefficient* from port j to port m, since on the real-frequency axis its

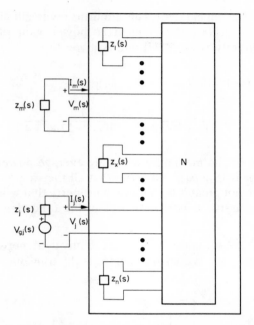

FIG. 2.13. A physical interpretation of the transmission coefficient $S_{mj}(s)$ of the n-port network N of Fig. 2.5.

squared magnitude denotes the fraction of the maximum available average power at the input port j that is transmitted to the load at the output port m.

EXAMPLE 2.6. Consider the lossless two-port network N of Fig. 2.8. The input impedance $Z_{11}(s)$ of N looking into port 1 when port 2 is terminated in its load, chosen to be its reference impedance, is given by

$$Z_{11}(s) = \frac{4s^4 + 12s^3 + 12s^2 + 6s + 2}{(s+1)(4s^4 + 8s^3 + 6s^2 + 6s + 1)}. \tag{2.115}$$

Substituting (2.115) in (2.112) yields the normalized reflection coefficient at the input port:

$$S_{11}(s) = \frac{-(4s^5 + 4s^4 + 2s^3 + 10s^2 + 3s + 2)}{4s^5 + 12s^4 + 18s^3 + 18s^2 + 7s + 2}, \tag{2.116}$$

where $h_1(s)$ is given in (2.96). This confirms the result given in (2.97). Appealing to (2.114), the transducer power gain of N can be computed directly from (2.97) and is given by

$$G_{21}(\omega^2) = |S_{21}(j\omega)|^2 = |S_{12}(j\omega)|^2$$

$$= \frac{8\omega^2(2\omega^2+1)}{16\omega^{10} - 52\omega^6 + 120\omega^4 - 23\omega^2 + 4}. \quad (2.117)$$

The fraction of the maximum available average power returned to the generator is $|S_{11}(j\omega)|^2$. Observe that all the squared magnitudes of $S_{ij}(j\omega)$ are not greater than unity, a property that will be shown to hold for all passive n-port networks in the following section.

EXAMPLE 2.7. Figure 2.14 is the equivalent network N of a transistor amplifier. We wish to compute the transducer power gain

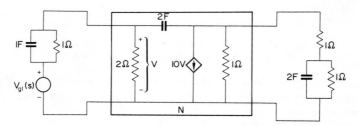

FIG. 2.14. The equivalent network of a transistor amplifier together with its loading.

through its scattering matrix normalizing to the load impedances as indicated in the figure.

We first compute the impedance matrix $Z(s)$ of N and the reference impedance matrix $z(s)$. The results are given by

$$z(s) = \begin{bmatrix} 1/(s+1) & 0 \\ 0 & (2s+2)/(2s+1) \end{bmatrix}, \quad (2.118a)$$

THE SCATTERING MATRIX

$$Z(s) = \frac{2}{46s+1} \begin{bmatrix} 2s+1 & 2s \\ 2s-10 & 2s+0.5 \end{bmatrix}. \qquad (2.118b)$$

The inverse of the matrix $Z(s) + z(s)$, which is also the admittance matrix $Y_{a1}(s)$ of the augmented two-port network N_{a1}, is obtained as

$$Y_{a1}(s) = [Z(s) + z(s)]^{-1}$$
$$= \frac{1}{q(s)} \begin{bmatrix} (s+1)(50s^2+50s+1.5) & -2s(s+1)(2s+1) \\ (s+1)(2s+1)(10-2s) & (2s+1)(2s^2+26s+1.5) \end{bmatrix}, \qquad (2.119a)$$

where

$$q(s) = 4s^3 + 62s^2 + 61s + 4.5. \qquad (2.119b)$$

Substituting these in (2.77) yields the current-based scattering matrix

$$S^I(s) = \frac{s+1}{q(s)}$$

$$\cdot \begin{bmatrix} \dfrac{4s^3+54s^2+45s-1.5}{s-1} & \dfrac{8s(2s^2-1)}{2s-1} \\ \dfrac{4(5-s)(2s+1)}{(s+1)(s-1)} & \dfrac{(2s+1)(-4s^3-42s^2+49s+1.5)}{(s+1)(2s-1)} \end{bmatrix}. \qquad (2.120)$$

The para-hermitian part of $z(s)$ is computed as

$$r(s) = \begin{bmatrix} 1/(1+s)(1-s) & 0 \\ 0 & 2(1-\sqrt{2}s)(1+\sqrt{2}s)/(1-2s)(1+2s) \end{bmatrix}$$
$$= h(s)h_*(s), \qquad (2.121a)$$

where

$$h(s) = \begin{bmatrix} 1/(1+s) & 0 \\ 0 & \sqrt{2}(1-\sqrt{2}s)/(1+2s) \end{bmatrix}. \qquad (2.121b)$$

Finally, using (2.89a) gives the normalized scattering matrix

$$S(s) = \frac{1}{q(s)}$$

$$\cdot \begin{bmatrix} -4s^3 - 54s^2 - 45s + 1.5 & 4\sqrt{2}s(1 - \sqrt{2}s) \\ 4\sqrt{2}(s-5)(1-\sqrt{2}s) & \dfrac{(1-\sqrt{2}s)(4s^3 + 42s^2 - 49s - 1.5)}{1+\sqrt{2}s} \end{bmatrix}. \quad (2.122)$$

Observe that since N is nonreciprocal, $S(s)$ is not symmetric. According to (2.114), the transducer power gain $G_{12}(\omega^2)$ from port 1 to port 2 is given by

$$G_{12}(\omega^2) = |S_{21}(j\omega)|^2 = \frac{32(2\omega^4 + 51\omega^2 + 25)}{16\omega^6 + 3356\omega^4 + 3163\omega^2 + 20.25}.$$

Likewise, the transducer power gain from port 2 to port 1 is $|S_{12}(j\omega)|^2$.

3.6. The normalized scattering matrix and passivity

In this section, we derive the power relations and the passivity criterion of an n-port network in terms of its normalized scattering matrix.

Using (2.91a) and (2.91b), we can express

$$\tfrac{1}{2}[V_*(s)I(s) + I_*(s)V(s)] = a_*(s)a(s) - b_*(s)b(s). \quad (2.124)$$

On the real-frequency axis, the average power absorbed by the n-port network is given by

$$P_{av} = \tfrac{1}{2}\text{Re}\,[V^*(j\omega)I(j\omega)] = \tfrac{1}{4}[V^*(j\omega)I(j\omega) + I^*(j\omega)V(j\omega)]$$
$$= \tfrac{1}{2}[a^*(j\omega)a(j\omega) - b^*(j\omega)b(j\omega)]$$
$$= \tfrac{1}{2}a^*(j\omega)[U_n - S^*(j\omega)S(j\omega)]a(j\omega). \quad (2.125)$$

For a passive n-port network, P_{av} is nonnegative for all $a(j\omega)$, showing that the hermitian matrix defined by the relation

$$Q(j\omega) = U_n - S^*(j\omega)S(j\omega) \qquad (2.126)$$

is nonnegative-definite. If, in addition, the n-port network is lossless, then $P_{av} = 0$ for all $a(j\omega)$, whence

$$S^*(j\omega)S(j\omega) = U_n = S(j\omega)S^*(j\omega). \qquad (2.127)$$

A matrix whose inverse equals to its transposed conjugate is called a *unitary matrix*. Thus, on the real-frequency axis, the normalized scattering matrix of a lossless n-port network is unitary. This unitary property of $S(j\omega)$ imposes additional constraints on its elements. For example, the absolute value of its determinant must be unity. In fact, a necessary and sufficient condition for $S(j\omega)$ to be unitary is that its columns (rows) be mutually orthogonal unit vectors. In terms of the elements of $S(j\omega)$, we have

$$\sum_{j=1}^{n} S_{ij}(j\omega)\bar{S}_{kj}(j\omega) = \delta_{ik}, \qquad (2.128a)$$

$$\sum_{j=1}^{n} S_{ji}(j\omega)\bar{S}_{jk}(j\omega) = \delta_{ik}, \qquad (2.128b)$$

δ_{ik} being the Kronecker delta.

In Example 2.4, it is straightforward to confirm that the normalized scattering matrix (2.97), when evaluated on the $j\omega$-axis, is unitary.

As stated in Theorem 1.2 of Chapter 1, a hermitian matrix is nonnegative definite if and only if all of its principal minors are nonnegative. Applying this to (2.126) for the first-order minors gives

$$1 - \sum_{j=1}^{n} \bar{S}_{ji}(j\omega)S_{ji}(j\omega) = 1 - \sum_{j=1}^{n} |S_{ji}(j\omega)|^2 \geq 0 \qquad (2.129)$$

for all ω, showing that for a passive n-port network, the magnitudes of its normalized reflection coefficients and transmission coefficients, when evaluated on the real-frequency axis, are bounded by unity, i.e.

$$|S_{ij}(j\omega)| \leq 1 \qquad (2.130)$$

for all ω. In §4 we shall show that this property together with (2.127) can be extended to the entire RHS.

3.7. The normalized scattering parameters of a lossless two-port network

Since lossless two-port networks are of considerable interest in practical applications, in this section we derive additional constraints imposed on their scattering parameters. Our starting point is (2.127) with $n=2$. This gives

$$|S_{11}(j\omega)|^2 + |S_{12}(j\omega)|^2 = 1, \qquad (2.131a)$$

$$|S_{21}(j\omega)|^2 + |S_{22}(j\omega)|^2 = 1, \qquad (2.131b)$$

$$|S_{11}(j\omega)|^2 + |S_{21}(j\omega)|^2 = 1, \qquad (2.131c)$$

$$|S_{12}(j\omega)|^2 + |S_{22}(j\omega)|^2 = 1. \qquad (2.131d)$$

Combining (2.131a) and (2.131d) or (2.131b) and (2.131c) yields

$$|S_{11}(j\omega)|^2 = |S_{22}(j\omega)|^2, \qquad (2.132)$$

showing that for a lossless two-port network, reciprocal or nonreciprocal, the magnitude of its reflection coefficient evaluated on the real-frequency axis at port 1 with port 2 terminating in its reference impedance is equal to that at port 2 with port 1 terminating in its reference impedance.

Also from (2.131) and (2.132), we get

$$|S_{12}(j\omega)|^2 = |S_{21}(j\omega)|^2, \qquad (2.133)$$

meaning that the magnitude of the transmission coefficient is the same in both directions, again for all real frequencies. Since $S(j\omega)$ is unitary, the off-diagonal elements of $S^*(j\omega)S(j\omega)$ or $S(j\omega)S^*(j\omega)$ are identically zero, resulting in

$$S_{11}(j\omega)\bar{S}_{12}(j\omega) = -S_{21}(j\omega)\bar{S}_{22}(j\omega), \qquad (2.134a)$$

$$S_{11}(j\omega)\bar{S}_{21}(j\omega) = -S_{12}(j\omega)\bar{S}_{22}(j\omega). \qquad (2.134b)$$

As an illustration, let us again consider the two-port network of Fig. 2.8. It is easy to confirm that the elements of (2.97), when evaluated on the $j\omega$-axis, satisfy the relations (2.131)–(2.134). An example of a lossless nonreciprocal two-port network possessing these properties is given in Problem 2.37.

4. The bounded-real scattering matrix

To investigate the global characteristics of the normalized scattering matrix in the entire complex-frequency plane, we first introduce the concept of a bounded-real matrix.

DEFINITION 2.1. *Bounded-real matrix.* A square matrix $A(s)$ is said to be *bounded-real* if it satisfies the following three conditions:

(i) $\bar{A}(s) = A(\bar{s})$ for all s in the open RHS,
(ii) (each of the elements of) $A(s)$ is analytic in the open RHS,
(iii) $U_n - A^*(s)A(s)$ is nonnegative-definite for all s in the open RHS.

DEFINITION 2.2. *Bounded-real function.* A 1×1 bounded-real matrix is called a *bounded-real function.*

The first condition in Definition 2.1 is equivalent to stating that each element of $A(s)$ is real when s is real, and, for rational $A(s)$, is always satisfied if all the coefficients of its elements are real. The third condition defines the boundedness of the quadratic form of $A(s)$ by virtue of the fact that, for all complex constant vectors x,

$$[A(s)x]^*[A(s)x] \leq x^*x, \qquad (2.135)$$

meaning that the sum of the squared magnitudes of the elements of the vector $A(s)x$ is bounded above by that of x for all x and all s in the open RHS. The second condition permits us to test the third one by using only $s = j\omega$ for rational $A(s)$, as will be demonstrated shortly. We note that the elements of $A(s)$ in the above definitions need not be rational functions. However, if they are, then not all of the three conditions are independent; for, in this case, conditions (i) and (iii) would require that $A(s)$ be analytic in the closed RHS. To see this, assume that the (i, j)-element of $A(s)$ has a pole in the closed RHS, and let x be the vector all of whose elements are zero except the jth one which is 1. It is clear that, in the neighborhood of this pole in the closed RHS, the inequality (2.135) is violated. Consequently, $A(s)$ must be analytic in the closed RHS.

We now state and prove a set of equivalent conditions for a rational matrix to be bounded-real. A generalization of this result to any matrix can be found in Chen (1973).

THEOREM 2.1. *A square matrix $A(s)$ of rational functions is bounded-real if and only if the following three conditions are satisfied:*

  (i) $\bar{A}(s) = A(\bar{s})$ *for all s in the open RHS,*
  (ii) $A(s)$ *is analytic in the closed RHS,*
  (iii) $U_n - A^*(j\omega)A(j\omega)$ *is nonnegative-definite for all real ω.*

Proof. Necessity. Let $A(s)$ be a bounded-real matrix of order n. Then condition (i) is certainly satisfied. The proof of condition (ii) was outlined above. Condition (iii) cannot be violated, since $x^*[U_n - A^*(s)A(s)]x$ is a continuous function of s and is nonnegative in the open RHS, and since $x^*[U_n - A^*(j\omega)A(j\omega)]x$ is a limit of this nonnegative function. Thus, all of the three conditions are necessary for $A(s)$ to be bounded-real.

Sufficiency. Let

$$M(s) = U_n - A(s). \qquad (2.136)$$

We first demonstrate that if $A(s)$ satisfies the three conditions stated in the theorem, the matrix $M(s)$ is a positive-real matrix. To this end, we first compute the hermitian part $M_h(s)$ of $M(s)$:

$$2M_h(s) = M(s) + M^*(s)$$
$$= M^*(s)M(s) + [U_n - A^*(s)A(s)], \qquad (2.137)$$

the matrix $M^*(s)M(s)$ being nonnegative definite for all s whenever $M(s)$ is defined (Problem 2.11). Since on the $j\omega$-axis the second term on the right-hand side of (2.137) is nonnegative-definite, it follows that $M_h(j\omega)$ is nonnegative-definite for all real ω which coupled with the fact that $M(s)$ is analytic in the closed RHS indicates that, after appealing to Theorem 1.3, $M(s)$ is a positive-real matrix. Two cases are considered.

Case 1. $\det M(s)$ is not identically zero. Then according to Theorem 1.4 the inverse $M^{-1}(s)$ is a positive-real matrix. Now we show that the matrix

$$Z(s) = 2M^{-1}(s) - U_n = [U_n + A(s)][U_n - A(s)]^{-1} \quad (2.138)$$

is also positive-real. For this purpose, we compute the hermitian part of $Z(s)$:

$$Z_h(s) = \tfrac{1}{2}[Z(s) + Z^*(s)] = M^{*-1}(s)[U_n - A^*(s)A(s)]M^{-1}(s). \quad (2.139)$$

Since, for all real ω, $U_n - A^*(j\omega)A(j\omega)$ is nonnegative-definite, it follows that $Z_h(j\omega)$ is also nonnegative-definite for all $j\omega$-axis points except those that are $j\omega$-axis poles of the elements of $M^{-1}(s)$.

Now consider a contour C_x enclosing the RHS, with its boundary being the $j\omega$-axis, except for arbitrarily small indentations into the RHS around the poles of $x^*M^{-1}(s)x$ on the $j\omega$-axis, as shown in Fig. 1.12. Note that C_x generally changes with x as $x^*M^{-1}(s)x$ does. For s on one of the indentations of C_x very near a finite $j\omega$-axis pole, the quadratic form of (2.138) becomes

$$x^*Z(s)x \approx 2x^*M^{-1}(s)x. \quad (2.140)$$

Since $M^{-1}(s)$ is positive-real, from Theorem 1.3 and its proof, we see that the hermitian part of $M^{-1}(s)$ is nonnegative-definite on these small indentations, and so is $Z_h(s)$. Thus, we conclude that

$$\operatorname{Re} x^*Z(s)x = x^*Z_h(s)x \geqq 0 \quad (2.141)$$

for all s on the contour C_x. Applying the maximum-modulus theorem to the function

$$f(s) = \exp[-x^*Z(s)x], \quad (2.142)$$

which is analytic inside and on C_x, we see that the maximum value of

$|f(s)|$, which corresponds to the minimum value of $x^*Z_h(s)x$, occurs on C_x. This means that (2.141) is satisfied for all s in the open RHS. Hence, $Z(s)$ is a positive-real matrix.

Finally, from (2.139) we have

$$U_n - A^*(s)A(s) = M^*(s)Z_h(s)M(s), \quad (2.143)$$

which is clearly nonnegative-definite for all s in the closed RHS.

Case 2. det $M(s)$ is identically zero. Let

$$\hat{M}(s) = U_n - DA(s), \quad (2.144)$$

where D is some diagonal matrix of entries 1 and -1 such that $\det[A(s) - D]$ is not identically zero (Problem 2.12). This would imply that $\hat{M}(s)$ is not identically singular, and following the arguments outlined in Case 1, we can show that $\hat{M}(s)$ and

$$\hat{Z}(s) = 2\hat{M}^{-1}(s) - U_n \quad (2.145)$$

are both positive-real. Since

$$U_n - A^*(s)A(s) = U_n - [DA(s)]^*DA(s) = \hat{M}^*(s)\hat{Z}_h(s)\hat{M}(s), \quad (2.146)$$

$\hat{Z}_h(s)$ being the hermitian part of $\hat{Z}(s)$, and since $\hat{Z}_h(s)$ is nonnegative-definite for all s in the closed RHS, we conclude that $U_n - A^*(s)A(s)$ is nonnegative-definite for all s in the closed RHS. So $A(s)$ is a bounded-real matrix. This completes the proof of the theorem.

Before proceeding to the discussion of the scattering matrix, we note that one might attempt to prove the nonnegative-definiteness of $U_n - A^*(s)A(s)$ in the open RHS by directly appealing to the maximum-modulus theorem and conditions (ii) and (iii) of Theorem 2.1. This argument, however, is not valid since the matrix $U_n - A^*(s)A(s)$ is not analytic in the open RHS.

We now summarize our discussions on the characterizations of a passive n-port network in terms of its normalized scattering matrix by the following fundamental theorem, which is perhaps the most penetrating result presently available in network theory.

THEOREM 2.2. *The necessary and sufficient conditions for an $n \times n$ matrix $S(s)$ to be the scattering matrix of a linear, lumped, time-invariant and passive n-port network, normalizing to the n minimum reactance functions (no poles on the real-frequency axis), are that*

(i) $S(s)$ *be rational and bounded-real;*
(ii) *the matrix*

$$Y_{a1}(s) = \tfrac{1}{2}h^{-1}(s)[d(s) - S(s)]h^{-1}(s) \qquad (2.147)$$

be analytic in the open RHS;

(iii) $Y_{a1}(s)$ *have at most simple poles on the real-frequency axis and the residue matrix evaluated at each of these poles be hermitian and nonnegative-definite;*

where $d(s) = h(s)h_*^{-1}(s)$, and $h(s)h_*(s)$ is the para-hermitian part of the reference impedance matrix $z(s)$ corresponding to the n minimum reactance functions.

Proof. Necessity. Condition (i) follows directly from (2.126), Theorem 2.1 and the fact that the normalized scattering matrix $S(s)$ of a passive n-port network N is analytic in the closed RHS. If N is passive, Y_{a1} represents the short-circuit admittance matrix of the augmented passive n-port network N_{a1}, as defined in § 3.3 and (2.896). Thus, $Y_{a1}(s)$ is positive-real, and by appealing to Theorem 1.3 conditions (ii) and (iii) must be satisfied.

Sufficiency. Two cases are considered. *Case* 1. $\det[d(s) - S(s)] \not\equiv 0$. Then the matrix

$$Z(s) = 2h(s)[d(s) - S(s)]^{-1}h(s) - z(s) \qquad (2.148)$$

is well defined, and by (2.89b) it is the impedance matrix of N. To complete the proof, it is sufficient to show that $Z(s)$ is positive-real. From (2.148), it is found that, after a little simplification, the para-hermitian part $Z_p(s)$ of $Z(s)$ is given by

$$Z_p(s) = h_*(s)[d_*(s) - S_*(s)]^{-1}[U_n - S_*(s)S(s)][d(s) - S(s)]^{-1}h(s). \qquad (2.149)$$

Since $S(s)$ is rational and bounded-real, by Theorem 2.1

$$U_n - S_*(j\omega)S(j\omega) = U_n - S^*(j\omega)S(j\omega) \qquad (2.150)$$

is nonnegative-definite for all real ω, from which it follows that, on the $j\omega$-axis, the hermitian part $Z_h(j\omega)$ of $Z(s)$, which is the same as $Z_p(j\omega)$, is also nonnegative-definite whenever $[d(j\omega) - S(j\omega)]^{-1}$ is defined. Under the stipulated hypotheses, $z(s)$ is analytic and its hermitian part is nonnegative definite in the closed RHS, and in particular on the real-frequency axis. Thus, we conclude from (2.148), which can also be expressed as

$$Z(s) = Y_{a1}^{-1}(s) - z(s), \qquad (2.151)$$

that the hermitian part of $Y_{a1}^{-1}(s)$ is nonnegative-definite on the $j\omega$-axis, and that the residue matrix at each of the $j\omega$-axis poles of $Z(s)$ is nonnegative-definite if and only if the corresponding residue matrix of $Y_{a1}^{-1}(s)$ is nonnegative-definite. Now the nonnegativity of the hermitian part of $Y_{a1}^{-1}(s)$ on the $j\omega$-axis implies the nonnegativity of the hermitian part of $Y_{a1}(s)$ on the $j\omega$-axis. This, together with conditions (ii) and (iii), would imply, after appealing to Theorem 1.3, that $Y_{a1}(s)$ is positive-real, and therefore so does its inverse $Y_{a1}^{-1}(s)$. This means that the $j\omega$-axis poles of $Y_{a1}^{-1}(s)$ are simple, and the residue matrix evaluated at each of these poles is hermitian and nonnegative-definite. In view of the above comments and again appealing to Theorem 1.3, $Z(s)$ is positive-real.

Case 2. $\det[d(s) - S(s)] \equiv 0$. Then as in the proof for Case 2 of Theorem 2.1, there exists a diagonal matrix D whose diagonal elements are 1 or -1, and such that

$$\det[d(s) - DS(s)] \not\equiv 0. \qquad (2.152)$$

Consider the matrix

$$Z(s) = 2h(s)[d(s) - DS(s)]^{-1}h(s) - z(s), \qquad (2.153)$$

which is well defined. Following Case 1, we can also prove this case. However, the details are omitted, Q.E.D.

It is of practical interest to single out an important special case where the impedance functions, to which the scattering matrix is normalized, are positive-real constants.

COROLLARY 2.1. *An $n \times n$ rational matrix is the scattering matrix of a linear, lumped, time-invariant and passive n-port network, normalizing to the n positive resistances, if and only if it is bounded-real.*

Proof. To prove the corollary, it suffices, under the stipulated hypothesis, to show that condition (i) of Theorem 2.2 would imply the positive realness of $Y_{a1}(s)$. From (2.147), it is clear that $Y_{a1}(s)$ is positive-real if and only if $U_n - S(s)$ is positive-real. But in the proof of Theorem 2.1, we showed that $U_n - S(s)$ is positive-real if $S(s)$ is rational and bounded-real. The conclusion of the corollary follows.

We now proceed to consider the lossless n-port networks. However, before we do this, we need the concept of the para-unitary matrix.

DEFINITION 2.3. *Para-unitary matrix.* An $n \times n$ matrix $A(s)$ is called *para-unitary* if

$$A_*(s)A(s) = U_n. \tag{2.154}$$

In the following, we show that, for a lossless n-port network, its scattering matrix normalizing to the n strictly passive impedances is para-unitary. To this end, let

$$Q(s) = U_n - S_*(s)S(s). \tag{2.155}$$

Since from (2.127), $Q(j\omega) = O_n$ for a lossless n-port network, and since $Q(s)$ represents the analytic continuation of $Q(j\omega)$, it follows that each element of $Q(s)$ must vanish identically for s with which $Q(s)$ is defined. Thus, we conclude that $S(s)$, in addition to being rational and bounded-real, is para-unitary for almost all the complex frequencies. These properties of $S(s)$ are all necessary conditions that $S(s)$ must possess, but they are not sufficient as demonstrated in Theorem 2.2. However, if the reference impedances, to which the scattering matrix is normalized, are positive-real constants, the above conditions become both necessary and sufficient.

COROLLARY 2.2. *An $n \times n$ rational matrix is the scattering matrix of a linear, lumped, time-invariant and lossless n-port network, normalizing to the n positive resistances, if and only if it is bounded-real and para-unitary whenever it is defined.*

The proof of the corollary is straightforward, and is left as an exercise (Problem 2.15).

We remark that from (2.126) one may attempt to conclude through analytic continuation that, for a lossless n-port network, each element of the matrix

$$U_n - S^*(s)S(s) \qquad (2.156)$$

must also be identically zero for almost all s. This is *not* valid since the analytic continuation of (2.126) is (2.155), not (2.156), the reason being that the elements of (2.156) are not analytic functions in general.

As an illustration, consider the lossless two-port network of Fig. 2.8 whose normalized scattering matrix is given in (2.97). It is easy to confirm that $S(s)S_*(s) = S_*(s)S(s) = U_n$, which reduces to $S^*(j\omega)S(j\omega) = S(j\omega)S^*(j\omega) = U_n$ on the real-frequency axis. However, we can easily verify that for $s \neq j\omega$

$$S^*(s)S(s) \neq U_n. \qquad (2.157)$$

5. Interconnection of multi-port networks

It is well known that simple multi-ports can be interconnected to make up a complex n-port network. The reasons for doing this are that from the designer's viewpoint it is much easier to design simple blocks and interconnect them than to design a complex network in one piece, and that each specific type of interconnection can be accompanied by a characterization which facilitates the matrix manipulation. For example, if the voltage–current relationships of the individual multi-ports remain unaltered when they are connected in parallel (series), the admittance (impedance) matrix of the composite network is the sum of the admittance (impedance) matrices of the individual networks. In this section, we shall present formulas relating the scattering matrix of an n-port network in terms of the scattering matrices of its component networks.

Let N be an n-port network which is an interconnection of an n_a-port N_a and an n_b-port N_b, as depicted symbolically in Fig. 2.15. The scattering equations in partitioned form for N, N_a and N_b, normalized with respect to the reference impedance matrices z, z_a

and z_b, respectively, are given by

$$\begin{bmatrix} b_1 \\ b_2 \end{bmatrix} = \begin{bmatrix} S_{11} & S_{12} \\ S_{21} & S_{22} \end{bmatrix} \begin{bmatrix} a_1 \\ a_2 \end{bmatrix}, \qquad (2.158a)$$

$$\begin{bmatrix} b_{1a} \\ b_{2a} \end{bmatrix} = \begin{bmatrix} S_{11a} & S_{12a} \\ S_{21a} & S_{22a} \end{bmatrix} \begin{bmatrix} a_{1a} \\ a_{2a} \end{bmatrix}, \qquad (2.158b)$$

$$\begin{bmatrix} b_{1b} \\ b_{2b} \end{bmatrix} = \begin{bmatrix} S_{11b} & S_{12b} \\ S_{21b} & S_{22b} \end{bmatrix} \begin{bmatrix} a_{1b} \\ a_{2b} \end{bmatrix} \qquad (2.158c)$$

with

$$z = \begin{bmatrix} z_1 & 0 \\ 0 & z_2 \end{bmatrix}, \qquad (2.159a)$$

$$z_a = \begin{bmatrix} z_{1a} & 0 \\ 0 & z_{2a} \end{bmatrix}, \qquad (2.159b)$$

$$z_b = \begin{bmatrix} z_{1b} & 0 \\ 0 & z_{2b} \end{bmatrix}. \qquad (2.159c)$$

In order to simplify the formula and to make the physical interpretation of the incident and reflected waves meaningful, we set

$$z_1 = z_{1a}, \qquad z_{2a} = (z_{1b})_* = z_{1b*}, \qquad z_{2b} = z_2. \qquad (2.160)$$

Then from (2.92) the incident and reflected waves for the networks N, N_a and N_b can be expressed in terms of their port voltages and currents, as indicated in Fig. 2.15 and again in partitioned form:

$$\begin{aligned} a_1 &= \tfrac{1}{2} h_1^{-1}(V_1 + z_1 I_1) \\ &= \tfrac{1}{2} h_{1a}^{-1}(V_1 + z_{1a} I_1) = a_{1a}, \end{aligned} \qquad (2.161a)$$

$$\begin{aligned} b_{1b} &= \tfrac{1}{2} h_{1b*}^{-1}(V_3 - z_{1b*} I_3) \\ &= \tfrac{1}{2} h_{2a}^{-1}[V_3 + z_{2a}(-I_3)] = a_{2a}, \end{aligned} \qquad (2.161b)$$

where h_1, h_{1a}, h_{1b} and h_{2a} are the decompositions of the parahermitian parts of z_1, z_{1a}, z_{1b} and z_{2a}, respectively. Similarly, we can show that

$$a_{2b} = a_2, \qquad (2.162a)$$

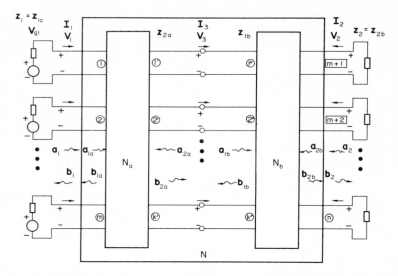

FIG 2.15. An n-port network composed of an interconnection of an n_a-port network N_a and an n_b-port network N_b.

$$b_{2a} = a_{1b}, \qquad (2.162b)$$

$$b_{1a} = b_1, \qquad (2.162c)$$

$$b_{2b} = b_2. \qquad (2.162d)$$

Using (2.161) and (2.162), the second equation of (2.158b) and the first of (2.158c) become

$$a_{1b} = S_{21a}a_1 + S_{22a}a_{2a}, \qquad (2.163a)$$

$$a_{2a} = S_{11b}a_{1b} + S_{12b}a_2. \qquad (2.163b)$$

From these two equations, we can solve for a_{1b} and a_{2a} in terms of a_1 and a_2:

$$a_{1b} = (U_k - S_{22a}S_{11b})^{-1}(S_{21a}a_1 + S_{22a}S_{12b}a_2), \qquad (2.164a)$$

$$a_{2a} = (U_k - S_{11b}S_{22a})^{-1}(S_{11b}S_{21a}a_1 + S_{12b}a_2), \qquad (2.164b)$$

assuming, of course, the existence of the required inverses. Substituting (2.164) in the first equation of (2.158b) and the second of

THE SCATTERING MATRIX

(2.158c) and applying (2.161) and (2.162) yield the desired formulas

$$S_{11} = S_{11a} + S_{12a}(U_k - S_{11b}S_{22a})^{-1}S_{11b}S_{21a}, \qquad (2.165a)$$

$$S_{12} = S_{12a}(U_k - S_{11b}S_{22a})^{-1}S_{12b}, \qquad (2.165b)$$

$$S_{21} = S_{21b}(U_k - S_{22a}S_{11b})^{-1}S_{21a}, \qquad (2.165c)$$

$$S_{22} = S_{22b} + S_{21b}(U_k - S_{22a}S_{11b})^{-1}S_{22a}S_{12b}. \qquad (2.165d)$$

We remark that the formulas are valid only under the assumption of (2.160). This is equivalent to saying that if the reference impedance matrix z_{2a} of the output ports of N_a is equal to z_{1b*}, z_{1b} being the reference impedance matrix for the input ports of N_b, then (2.165) represents the submatrices of the scattering matrix of the composite n-port network N, normalizing to the reference impedance matrix

$$z = \begin{bmatrix} z_{1a} & 0 \\ 0 & z_{2b} \end{bmatrix}, \qquad (2.166)$$

z_{1a} and z_{2b} being the reference impedance matrices of the input and output ports of N_a and N_b, respectively.

It is significant to point out that, under the assumed constraints (2.160), the identities (2.161) and (2.162) can also be obtained intuitively from the physical argument of waves as symbolically depicted in Fig. 2.15. At the k ports of interconnection of N_a and N_b, the wave incident to one network must be that reflected from the other since there are no independent sources at these junctions.

Finally, we mention that in Fig. 2.15 if N_a is an $(n + k)$-port and N_b a k-port, the scattering matrix of the resulting n-port N can be greatly simplified, and is given by

$$S = S_{11a} + S_{12a}(U_k - S_b S_{22a})^{-1} S_b S_{21a}, \qquad (2.167)$$

where S_b is the scattering matrix of N_b. Putting it differently, (2.167) indicates that the network N_b of scattering matrix S_b is transformed by the network N_a of scattering matrix S_a to yield the n-port network N of scattering matrix S.

We shall illustrate the use of the above formulas by the following example.

EXAMPLE 2.8. Consider the network N of Fig. 2.16, which can be viewed as an interconnection of a four-port ideal transformer N_a and a six-port N_b made up of three two-ports, as indicated in Fig. 2.16. For simplicity, let

$$k_3^2 = \frac{1}{2R} \quad \text{and} \quad k_1^2 = 2R_1 k_2^2. \tag{2.168}$$

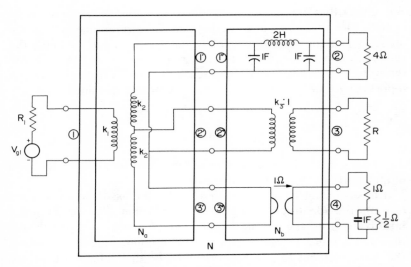

FIG. 2.16. A four-port network which can be viewed as an interconnection of a four-port ideal transformer N_a and a six-port network N_b composed of three two-port networks.

It is straightforward to show that the scattering matrix S_a of the four-port N_a, normalizing to the reference impedance matrix

$$z_a = \begin{bmatrix} R_1 & 0 & 0 & 0 \\ 0 & 1 & 0 & 0 \\ 0 & 0 & \frac{1}{2} & 0 \\ 0 & 0 & 0 & 1 \end{bmatrix} = \begin{bmatrix} z_{1a} & \mathbf{0} \\ \mathbf{0} & z_{1b} \end{bmatrix}, \tag{2.169}$$

is given by

$$S_a = \frac{1}{\sqrt{2}} \begin{bmatrix} 0 & 1 & 0 & 1 \\ 1 & 0 & 1 & 0 \\ 0 & 1 & 0 & -1 \\ 1 & 0 & -1 & 0 \end{bmatrix} = \begin{bmatrix} S_{11a} & S_{12a} \\ S_{21a} & S_{22a} \end{bmatrix}. \quad (2.170)$$

From Problems 2.4, 2.6 and 2.8, the scattering matrix S_b of the six-port N_b, normalizing to the reference impedance matrix

$$z_b = \begin{bmatrix} 1 & 0 & 0 & 0 & 0 & 0 \\ 0 & \tfrac{1}{2} & 0 & 0 & 0 & 0 \\ 0 & 0 & 1 & 0 & 0 & 0 \\ 0 & 0 & 0 & 4 & 0 & 0 \\ 0 & 0 & 0 & 0 & R & 0 \\ 0 & 0 & 0 & 0 & 0 & \dfrac{s+3}{s+2} \end{bmatrix} = \begin{bmatrix} z_{1b} & 0 \\ 0 & z_{2b} \end{bmatrix}, \quad (2.171)$$

is obtained as

$$S_b = \begin{bmatrix} \dfrac{w(s)}{q(s)} & 0 & 0 & \dfrac{4}{q(s)} & 0 & 0 \\ 0 & 0 & 0 & 0 & 1 & 0 \\ 0 & 0 & \dfrac{-1}{2s+5} & 0 & 0 & \dfrac{-2(s-\sqrt{6})}{2s+5} \\ \dfrac{4}{q(s)} & 0 & 0 & \dfrac{-w(-s)}{q(s)} & 0 & 0 \\ 0 & 1 & 0 & 0 & 0 & 0 \\ 0 & 0 & \dfrac{2(s-\sqrt{6})}{2s+5} & 0 & 0 & \dfrac{(s-\sqrt{6})}{(2s+5)(s+\sqrt{6})} \end{bmatrix}$$

$$\quad (2.172a)$$

$$= \begin{bmatrix} S_{11b} & S_{12b} \\ S_{21b} & S_{22b} \end{bmatrix}, \quad (2.172b)$$

where
$$w(s) = -8s^3 + 6s^2 - 6s + 3, \tag{2.173a}$$
$$q(s) = 8s^3 + 10s^2 + 10s + 5. \tag{2.173b}$$

We next compute the matrix

$$U_3 - S_{11b}S_{22a} = \begin{bmatrix} 1 & 0 & 0 \\ 0 & 1 & 0 \\ 0 & 0 & 1 \end{bmatrix} - 2^{-1/2} \begin{bmatrix} \dfrac{w(s)}{q(s)} & 0 & 0 \\ 0 & 0 & 0 \\ 0 & 0 & \dfrac{-1}{2s+5} \end{bmatrix} \begin{bmatrix} 0 & 1 & 0 \\ 1 & 0 & -1 \\ 0 & -1 & 0 \end{bmatrix}$$

$$= \begin{bmatrix} 1 & -2^{-1/2}\dfrac{w(s)}{q(s)} & 0 \\ 0 & 1 & 0 \\ 0 & \dfrac{-2^{-1/2}}{2s+5} & 1 \end{bmatrix}, \tag{2.174}$$

whose inverse is

$$(U_3 - S_{11b}S_{22a})^{-1} = \begin{bmatrix} 1 & 2^{-1/2}\dfrac{w(s)}{q(s)} & 0 \\ 0 & 1 & 0 \\ 0 & \dfrac{2^{-1/2}}{2s+5} & 1 \end{bmatrix}. \tag{2.175}$$

Substituting these in (2.165a) and (2.165b) yields

$$S_{11} = S_{11a} + S_{12a}(U_3 - S_{11b}S_{22a})^{-1}S_{11b}S_{21a}$$

$$= 0 + \tfrac{1}{2}[1 \quad 0 \quad 1] \begin{bmatrix} 1 & 2^{-1/2}\dfrac{w(s)}{q(s)} & 0 \\ 0 & 1 & 0 \\ 0 & \dfrac{2^{-1/2}}{2s+5} & 1 \end{bmatrix} \begin{bmatrix} \dfrac{w(s)}{q(s)} & 0 & 0 \\ 0 & 0 & 0 \\ 0 & 0 & \dfrac{-1}{2s+5} \end{bmatrix} \begin{bmatrix} 1 \\ 0 \\ 1 \end{bmatrix}$$

$$= \frac{1}{2}\left(\frac{w(s)}{q(s)} - \frac{1}{2s+5}\right), \tag{2.176}$$

THE SCATTERING MATRIX

$S_{12} = S_{12a}(U_3 - S_{11b}S_{22a})^{-1}S_{12b}$

$$= 2^{-1/2}[1 \quad 0 \quad 1]\begin{bmatrix} 1 & 2^{-1/2}\dfrac{w(s)}{q(s)} & 0 \\ 0 & 1 & 0 \\ 0 & \dfrac{2^{-1/2}}{2s+5} & 1 \end{bmatrix}\begin{bmatrix} \dfrac{4}{q(s)} & 0 & 0 \\ 0 & 1 & 0 \\ 0 & 0 & \dfrac{-2(s-\sqrt{6})}{2s+5} \end{bmatrix}$$

$$= 2^{-1/2}\left[\dfrac{4}{q(s)} \quad 2^{-1/2}\left(\dfrac{w(s)}{q(s)} + \dfrac{1}{2s+5}\right) \quad \dfrac{-2(s-\sqrt{6})}{2s+5}\right]. \quad (2.177)$$

Similarly, we compute the inverse of $(U_3 - S_{22a}S_{11b})$, which is

$$(U_3 - S_{22a}S_{11b})^{-1} = \begin{bmatrix} 1 & 0 & 0 \\ 2^{-1/2}\dfrac{w(s)}{q(s)} & 1 & \dfrac{2^{-1/2}}{2s+5} \\ 0 & 0 & 1 \end{bmatrix}, \quad (2.178)$$

and from (2.165c) and (2.165d) we obtain

$$S_{21} = \dfrac{1}{2}\begin{bmatrix} \dfrac{4\sqrt{2}}{q(s)} \\ \dfrac{w(s)}{q(s)} + \dfrac{1}{2s+5} \\ \dfrac{2\sqrt{2}(s-\sqrt{6})}{2s+5} \end{bmatrix}, \quad (2.179)$$

$$S_{22} = \begin{bmatrix} -\dfrac{w(-s)}{q(s)} & \dfrac{2\sqrt{2}}{q(s)} & 0 \\ \dfrac{2\sqrt{2}}{q(s)} & \dfrac{w(s)}{2q(s)} - \dfrac{1}{2(2s+5)} & \dfrac{\sqrt{2}(s-\sqrt{6})}{2s+5} \\ 0 & \dfrac{-\sqrt{2}(s-\sqrt{6})}{2s+5} & \dfrac{s-\sqrt{6}}{(2s+5)(s+\sqrt{6})} \end{bmatrix}. \quad (2.180)$$

Combining these submatrices yields the scattering matrix $S(s)$ of the composite four-port network N, normalizing to the reference impedance matrix

$$z = \begin{bmatrix} R_1 & 0 & 0 & 0 \\ 0 & 4 & 0 & 0 \\ 0 & 0 & R & 0 \\ 0 & 0 & 0 & \dfrac{s+3}{s+2} \end{bmatrix}. \quad (2.181)$$

The resulting matrix is given by

$$S(s) =$$

$$\begin{bmatrix} \dfrac{w(s)}{2q(s)} - \dfrac{1}{2(2s+5)} & \dfrac{2\sqrt{2}}{q(s)} & \dfrac{w(s)}{2q(s)} + \dfrac{1}{2(2s+5)} & \dfrac{-\sqrt{2}(s-\sqrt{6})}{2s+5} \\ \dfrac{2\sqrt{2}}{q(s)} & \dfrac{-w(-s)}{q(s)} & \dfrac{2\sqrt{2}}{q(s)} & 0 \\ \dfrac{w(s)}{2q(s)} + \dfrac{1}{2(2s+5)} & \dfrac{2\sqrt{2}}{q(s)} & \dfrac{w(s)}{2q(s)} - \dfrac{1}{2(2s+5)} & \dfrac{\sqrt{2}(s-\sqrt{6})}{2s+5} \\ \dfrac{\sqrt{2}(s-\sqrt{6})}{2s+5} & 0 & \dfrac{-\sqrt{2}(s-\sqrt{6})}{2s+5} & \dfrac{s-\sqrt{6}}{(2s+5)(s+\sqrt{6})} \end{bmatrix}$$

$$(2.182)$$

We remark that since N is lossless but not reciprocal, according to Theorem 2.2 and the paragraph following (2.155), $S(s)$ must be a bounded-real and para-unitary matrix whenever it is defined, but it is not symmetric. It is interesting to note that the submatrix formed by the first three rows and columns is symmetric since it corresponds to the reciprocal part of N. We leave to the reader to confirm that

$$S(s)S_*(s) = S_*(s)S(s) = U_4. \quad (2.183)$$

6. Conclusions

In this chapter we began our discussion on the scattering matrix of a one-port network based on the concepts from the transmission-line theory, and then extended it to an n-port network. Much of the manipulation of the first three sections could have been avoided if we had at the outset defined the normalized waves in terms of the actual port voltages and currents. This would be at the expense of the intuitive physical interpretation and motivation.

To summarize our results, in this chapter we showed that to any linear, lumped, passive and time-invariant n-port network and any prescribed set of n strictly passive impedances, we can define a basis-independent normalized scattering matrix $S(s)$ that enjoys the following characteristics:

(i) $S(s)$ is analytic in the open RHS;
(ii) $U_n - S^*(s)S(s)$ is nonnegative-definite for all s in the closed RHS;
(iii) $S(s)$ is para-unitary if the n-port network is lossless.
(iv) $S(s)$ is symmetric when both the n-port network and its normalizing load impedances are reciprocal;
(v) $|S_{ij}(j\omega)|^2$ is the transducer power gain from port j to port i.

Since a complex n-port network is usually made up of several simple multi-ports, to simplify the computation of its scattering matrix, in §5 we presented formulas relating the scattering matrix of the n-port network in terms of the scattering matrices of the component multi-port networks.

The application of the scattering matrix to the synthesis of multi-ports is beyond the scope of this book, and we refer to Belevitch (1968) and Newcomb (1966) for a detailed treatment. However, its applications to the design of optimum equalizers in the theory of broadband matching and other related topics will be discussed thoroughly in the subsequent chapters.

Problems

2.1. A one-port network N containing sinusoidal sources can be represented by its Thévenin equivalent consisting of a voltage source and an impedance

$$Z_0(j\omega) = |Z_0|e^{j\theta_0}. \tag{2.184}$$

Let

$$Z(j\omega) = |Z|e^{j\theta} \tag{2.185}$$

be the load of N. Assume that $|Z|$ and θ can be independently varied.

108 BROADBAND MATCHING NETWORKS

(i) If θ is held constant, show that the real power entering the load will be maximum if $|Z| = |Z_0|$.

(ii) If $|Z|$ is held constant, find the condition under which maximum power will be transferred to the load.

(iii) Under the condition of maximum power transfer of (ii), show that the magnitude of θ will be largest when $|Z| = |Z_0|$.

(iv) If $|Z|$ and θ can both be varied, show that maximum power transfer will occur when $Z = \bar{Z}_0$.

2.2. Derive the identity (2.80).

2.3. Consider the ideal transformer N of Fig. 2.7. Determine its scattering matrix normalized to the reference impedance matrix as given in (2.118a).

2.4. Consider the ideal transformer of Fig. 2.7. Show that its scattering matrix normalizing to the load resistances R_1 and R_2 is given by

$$S = \frac{1}{R_1 + m^2 R_2} \begin{bmatrix} m^2 R_2 - R_1 & 2mR_1^{1/2} R_2^{1/2} \\ 2mR_1^{1/2} R_2^{1/2} & R_1 - m^2 R_2 \end{bmatrix}. \quad (2.186)$$

2.5. Show that the admittance matrix Y_{a1} and impedance matrix Z_{a2} of the augmented two-port networks N_{a1} and N_{a2} of the ideal transformer of Fig. 2.7 are given, respectively, by

$$Y_{a1} = \frac{1}{R_1 + m^2 R_2} \begin{bmatrix} 1 & -m \\ -m & m^2 \end{bmatrix}, \quad Z_{a2} = \frac{R_1 R_2}{R_1 + m^2 R_2} \begin{bmatrix} m^2 & m \\ m & 1 \end{bmatrix}. \quad (2.187)$$

2.6. A nonreciprocal two-port network N is composed of an ideal gyrator of gyration resistance r in series with an R-Ω resistor. The impedance matrix of the resulting two-port network N is given by

$$Z = \begin{bmatrix} R & -r \\ r & 0 \end{bmatrix}. \quad (2.188)$$

Show that the scattering matrix of N normalizing to the resistances R_1 and R_2 is given by

$$S = \frac{1}{R_2(R + R_1) + r^2} \begin{bmatrix} R_2(R - R_1) + r^2 & -2R_1^{1/2} R_2^{1/2} r \\ 2R_1^{1/2} R_2^{1/2} r & -R_2(R + R_1) + r^2 \end{bmatrix}. \quad (2.189)$$

2.7. Show that on the real-frequency axis, a linear time-invariant n-port network is lossless if and only if its normalized scattering matrix is unitary.

2.8. Confirm that the scattering matrix of the lossless two-port network of Fig. 2.8, normalizing to the resistances $z_1 = 4\Omega$ and $z_2 = 1\Omega$, is given by

$$S(s) = \frac{1}{q(s)} \begin{bmatrix} -8s^3 - 6s^2 - 6s - 3 & 4 \\ 4 & -8s^3 + 6s^2 - 6s + 3 \end{bmatrix}, \quad (2.190a)$$

where

$$q(s) = 8s^3 + 10s^2 + 10s + 5. \quad (2.190b)$$

THE SCATTERING MATRIX 109

2.9. Show that the matrix (2.190a) is para-unitary whenever it is defined.

2.10. Confirm that (2.35) is an all-pass function.

2.11. In (2.137), prove that the matrix $M^*(s)M(s)$ is nonnegative-definite for all s whenever $M(s)$ is defined.

2.12. In (2.144) show that there exists a diagonal matrix D whose diagonal entries are 1 or -1 such that $A(s) - D$ is not identically singular.

2.13. Using (2.104c), compute the scattering matrix of an ideal gyrator with gyration resistance of 1 Ω, normalized to the load impedances as given in (2.118a).

2.14. In Theorem 2.2, show that condition (iii) is always satisfied if $h^{-1}(s)$ is analytic on the real-frequency axis, which is equivalent to the statement that the even parts of the reference impedances are devoid of zeros on the entire real-frequency axis.

2.15. Prove Corollary 2.2.

2.16. Show that the characterization of a lossless n-port network is the same for all three scattering matrices:

$$S_*(s)S(s) = S_*^I(s)S^I(s) = S_*^V(s)S^V(s) = U_n. \qquad (2.191)$$

2.17. Show that the scattering matrix (2.182) is para-unitary whenever it is defined.

2.18. Using Theorem 2.1, show that the scattering matrix (2.182) is bounded-real.

2.19. Assume that the reference impedances of an n-port network to which its scattering matrix is normalized are all real and positive. Show that (2.89) and (2.105) can be simplified to

$$S(s) = [Z_n(s) + U_n]^{-1}[Z_n(s) - U_n] = [Z_n(s) - U_n][Z_n(s) + U_n]^{-1}, \qquad (2.192a)$$
$$= [U_n + Y_n(s)]^{-1}[U_n - Y_n(s)] = [U_n - Y_n(s)][U_n + Y_n(s)]^{-1}, \qquad (2.192b)$$

in which the factors commute, where

$$Z_n(s) = r^{-1/2}Z(s)r^{-1/2}, \qquad (2.193a)$$
$$Y_n(s) = r^{1/2}Y(s)r^{1/2}, \qquad (2.193b)$$

and $r^{1/2}$ and $r^{-1/2}$ denote the diagonal matrices whose diagonal elements are $r_k^{1/2}$ and $r_k^{-1/2}$, respectively, r_k being the reference resistances.

2.20. Consider the three-port network of Fig. 2.17 which is formed with an ideal gyrator of gyration resistance $r = 1\Omega$ by making an extra port as illustrated in the figure. Show that its scattering matrix normalizing to the 1-Ω resistance is given by

$$S = \begin{bmatrix} 0 & 1 & 0 \\ 0 & 0 & -1 \\ 1 & 0 & 0 \end{bmatrix}. \qquad (2.194)$$

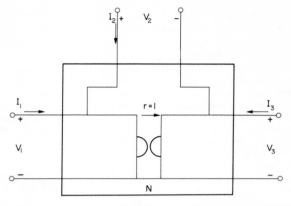

FIG. 2.17. A three-port network obtained from an ideal gyrator by making an extra port as indicated.

2.21. Let $S(s)$ be the scattering matrix of a two-port network normalizing to the 1-Ω resistance. Let h_{ij} be the hybrid parameters of the two-port network. Show that

$$S(s) = \begin{bmatrix} h_{11}+1 & -h_{12} \\ h_{21} & -h_{22}-1 \end{bmatrix}^{-1} \begin{bmatrix} h_{11}-1 & h_{12} \\ h_{21} & h_{22}-1 \end{bmatrix}. \quad (2.195)$$

2.22. A three-port network is characterized by its normalized scattering matrix

$$S = \begin{bmatrix} \alpha_1 & \beta_1 & 0 \\ \beta_4 & \alpha_2 & \beta_2 \\ \beta_3 & 0 & \alpha_3 \end{bmatrix}, \quad (2.196)$$

where β's are nonzero. Show that if the three-port network is lossless, then S can be written as

$$S = \begin{bmatrix} 0 & 1 & 0 \\ \sin\theta & 0 & -\cos\theta \\ \cos\theta & 0 & \sin\theta \end{bmatrix}, \quad (2.197)$$

where $\cos\theta = \beta_3$.

2.23. In Problem 2.22, set $\beta_4 = 0$. Show that if the three-port network is lossless, then $\alpha_1 = \alpha_2 = \alpha_3 = \beta_4 = 0$ and $\beta_x = e^{j\theta_x}$ ($x = 1, 2, 3$) or $S = U_3$. Compare this with (2.194).

2.24. A lossless reciprocal three-port network is characterized by its normalized scattering matrix:

$$S = \begin{bmatrix} \alpha & \beta & \beta \\ \beta & \alpha & \beta \\ \beta & \beta & \alpha \end{bmatrix}, \quad (2.198)$$

α being real. Prove that $\alpha \neq 0$, and determine the minimum α.

2.25. Show that the product of two bounded-real matrices is bounded-real. [*Hint*. Let $S(s) = S_1(s)S_2(s)$. Then

$$U_n - S^*(s)S(s) = S_2^*(s)[U_n - S_1^*(s)S_1(s)]S_2(s) + [U_n - S_2^*(s)S_2(s)].] \quad (2.199)$$

2.26. Show that (2.195) can be expressed as

$$\begin{bmatrix} h_{11} & h_{12} \\ h_{21} & h_{22} \end{bmatrix} = -\begin{bmatrix} S_{11}-1 & S_{12} \\ S_{21} & S_{22}+1 \end{bmatrix}^{-1}\begin{bmatrix} S_{11}+1 & S_{12} \\ S_{21} & S_{22}-1 \end{bmatrix}, \quad (2.200)$$

where S_{ij} are the elements of S.

2.27. Let S be the scattering matrix of a two-port network N normalizing to the 1-Ω resistance. Let A, B, C and D be the transmission parameters of N. Show that the scattering parameters S_{ij} and the transmission parameters are related by the equations

$$S = \begin{bmatrix} 1 & -A-B \\ -1 & -C-D \end{bmatrix}^{-1}\begin{bmatrix} -1 & A-B \\ -1 & C-D \end{bmatrix}, \quad (2.201)$$

$$\begin{bmatrix} A & B \\ C & D \end{bmatrix} = \begin{bmatrix} S_{11}-1 & S_{11}+1 \\ S_{21} & S_{21} \end{bmatrix}^{-1}\begin{bmatrix} -S_{12} & S_{12} \\ 1-S_{22} & S_{22}+1 \end{bmatrix}. \quad (2.202)$$

2.28. Show that the determinant of the transmission matrix (2.202) is equal to S_{12}/S_{21}, thus being one for reciprocal two-port networks.

2.29. From Problem 2.19, show that

$$Z_n(s) = [U_n + S(s)][U_n - S(s)]^{-1} = [U_n - S(s)]^{-1}[U_n + S(s)] \quad (2.203a)$$

$$= 2[U_n - S(s)]^{-1} - U_n, \quad (2.203b)$$

$$Y_n(s) = [U_n - S(s)][U_n + S(s)]^{-1} = [U_n + S(s)]^{-1}[U_n - S(s)] \quad (2.204a)$$

$$= 2[U_n + S(s)]^{-1} - U_n, \quad (2.204b)$$

indicating that the factors in (2.203a) and (2.204a) commute.

2.30. Confirm that the normalized scattering matrix $S_a(s)$ of the subnetwork N_a in Fig. 2.16 is given by (2.170).

2.31. Show that a lossless reciprocal three-port network that is matched at each port ($S_{ii} = 0$) is not a physically realizable passive network.

2.32. Let $S(s)$ be the scattering matrix of a two-port network N normalizing to the 1-Ω resistance. Let $Z(s)$ be the impedance matrix of N. Justify each of the

following statements:
(i) The functional relationship between $S(s)$ and $Z(s)$ remains invariant under the similarity transformation.
(ii) If $S_{11} = -S_{22}$, then det $Z(s) = 1$ where $S(s) = [S_{ij}]$.

2.33. In Fig. 2.15, let N_b be a unilateral amplifier of gain G whose normalized scattering matrix is given by

$$S_b = \begin{bmatrix} 0 & 0 \\ G & 0 \end{bmatrix}. \tag{2.205}$$

Also let N_a be a four-port network with $m = k = 2$. The overall two-port network is thus a general representation of a feedback amplifier. Show that the normalized scattering matrix of the complete amplifier N will be

$$S = \begin{bmatrix} S_{11a} & S_{12a} \\ S_{21a} & S_{22a} \end{bmatrix} + \frac{G}{1 - S_{34a}G} \begin{bmatrix} S_{14a}S_{31a} & S_{14a}S_{32a} \\ S_{24a}S_{31a} & S_{24a}S_{32a} \end{bmatrix}, \tag{2.206}$$

where $S_a(s) = [S_{ija}]$. The term $G(1 - S_{34a}G)^{-1}$ is recognized as an extension of the usual $\mu(1 - \mu\beta)^{-1}$ in the feedback amplifier theory.

2.34. Apply the infinite series expansion of the matrix

$$(U_n - M)^{-1} = U_n + M + M^2 + M^3 + \cdots, \tag{2.207}$$

whose validity can be checked heuristically by multiplying through by $U_n - M$, to (2.165) and show that the resulting series expansions of these identities can be interpreted physically as a series of waves being reflected back and forth. Thus, if the original network is nearly matched, the reflected waves grow successively weaker in each reflection, and the infinite series converges rapidly. This means that the first few terms are a good approximation in practical design. Note that (2.207) is not valid in general. However, it converges if every eigenvalue of M lies inside the unit circle, which is the case for almost all the nonpathological networks.

2.35. Let N be an n-port network composed only of ideal transformers. Let S be its scattering matrix normalizing to a set of positive resistances. Show that if n is odd, then not all of the ports can be matched simultaneously. [*Hint.* Use property (i) of Problem 2.32.]

2.36. Consider the subnetwork N_a of Fig. 2.16. Show that the scattering matrix $S_a(s)$ of the four-port network N_a normalizing to the reference resistances R_1, R_2, R_3 and R_4 is given by

$$S_a(s) = \frac{1}{R_3 + 2R_2} \begin{bmatrix} R_3 - 2R_2 & 2R_2^{1/2}R_3^{1/2} & 0 & 2R_2^{1/2}R_3^{1/2} \\ 2R_2^{1/2}R_3^{1/2} & 2R_2 - R_3 & 2R_2^{1/2}R_3^{1/2} & 0 \\ 0 & 2R_2^{1/2}R_3^{1/2} & R_3 - 2R_2 & -2R_2^{1/2}R_3^{1/2} \\ 2R_2^{1/2}R_3^{1/2} & 0 & -2R_2^{1/2}R_3^{1/2} & 2R_2 - R_3 \end{bmatrix}, \tag{2.208}$$

where $R_3 = R_4$ and $k_1^2 R_2 = k_2^2 R_1$.

2.37. In Problem 2.6, set $R = 0$. Then (2.189) becomes the scattering matrix of an ideal gyrator normalizing to the resistances R_1 and R_2. Since an ideal gyrator is lossless, show that the elements of its normalized scattering matrix S satisfy the relations (2.131)–(2.134).

2.38. Show that the most general orthogonal matrix of order 2 is of the form

$$\begin{bmatrix} \alpha & \pm\beta \\ \beta & \mp\alpha \end{bmatrix} \quad (2.209)$$

with $\alpha^2 + \beta^2 = 1$.

References

1. Belevitch, V. (1956) Elementary applications of the scattering formalism in network design. *IRE Trans. Circuit Theory*, vol. CT-3, no. 2, pp. 97–104.
2. Belevitch, V. (1963) Factorization of scattering matrices with applications to passive-network synthesis. *Philips Res. Rept.*, vol. 18, no. 4, pp. 275–317.
3. Belevitch, V. (1968) *Classical Network Theory*, San Francisco, Calif.: Holden-Day.
4. Bodharamik, P., Besser, L. and Newcomb, R. W. (1971) Two scattering matrix programs for active circuit analysis. *IEEE Trans. Circuit Theory*, vol. CT-18, no. 6, pp. 610–619.
5. Carlin, H. J. (1956) The scattering matrix in network theory. *IRE Trans. Circuit Theory*, vol. CT-3, no. 2, pp. 88–97.
6. Carlin, H. J. and Giordano, A. B. (1964) *Network Theory: An Introduction to Reciprocal and Nonreciprocal Circuit*, Englewood Cliffs, N.J.: Prentice-Hall.
7. Chen, W. K. (1973) The scattering matrix and the passivity condition. *Matrix Tensor Quart.*, vol. 24, nos. 1 and 2, pp. 30–32 and 74–75.
8. Chen, W. K. (1975) Relationships between scattering matrix and other matrix representations of linear two-port networks. *Int. J. Electronics*, vol. 38, no. 4, pp. 433–441.
9. Kuh, E. S. and Rohrer, R. A. (1967) *Theory of Linear Active Networks*, San Francisco, Calif.: Holden-Day.
10. Newcomb, R. W. (1966) *Linear Multiport Synthesis*, New York: McGraw-Hill.
11. Rohrer, R. A. (1965) The scattering matrix: normalized to complex n-port load networks. *IEEE Trans. Circuit Theory*, vol. CT-12, no. 2, pp. 223–230.
12. Rohrer, R. A. (1968) Optimal matching: a new approach to the matching problem for real linear time-invariant one-port networks. *IEEE Trans. Circuit Theory*, vol. 15, no. 2, pp. 118–124.
13. Saeks, R. (1972) *Generalized Networks*, New York: Holt, Rinehart & Winston.
14. Wohlers, M. R. (1965) Complex normalization of scattering matrices and the problem of compatible impedances. *IEEE Trans. Circuit Theory*, vol. CT-12, no. 4, pp. 528–535.
15. Wohlers, M. R. (1969) *Lumped and Distributed Passive Networks: A Generalized and Advanced Viewpoint*, New York: Academic Press.
16. Wohlers, M. R. (1969) On scattering matrices normalized to active n-ports at real frequencies. *IEEE Trans. Circuit Theory*, vol. CT-16, no. 2, pp. 254–256.

17. Wohlers, M. R. (1970) A quantitative measure of the passivity and activity of networks. *IEEE Trans. Circuit Theory*, vol. CT-17, no. 1, pp. 46–54.
18. Youla, D. C. (1961) On scattering matrices normalized to complex port numbers. *Proc. IRE*, vol. 49, no. 7, p. 1221.
19. Youla, D. C. (1964) An extension of the concept of scattering matrix. *IEEE Trans. Circuit Theory*, vol. CT-11, no. 2, pp. 310–312.
20. Youla, D. C., Castriota L. J. and Carlin H. J. (1959) Bounded real scattering matrices and the foundations of linear passive network theory. *IRE Trans. Circuit Theory*, vol. CT-6, no. 1, pp. 102–124.

CHAPTER 3

Approximation and Ladder Realization

IN CHAPTERS 4 and 5, we shall apply the concept of the scattering parameters defined in the foregoing to the design of a coupling network that matches a given load impedance to a resistive generator and that achieves a preassigned transducer power-gain characteristic over a frequency band of interest. Ideally, we hope that we can design such a coupling network having any desired gain characteristic, such as the ideal *brick-wall* type of response shown in Fig. 3.1, which is constant from $\omega = 0$ to $\omega = \omega_c$ and zero for all ω

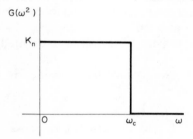

FIG. 3.1. The ideal brick-wall type of low-pass response.

greater than ω_c. However, such niceties cannot be achieved with a finite number of network elements. What then can be done in order to obtain a desired gain characteristic? Instead of seeking an overly idealistic performance criteria, we specify the maximum permissible loss or maximum permissible reflection coefficient over a given frequency band of interest called the *passband*, the minimum allowable loss or reflection coefficient over another frequency band called the *stopband*, and a statement about the selectivity or the tolerable interval between these two bands. We then seek a rational function that meets all the specifications and at the same time it must be realizable for the class of networks desired. This is known as the *approximation problem*.

To introduce this concept along with a discussion of the approximating functions, we consider the ideal low-pass brick-wall type of gain response of Fig. 3.1, and show how it can be approximated by three popular rational function approximation schemes: the maximally-flat (Butterworth) response, the equiripple (Chebyshev) response, and the elliptic (Cauer-parameter) response. We then present the corresponding ladder network realizations, which are attractive from an engineering viewpoint in that they are unbalanced and contain no coupling coils. Explicit formulas for element values of these ladder networks with Butterworth or Chebyshev gain characteristic will be given, which reduce the design of these networks to simple arithmetic.

Confining attention to the low-pass gain characteristic is not to be deemed as restrictive as it may appear. We shall demonstrate this by considering frequency transformations that permit low-pass characteristic to be converted to a high-pass, band-pass or band-elimination characteristic. This will be presented at the end of the chapter.

We begin our discussion by considering the Butterworth response.

1. The Butterworth response

The transducer power-gain characteristic (Problem 3.3)

$$G(\omega^2) = \frac{K_n}{1+(\omega/\omega_c)^{2n}}, \qquad K_n \geq 0, \tag{3.1}$$

ω_c being the 3-dB radian bandwidth, is known as the nth-order *Butterworth* or *maximally-flat* low-pass response, and was first suggested by Butterworth (1930). The constant K_n is the dc gain, which may be greater, equal to or less than unity. In Chapter 4, when we discuss passive matching, K_n is bounded above by unity, while in

Chapter 5, when we are concerned with negative-resistance amplifiers, K_n must be larger than unity.

The term "maximally flat" was coined by Landon (1941), the reason being that the first $2n - 1$ derivatives of $G(\omega^2)$ are zero at $\omega = 0$. To see this, we apply the binomial series expansion

$$(1+x)^{-1} = 1 - x + x^2 - x^3 + x^4 - x^5 + \cdots, \qquad x^2 < 1, \quad (3.2)$$

to (3.1), giving

$$G(\omega^2) = K_n[1 - (\omega/\omega_c)^{2n} + (\omega/\omega_c)^{4n} - (\omega/\omega_c)^{6n} + \cdots], \qquad \omega < \omega_c. \quad (3.3)$$

From this expression, it is clear that the first $2n - 1$ derivatives are zero at $\omega = 0$, which yields a maximally-flat gain characteristic at dc. Equation (3.3) is known as the *Maclaurin* series expansion of $G(\omega^2)$. We remark that in $G(\omega^2)$ if we replace ω by $1/\omega$ and then derive the Maclaurin series expansion for the new function, we obtain a series which begins with a term $K_n\omega_c^{2n}\omega^{2n}$, indicating again that the first $2n - 1$ derivatives of the new function are zero at $\omega = 0$. This means that the Butterworth response automatically gives a maximally-flat gain characteristic in the stopband at infinity.

Figure 3.2 illustrates the Butterworth response of several orders

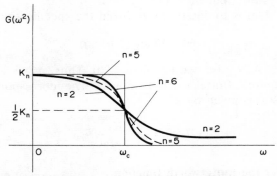

FIG. 3.2. The Butterworth response of several orders as well as the ideal brick-wall type of low-pass response.

as well as the ideal brick-wall type of response, which corresponds to the limiting case as n approaches to infinity. Observe that all the curves intersect the line $G(\omega^2) = \frac{1}{2}K_n$ at $\omega = \omega_c$, showing a 3-dB attenuation at the radian cutoff frequency ω_c. For frequencies far above ω_c, the gain characteristic becomes

$$G(\omega^2) = K_n \omega_c^{2n}/\omega^{2n}. \tag{3.4}$$

By expressing the gain in decibels, we have

$$\alpha = 10 \log G(\omega^2) = 10 \log K_n - 20n \log (\omega/\omega_c), \tag{3.5}$$

yielding an asymptotic slope of $-20n$ dB/decade or, equivalently, $-6n$ dB/octave.

Apart from the dc gain K_n, the Butterworth response is specified by a single parameter n, which can be chosen from the given specifications. The constant K_n will be determined by the conditions derived from the physical realizability of the networks, to be presented in the following chapters.

We shall henceforth assume that the quantities K_n and n are all known in accordance with the preceding discussion.

EXAMPLE 3.1. It is desired to have a Butterworth gain characteristic that gives at least 60 dB attenuation at the frequency five times the cutoff frequency and beyond.

The problem is to determine n. From the specifications, we can write

$$60 \leqq 10 \log (1 + 5^{2n}), \tag{3.6}$$

yielding $n \geqq 4.29$ (also see Problem 3.20). Thus, $n = 5$ is the order of the required Butterworth response. The corresponding gain characteristic is given by

$$G(\omega^2) = \frac{K_5}{1 + (\omega/\omega_c)^{10}}. \tag{3.7}$$

1.1. Poles of the Butterworth function

Once the order of the Butterworth response is chosen, the next step is to determine its corresponding pole locations. To this end, we appeal to the theorem on the uniqueness of analytic continuation in the theory of analytic functions of a complex variable by substituting ω by $-js$ in (3.1), resulting in

$$G(-s^2) = \frac{K_n}{1+(-1)^n y^{2n}}, \quad (3.8a)$$

where

$$y = s/\omega_c. \quad (3.8b)$$

Clearly, the poles of this function are defined by the zeros of the polynomial

$$1+(-1)^n y^{2n} = 0, \quad (3.9)$$

which are located on a circle of radius ω_c in the s-plane and are given by (Problem 3.2)

$$y_{k+1} = s_{k+1}/\omega_c = \exp[j(2k+n+1)\pi/2n] \quad (3.10)$$

for $k = 0, 1, 2, \ldots, 2n-1$. Pole locations for $n = 5$ and $n = 6$ are presented in Fig. 3.3. It is seen that these poles are located

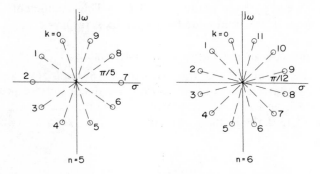

FIG. 3.3. The pole locations of the fifth-order and sixth-order Butterworth responses.

symmetrically with respect to both the real and the imaginary axes with quadrantal symmetry. For n odd, a pair of poles are located on the real axis, but they never lie on the imaginary axis for any n.

These follow directly from (3.10) and the fact that the poles are separated by π/n radians.

For reasons to be given shortly, we decompose the left-hand side of (3.9) into the form

$$1 + (-1)^n y^{2n} = q(y)q(-y), \qquad (3.11)$$

where

$$q(y) = a_0 + a_1 y + \cdots + a_{n-1} y^{n-1} + a_n y^n = \sum_{m=0}^{n} a_m y^m \qquad (3.12)$$

with $a_n = 1$, is the Hurwitz polynomial of degree n formed by the LHS zeros of (3.10), which are given by

$$y_{k+1} = \exp[j(2k + n + 1)\pi/2n], \qquad k = 0, 1, 2, \ldots, n-1, \qquad (3.13a)$$

or in trigonometric form by

$$y_{k+1} = -\sin[(2k+1)\pi/2n] + j\cos[(2k+1)\pi/2n], \qquad (3.13b)$$
$$k = 0, 1, 2, \ldots, n-1.$$

Though the Hurwitz polynomials $q(y)$ can easily be calculated, it is useful to have them readily available. The coefficients of the polynomials $q(y)$ are tabulated in Appendix A for $n = 2, 3, \ldots, 10$.

1.2. Coefficients of the Butterworth polynomials

The coefficients of the *Butterworth polynomial* $q(y)$ defined in (3.12) are interrelated. In this section, we discuss these properties and present an explicit formula for their computation.

From (3.13a), the Butterworth polynomial $q(y)$ can be written as

$$q(y) = \prod_{k=0}^{n-1} (y - \beta^{2k+n+1}), \qquad (3.14a)$$

where

$$\beta = \exp(j\pi/2n). \qquad (3.14b)$$

The remainder of the derivations amounts to the manipulation of the summation index k. From (3.14a), we have

$$q(y) = \frac{y - \beta^{n+1}}{y - \beta^{3n+1}} \prod_{k=1}^{n} (y - \beta^{2k+n+1})$$

$$= \frac{y - j\beta}{y + j\beta} \prod_{k=0}^{n-1} (y - \beta^{2k+n+3})$$

$$= \frac{y - j\beta}{y + j\beta} \beta^{2n} \prod_{k=0}^{n-1} (\beta^{-2}y - \beta^{2k+n+1})$$

$$= -\frac{y - j\beta}{y + j\beta} q(\beta^{-2}y). \tag{3.15}$$

Substituting (3.12) in (3.15) and equating coefficients of each power of y yield

$$a_k + j\beta a_{k+1} + \beta^{-2k} a_k - j\beta^{-2k-1} a_{k+1} = 0, \tag{3.16}$$

which gives the recursion formula

$$\frac{a_{k+1}}{a_k} = \frac{\cos(k\pi/2n)}{\sin(k+1)\pi/2n}, \qquad k = 0, 1, 2, \ldots, n-1. \tag{3.17}$$

Thus, we can calculate the coefficients starting from the first a_0 or the last a_n, which is known to be unity. Forming the product of the terms given in (3.17) yields an explicit formula for the coefficients a_k as

$$a_k = \prod_{u=1}^{k} \frac{\cos(u-1)\pi/2n}{\sin(u\pi/2n)}, \tag{3.18}$$

where $a_0 = 1$ and some a_k are explicitly listed in Problem 3.12. Finally, since

$$\cos(u\pi/2n) = \sin[(n-u)\pi/2n], \tag{3.19}$$

it follows that

$$a_k = a_{n-k}, \tag{3.20}$$

indicating that the coefficients of the Butterworth polynomial are symmetric from its two ends.

We shall illustrate this by the following example.

EXAMPLE 3.2. Consider the fifth-order Butterworth gain response shown in (3.7), whose corresponding Hurwitz polynomial is given by

$$q(y) = 1 + a_1 y + a_2 y^2 + a_3 y^3 + a_4 y^4 + y^5. \qquad (3.21)$$

The coefficients a_k can be computed recurrently by the formula (3.17), as follows:

$a_1 = a_0(\cos 0)/(\sin 18°) = 1/0.30902 = 3.2360365,$

$a_2 = a_1(\cos 18°)/(\sin 36°) = (3.2360365)(0.95106)/0.58779 = 5.2359938,$

$a_3 = a_2(\cos 36°)/(\sin 54°) = (5.2359938)(0.80902)/0.80902 = 5.2359938,$

$a_4 = a_3(\cos 54°)/(\sin 72°) = (5.2359938)(0.58779)/0.95106 = 3.2360365,$

which confirm the numbers given in Table A of the Appendix A.†
Alternatively, we can compute these coefficients from the other end, as follows:

$a_4 = a_5(\sin 90°)/(\cos 72°) = 1/0.30902 = 3.2360365,$

$a_3 = a_4(\sin 72°)/(\cos 54°) = (3.2360365)(0.95106)/0.58779 = 5.2359938,$

$a_2 = a_3(\sin 54°)/(\cos 36°) = (5.2359938)(0.80902)/0.80902 = 5.2359938,$

$a_1 = a_2(\sin 36°)/(\cos 18°) = (5.2359938)(0.58779)/0.95106 = 3.2360365.$

This also confirms the symmetric property of (3.20).

1.3. Butterworth networks

The problem considered in this section is to derive formulas pertinent to the design of a lossless two-port network operating between a resistive generator with internal resistance R_1 and a resistive load with resistance R_2 and having a preassigned Butterworth transducer power-gain characteristic, as depicted in Fig. 3.4. The general problem of matching an arbitrary load

---

†There is a slight discrepancy due to the fact that we carry only five significant digits in computing the trigonometric functions.

APPROXIMATION AND LADDER REALIZATION

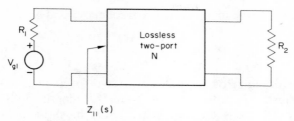

FIG. 3.4. A lossless two-port network with a preassigned Butterworth transducer power-gain characteristic.

impedance, which may be passive or active, to a resistive generator having a prescribed transducer power-gain characteristic will be considered in the next two chapters.

Referring to Fig. 3.4, let

$$S(s) = [S_{ij}] \tag{3.22}$$

be the scattering matrix of the lossless two-port network N normalizing to the resistances R_1 and R_2, which has an nth-order Butterworth transducer power-gain characteristic. From (2.114) and (3.1), we get

$$|S_{21}(j\omega)|^2 = G(\omega^2) = \frac{K_n}{1 + (\omega/\omega_c)^{2n}}. \tag{3.23}$$

Since $|S_{21}(j\omega)|$ is bounded between 0 and 1 for a passive two-port, the dc gain K_n is restricted by

$$0 \leqq K_n \leqq 1. \tag{3.24}$$

For a lossless two-port network N, its scattering matrix $S(j\omega)$ is unitary, whose first column elements according to (2.131c) are related by

$$|S_{11}(j\omega)|^2 = 1 - |S_{21}(j\omega)|^2 = \frac{1 - K_n + (\omega/\omega_c)^{2n}}{1 + (\omega/\omega_c)^{2n}}, \tag{3.25}$$

which after analytic continuation becomes

$$S_{11}(s)S_{11}(-s) = \delta^{2n}\frac{1 + (-1)^n x^{2n}}{1 + (-1)^n y^{2n}}, \tag{3.26a}$$

where y, the normalized complex frequency, is defined in (3.8b),

$$\delta = (1 - K_n)^{1/2n}, \qquad (3.26b)$$

$$x = y/\delta. \qquad (3.26c)$$

Now with the zeros and poles of (3.26a) specified as in (3.10), the input reflection coefficient $S_{11}(s)$ can be identified. Since from Corollary 2.2 the scattering matrix $S(s)$ must be bounded-real, $S_{11}(s)$ can have no poles in the closed RHS. Thus, we must assign all of the LHS poles of (3.26a) to $S_{11}(s)$, resulting in a unique decomposition of the denominator polynominal of (3.26a). However, the zeros of $S_{11}(s)$ may lie in the RHS, so that in general a number of different numerators are possible for $S_{11}(s)$. For reasons to be given in the following chapters, we choose only the LHS zeros for $S_{11}(s)$. Define a *minimum-phase reflection coefficient* as one that is devoid of zeros in the open RHS. Then a minimum-phase decomposition of (3.26a) is given by

$$S_{11}(s) = \pm \delta^n \frac{q(x)}{q(y)}. \qquad (3.27)$$

From (2.112), the impedance $Z_{11}(s)$ looking into the input port with the output port terminating in its reference impedance R_2, as indicated in Fig. 3.4, is given by

$$Z_{11}(s) = R_1 \frac{1 + S_{11}(s)}{1 - S_{11}(s)}. \qquad (3.28)$$

Substituting (3.27) in (3.28) yields

$$Z_{11}(s) = R_1 \frac{q(y) \pm \delta^n q(x)}{q(y) \mp \delta^n q(x)}. \qquad (3.29)$$

For some design applications, it may be desirable to specify both R_1 and R_2 in advance. In this case, the dc gain K_n cannot be chosen arbitrarily subject only to the constraint (3.24). In fact, substituting $s = 0$ in (3.29) gives the desired relationship for an LC two-port network with $K_n \neq 0$:

$$\frac{R_2}{R_1} = \left[\frac{1 + \delta^n}{1 - \delta^n}\right]^{\pm 1} \qquad (3.30)$$

APPROXIMATION AND LADDER REALIZATION 125

the $\pm$ signs being determined respectively according to $R_2 \gtreqqless R_1$ and $R_2 \lesseqgtr R_1$. Thus, if any two of the three quantities R_1, R_2 and K_n are specified, the third one is determined.

1.4. Butterworth *LC* ladder networks

In this section, we shall concentrate on the realization of a special class of lossless two-port networks known as the *LC ladders* that possess the properties discussed in the preceding section. The ladder networks are attractive from an engineering viewpoint in that they are unbalanced and contain no coupling coils. Also, explicit formulas for their element values can be obtained, which reduce the design problem to simple arithmetic.

Depending upon the choice of the plus and minus signs in (3.27), two cases are considered, each being presented in a separate section.

A. $S_{11}(0) \geqq 0$

With the choice of the plus sign in (3.27), the input impedance becomes

$$Z_{11}(s) = R_1 \frac{q(y) + \delta^n q(y/\delta)}{q(y) - \delta^n q(y/\delta)} = R_1 \frac{\sum_{m=0}^{n} a_m(1 + \delta^{n-m})y^m}{\sum_{m=0}^{n} a_m(1 - \delta^{n-m})y^m}. \quad (3.31)$$

According to (1.105) and (1.106), $Z_{11}(s)$ is a positive-real function. Its continued-fraction expansion about infinity yields a lossless ladder network terminated in a resistor. Hence we can write

$$Z_{11}(s) = L_1 s + \cfrac{1}{C_2 s + \cfrac{1}{L_3 s + \cfrac{1}{\ddots + \cfrac{1}{H}}}}, \quad (3.32)$$

H being a constant representing either a resistance or conductance. Depending upon whether n is odd or even, the *LC* ladder network

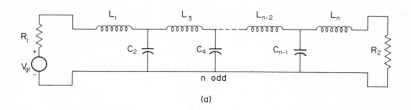

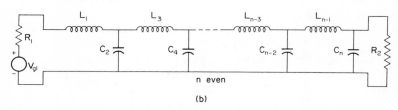

FIG. 3.5. The LC ladders with a preassigned Butterworth transducer power-gain characteristic together with its terminations, $S_{11}(0) \geqq 0$.

has the configuration as shown in Fig. 3.5. The first element L_1 can easily be determined from (3.31) in conjunction with (3.17), giving

$$L_1 = \frac{2R_1}{(1-\delta)a_{n-1}\omega_c} = \frac{2R_1 \sin(\pi/2n)}{(1-\delta)\omega_c}. \quad (3.33)$$

Moreover, it can be shown that the values of other elements can be computed by the recurrence formulas

$$L_{2m-1}C_{2m} = \omega_c^{-2} \frac{4 \sin \gamma_{4m-3} \sin \gamma_{4m-1}}{1 - 2\delta \cos \gamma_{4m-2} + \delta^2}, \quad (3.34a)$$

$$L_{2m+1}C_{2m} = \omega_c^{-2} \frac{4 \sin \gamma_{4m-1} \sin \gamma_{4m+1}}{1 - 2\delta \cos \gamma_{4m} + \delta^2} \quad (3.34b)$$

for $m = 1, 2, \ldots, [\tfrac{1}{2}n]$, where

$$\gamma_m = m\pi/2n, \quad (3.34c)$$

and $[\tfrac{1}{2}n]$ denotes the largest integer not greater than $\tfrac{1}{2}n$. (This symbol will be used throughout the remainder of the chapter.) The ladder is terminated in L_n as shown in Fig. 3.5(a) if n is odd, or C_n of Fig. 3.5(b) if n is even. In addition, the final elements are related to R_2 by

$$L_n = \frac{2R_2 \sin \gamma_1}{(1+\delta)\omega_c} \tag{3.35a}$$

for n odd, and

$$C_n = \frac{2 \sin \gamma_1}{R_2(1+\delta)\omega_c} \tag{3.35b}$$

for n even. The above formulas can be derived deductively by carrying out the calculations in detail for the cases of low order and then guessing the final result. A formal complete proof was first given by Bossé (1951). Hence we can calculate the element values starting from either the first or the last element (also see Problem 3.26).

In particular, for $R_1 = R_2$ we have $\delta = 0$ and the above formulas are amazingly simple, and are given by (Problem 3.5)

$$L_{2m-1} = \frac{2R_1}{\omega_c} \sin \gamma_{4m-3}, \tag{3.36a}$$

$$C_{2m} = \frac{2}{R_1 \omega_c} \sin \gamma_{4m-1}. \tag{3.36b}$$

We shall illustrate the above formulas by the following example.

EXAMPLE 3.3. Suppose that we wish to design a low-pass filter that has a maximally-flat transducer power-gain characteristic. The filter is to be operated between a resistive generator of internal resistance 100 Ω and a 200-Ω load and such that it gives at least 60 dB attenuation in gain at the frequency five times the radian cutoff frequency $\omega_c = 10^4$ rad/s and beyond.

From specifications, we have

$$\begin{matrix} R_1 = 100 \ \Omega, & R_2 = 200 \ \Omega, \\ \omega_c = 10^4 \text{ rad/s}, & n = 5. \end{matrix} \tag{3.37}$$

Since R_1 and R_2 are both specified, the dc gain K_5 cannot be chosen arbitrarily. For $R_2 > R_1$, we must select the plus sign in (3.27) and (3.30). From (3.30), we get

$$\delta^5 = 1/3,$$

which gives $\delta = 0.80274$. Thus from (3.33) we compute

$$L_1 = \frac{2 \times 100 \times \sin 18°}{(1 - 0.8027)10^4} = 31.331 \text{ mH}. \tag{3.38a}$$

Using (3.34a) for $m = 1$ in conjunction with (3.38a) yields

$$C_2 = \frac{10^{-6} \times 4 \times \sin 18° \sin 54°}{3.1331 \times (1 - 1.6055 \times \cos 36° + 0.6444)} = 0.9237 \text{ }\mu\text{F}. \tag{3.38b}$$

Now using (3.34b) with $m = 1$ and (3.38b) gives

$$L_3 = \frac{10^{-2} \times 4 \times \sin 54° \sin 90°}{0.9237 \times (1 - 1.6055 \times \cos 72° + 0.6444)} = 30.510 \text{ mH}. \tag{3.38c}$$

Repeat the above process for $m = 2$, and we obtain

$$C_4 = \frac{10^{-6} \times 4 \times \sin 90° \sin 126°}{3.051 \times (1 - 1.6055 \times \cos 108° + 0.6444)} = 0.4955 \text{ }\mu\text{F}, \tag{3.38d}$$

$$L_5 = \frac{10^{-2} \times 4 \times \sin 126° \sin 162°}{0.4955 \times (1 - 1.6055 \times \cos 144° + 0.6444)} = 6.857 \text{ mH}. \tag{3.38e}$$

Alternatively, L_5 can be computed directly from (3.35a), yielding

$$L_5 = \frac{2 \times 200 \times \sin 18°}{(1 + 0.8027)10^4} = 6.857 \text{ mH}, \tag{3.39}$$

which coincides with (3.38e). The ladder network together with its terminations is presented in Fig. 3.6.

Now suppose that we let $R_1 = R_2 = 100 \text{ }\Omega$. Then from (3.36) the

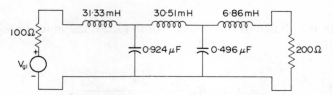

FIG. 3.6. A low-pass filter possessing the fifth-order Butterworth transducer power-gain characteristic.

element values for the desired ladder network are given by

$$L_1 = 200 \times 10^{-4} \times \sin 18° = 6.1804 \text{ mH},$$
$$C_2 = 2 \times 10^{-6} \times \sin 54° = 1.618 \text{ }\mu\text{F},$$
$$L_3 = 200 \times 10^{-4} \times \sin 90° = 20 \text{ mH},$$
$$C_4 = 2 \times 10^{-6} \times \sin 126° = 1.618 \text{ }\mu\text{F},$$
$$L_5 = 200 \times 10^{-4} \times \sin 162° = 6.1804 \text{ mH}.$$

The corresponding ladder network is shown in Fig. 3.7. The ladder of Fig. 3.6 yields a dc gain

$$K_5 = 1 - \delta^{10} = 1 - 1/9 = 8/9,$$

while the one in Fig. 3.7 gives $K_5 = 1$.

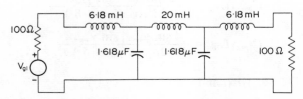

FIG. 3.7. A symmetric low-pass filter possessing the fifth-order Butterworth transducer power-gain characteristic.

B. $S_{11}(0) < 0$

With the choice of the minus sign in (3.27), the input impedance becomes

$$Z_{11}(s) = R_1 \frac{\sum_{m=0}^{n} a_m (1 - \delta^{n-m}) y^m}{\sum_{m=0}^{n} a_m (1 + \delta^{n-m}) y^m}, \qquad (3.40)$$

which, aside from the constant R_1, is the reciprocal of that given in (3.31). Hence (3.40) can be written as

$$\frac{1}{Z_{11}(s)} = C_1 s + \cfrac{1}{L_2 s + \cfrac{1}{C_3 s + \cfrac{1}{\ddots + \cfrac{1}{H}}}}, \qquad (3.41)$$

H being a constant representing either a resistance or conductance. Depending upon whether n is even or odd, the LC ladder network has the configuration as shown in Fig. 3.8. Formulas for the element values are the same as those given in (3.33)–(3.35) except that the roles of C's and L's are interchanged and that R_1 and R_2 are replaced by their reciprocals:

$$C_1 = \frac{2 \sin \gamma_1}{R_1 (1 - \delta) \omega_c}, \qquad (3.42a)$$

$$C_{2m-1} L_{2m} = \frac{4 \sin \gamma_{4m-3} \sin \gamma_{4m-1}}{(1 - 2\delta \cos \gamma_{4m-2} + \delta^2) \omega_c^2}, \qquad (3.42b)$$

$$C_{2m+1} L_{2m} = \frac{4 \sin \gamma_{4m-1} \sin \gamma_{4m+1}}{(1 - 2\delta \cos \gamma_{4m} + \delta^2) \omega_c^2} \qquad (3.42c)$$

for $m = 1, 2, \ldots, [\tfrac{1}{2} n]$, and

$$C_n = \frac{2 \sin \gamma_1}{R_2 (1 + \delta) \omega_c} \qquad (3.43a)$$

APPROXIMATION AND LADDER REALIZATION

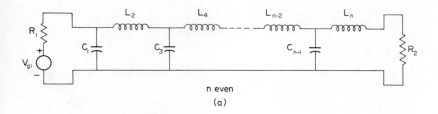

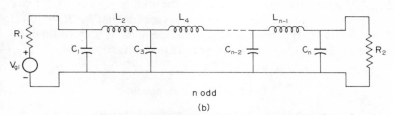

FIG. 3.8. The LC latters with a preassigned Butterworth transducer power-gain characteristic together with its terminations, $S_{11}(0) < 0$.

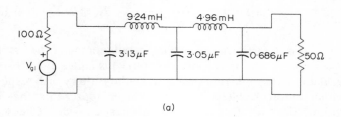

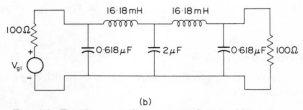

FIG. 3.9. Two low-pass filters possessing the fifth-order Butterworth transducer power-gain characteristic.

for n odd, and
$$L_n = \frac{2R_2 \sin \gamma_1}{(1+\delta)\omega_c} \qquad (3.43b)$$
for n even.

As an illustration, in Example 3.3 if we choose the minus sign in (3.27) rather than the plus sign, we obtain two ladders as shown in Fig. 3.9, whose normalized input impedances $Z_{11}(s)/R_1$ are reciprocals of those obtained for the corresponding ladders of Figs. 3.6 and 3.7. We remark that in selecting the minus sign in (3.27), we must choose the minus sign in (3.30). This means that for $R_1 = 100 \, \Omega$ and $\delta = 0.80274$, the terminating resistance R_2 is determined by (3.30) to be $50 \, \Omega$ and cannot be specified arbitrarily.

2. The Chebyshev response

Another useful transducer power-gain characteristic that approximates the ideal low-pass characteristic of Fig. 3.1 is given by (Problem 3.9)

$$G(\omega^2) = \frac{K_n}{1 + \epsilon^2 C_n^2(\omega/\omega_c)}, \qquad K_n \geqq 0, \qquad (3.44)$$

where $C_n(\omega)$ is the nth-order Chebyshev polynomial of the first kind and $\epsilon^2 \leqq 1$ and K_n are real constants. $G(\omega^2)$ is called the *nth-order Chebyshev* or *equiripple response*. Unlike the Butterworth response, the radian cutoff frequency ω_c no longer represents the 3-dB bandwidth. Since in this book we always consider Chebyshev polynomials of the first kind, we shall use a more concise designation by dropping the words *of the first kind*. We begin our discussion by considering the properties of the Chebyshev polynomials.

2.1. Chebyshev polynomials

It is convenient to define the *Chebyshev polynomial of order n* by the equations

$$C_n(\omega) = \cos(n \cos^{-1} \omega), \qquad 0 \leqq \omega \leqq 1 \qquad (3.45a)$$
$$= \cosh(n \cosh^{-1} \omega), \qquad \omega > 1. \qquad (3.45b)$$

APPROXIMATION AND LADDER REALIZATION   133

In fact, these two expressions are completely equivalent, each being valid for all ω (Problem 3.75). To show that the transcendental function (3.45) is indeed a polynomial, it is sufficient to consider (3.45a) and let

$$w = \cos^{-1}\omega. \tag{3.46}$$

Substituting it in (3.45a) gives

$$C_n(w) = \cos nw. \tag{3.47}$$

Then using the trigonometric identity

$$\cos(n+1)w = \cos nw \cos w - \sin nw \sin w$$
$$= \cos nw \cos w + \tfrac{1}{2}\cos(n+1)w - \tfrac{1}{2}\cos(n-1)w \tag{3.48}$$

yields

$$\cos(n+1)w = 2\cos nw \cos w - \cos(n-1)w. \tag{3.49}$$

Using (3.46) gives the desired recurrence formula

$$C_{n+1}(\omega) = 2\omega C_n(\omega) - C_{n-1}(\omega). \tag{3.50}$$

Since the Chebyshev polynomials of lower orders are known,

$$C_0(\omega) = 1 \quad \text{and} \quad C_1(\omega) = \omega,$$

the higher orders can be computed recurrently by means of (3.50), giving

$$\begin{aligned}C_2(\omega) &= 2\omega^2 - 1, \\ C_3(\omega) &= 4\omega^3 - 3\omega, \\ C_4(\omega) &= 8\omega^4 - 8\omega^2 + 1,\end{aligned} \tag{3.51}$$

and others are given in Table B.1 of Appendix B.

From this discussion, some characteristics of the Chebyshev polynomials are obvious. We shall now proceed to discuss some of these properties briefly as follows:

(i) $C_n(\omega)$ is either an even or an odd function depending upon whether n is even or odd. More specifically, we can write

$$C_n(-\omega) = C_n(\omega) \qquad \text{for } n \text{ even} \tag{3.52a}$$

$$C_n(-\omega) = -C_n(\omega) \qquad \text{for } n \text{ odd.} \tag{3.52b}$$

(ii) Every coefficient of $C_n(\omega)$ is an integer, the one associated with ω^n being 2^{n-1}. Thus, in the limit as ω approaches to infinity,

$$C_n(\omega) \to 2^{n-1}\omega^n. \tag{3.53}$$

(iii) In the range $-1 \leq \omega \leq 1$, all of the Chebyshev polynomials have the equal-ripple property, varying between a maximum of 1 and a minimum of -1. Outside of this interval, their magnitude increases monotonically as ω is increased, and approaches to infinity in accordance with (3.53). Sketches of the polynomials for $n = 4$ and $n = 5$ are shown in Fig. 3.10.

(iv) As indicated in Fig. 3.10, the polynomials possess special values at $\omega = 0, 1$ or -1:

$$\begin{aligned} C_n(0) &= (-1)^{n/2}, & n \text{ even} \\ &= 0, & n \text{ odd}; \end{aligned} \tag{3.54}$$

$$\begin{aligned} C_n(\pm 1) &= 1, & n \text{ even} \\ &= \pm 1, & n \text{ odd.} \end{aligned} \tag{3.55}$$

2.2. Equiripple characteristic

We now turn our attention back to (3.44) and examine the manner in which the Chebyshev polynomials as used in the transducer power-gain characteristic approximate the ideal response of Fig. 3.1. Apart from K_n, (3.44) dictates that we square $C_n(\omega/\omega_c)$, multiply it by the constant ϵ^2, not greater than unity, add unity, and form the reciprocal. If we carry out all these steps, the response that results has equal maxima and equal minima in the passband. Two typical plots corresponding to $n = 4$ and $n = 5$ are presented in Fig. 3.11. From these plots, it is clear that the total number of troughs and peaks for positive ω is equal to n, all lying within the passband, and

APPROXIMATION AND LADDER REALIZATION

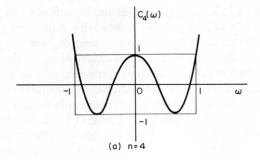

(a) n = 4

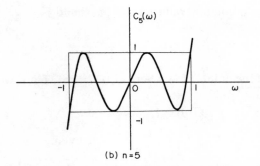

(b) n = 5

FIG. 3.10. Sketches of the fourth-order and fifth-order Chebyshev polynomials as a function of ω.

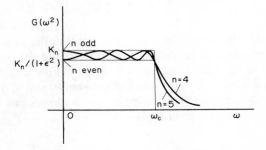

FIG. 3.11. The fourth-order and fifth-order Chebyshev responses.

outside the band the gain decreases monotonically. At the edge of the passband $\omega = \omega_c$, the gain goes through a minimum point. This is in contrast to the Butterworth approximation where at the cutoff frequency the gain is attenuated by 3 dB from its maximum value at $\omega = 0$. Because of the equal-ripple property in the passband, the Chebyshev gain response is also called the *equiripple* gain response. We shall use these two terms interchangeably in the discussion.

With the aid of the sketches of Fig. 3.11, it is clear that the maximum value of the Chebyshev response occurs at the points of ω where $C_n(\omega)$ vanishes or at the zeros of $C_n(\omega)$, giving

$$G(\omega^2)_{\max} = K_n, \qquad (3.56)$$

and that the minimum value in the passband is

$$G(\omega^2)_{\min} = \frac{K_n}{1+\epsilon^2}, \qquad (3.57)$$

occurring at the points of ω where $C_n(\omega) = \pm 1$. Thus, we see that the dc gain is

$$G(0) = K_n, \qquad n \text{ odd}$$
$$= \frac{K_n}{1+\epsilon^2}, \qquad n \text{ even.} \qquad (3.58)$$

At the edge of the passband, the gain is

$$G(\omega_c^2) = \frac{K_n}{1+\epsilon^2} \qquad (3.59)$$

for all n, as illustrated in Fig. 3.11. Observe that the quantity ϵ plays an important role in determining the maxima and the minima of the ripple, and it is called the *ripple factor*. For a fixed K_n, the peak-to-peak ripple in the passband, usually stated in terms of decibels, is determined by the ripple factor alone. For frequencies far above ω_c, the gain approaches to

$$G(\omega^2) \to \frac{K_n}{2^{2n-2}\epsilon^2(\omega/\omega_c)^{2n}}. \qquad (3.60)$$

APPROXIMATION AND LADDER REALIZATION

In terms of decibels, we have the attenuation

$$\alpha = 10 \log K_n - 6(n-1) - 20 \log \epsilon - 20n \log(\omega/\omega_c), \quad (3.61)$$

yielding an asymptotic slope of $-20n$ dB/decade or $-6n$ dB/octave, which is the same as that given for the Butterworth response except that it is offset by a value depending on both n and ϵ. Consideration of two limiting cases of the ripple factor is helpful in interpreting the difference, and the results are sketched in Fig. 3.12. Observe that for

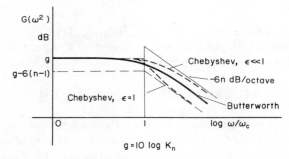

FIG. 3.12. A comparison of the Butterworth and Chebyshev responses.

$\epsilon = 1$, the cutoff frequency ω_c corresponds to a point having 3-dB attenuation from its maximum value, a convenient value for comparison with all the Butterworth response. Because of the equal ripple property, this also gives a 3-dB ripple in the passband.

The above discussion provides the procedure to be used in selecting a Chebyshev response function to match a particular set of specifications. The permissible ripple in the passband fixes the ripple factor ϵ. The rate of attenuation in the stopband determines the order of the Chebyshev polynomial to be used. Hence, we shall assume that the quantities K_n, ϵ and n are all known in accordance with the specifications (Problem 3.20).

EXAMPLE 3.4. It is desired to have a Chebyshev gain characteristic that gives at least 60-dB attenuation at the frequency five times the cutoff frequency and beyond, and that the peak-to-peak ripple in the passband must not exceed 1 dB.

The problem is to determine n and ϵ. From specifications, we can write

$$10 \log (1 + \epsilon^2) = 1, \qquad (3.62)$$

yielding $\epsilon^2 = 0.259$ or $\epsilon = 0.509$, and

$$10 \log [1 + \epsilon^2 C_n^2(5)] \geq 60, \qquad (3.63a)$$

or

$$C_n(5) \geq 1964.64. \qquad (3.63b)$$

For $n = 3$ and $n = 4$, we have from (3.51), $C_3(5) = 485$ and $C_4(5) = 4801$, which indicate that n should be 4. Alternatively, from (3.296) of Problem 3.20 we have

$$n \geq \tfrac{1}{2}[\ln (4 \times 10^6/0.509)]/\cosh^{-1} 5 = 3.463. \qquad (3.64)$$

Hence $n = 4$ is the required degree of the Chebyshev polynomial. The corresponding Chebyshev transducer power-gain characteristic is given by

$$G(\omega^2) = \frac{K_n}{1 + 0.259 C_4^2(\omega/\omega_c)}, \qquad (3.65)$$

in which K_n will be determined by other considerations such as the terminations of the desired two-port network.

2.3. Poles of the Chebyshev function

Like the Butterworth case, our next task is to determine the locations of the poles of the Chebyshev response. To this end, we again appeal to the theory of analytic continuation by replacing ω by $-js$ in (3.44), resulting in

$$G(-s^2) = \frac{K_n}{1 + \epsilon^2 C_n^2(-jy)}, \qquad (3.66)$$

y, the normalized complex frequency, being defined in (3.8b) as for the Butterworth response. Clearly, the poles of this function are given by the zeros of the polynomial

$$1 + \epsilon^2 C_n^2(-jy) = 0 \qquad (3.67)$$

with the *generalized Chebyshev polynomial* defined by the relation

$$C_n(-jy) = \cosh[n \cosh^{-1}(-jy)]. \qquad (3.68)$$

To put this in a more convenient form, write

$$\cosh^{-1}(-jy) = u + jv. \qquad (3.69)$$

Substituting it in (3.67) and expanding the resulting hyperbolic cosine yield

$$C_n(-jy) = \cosh nu \cosh jnv + \sinh nu \sinh jnv = \pm j/\epsilon. \qquad (3.70)$$

Applying the relations $\cosh ju = \cos u$ and $\sinh ju = j \sin u$ to (3.70) and equating the real and imaginary parts on both sides result in

$$\cosh nu \cos nv = 0, \qquad (3.71a)$$

$$\sinh nu \sin nv = \pm 1/\epsilon. \qquad (3.71b)$$

Since $\cosh nu \neq 0$, (3.71a) can be satisfied only if $\cos nv = 0$ or

$$v_k = (2k+1)\pi/2n, \quad k = 0, 1, 2, \ldots, 2n-1, \qquad (3.72)$$

giving $2n$ distinct solutions. At these values of v, $\sin nv = \pm 1$, so that from (3.71b) we obtain

$$u_k = \pm \frac{1}{n} \sinh^{-1} \frac{1}{\epsilon}. \qquad (3.73)$$

It is convenient to consider u_k positive and show the sign explicitly in the values of the roots of (3.67). Thus we choose and write

$$a = \frac{1}{n} \sinh^{-1} \frac{1}{\epsilon}. \qquad (3.74)$$

Substituting (3.72) and (3.73) in (3.69) and expanding the resulting hyperbolic cosine function give the desired locations of the poles of the Chebyshev response:

$$y_k = s_k/\omega_c = \sigma_k + j\omega_k$$
$$= -\sinh a \sin(2k+1)\pi/2n + j \cosh a \cos(2k+1)\pi/2n, \qquad (3.75)$$
$$k = 0, 1, 2, \ldots, 2n-1.$$

To find the locus of these roots, we square the real and imaginary parts of y_k and add. This leads to

$$\frac{\sigma_k^2}{\sinh^2 a} + \frac{\omega_k^2}{\cosh^2 a} = 1, \qquad (3.76)$$

which is the equation of an ellipse whose major semi-axis is $\cosh a$ and whose minor semi-axis is $\sinh a$. For $n = 5$ and $n = 6$, pole locations are presented in Fig. 3.13. Observe that these poles, like the Butterworth case, also possess quadrantal symmetry, being symmetric with respect to both the real and the imaginary axes. Hence, the left-hand side of (3.67) can be decomposed into the form

$$1 + \epsilon^2 C_n^2(-jy) = \epsilon^2 2^{2n-2} p(y) p(-y), \qquad (3.77)$$

where

$$p(y) = b_0 + b_1 y + \cdots + b_{n-1} y^{n-1} + b_n y^n = \sum_{m=0}^{n} b_m y^m \qquad (3.78)$$

with $b_n = 1$, is the Hurwitz polynomial of degree n formed by the LHS roots of (3.67), which are given by

$$y_{k+1} = -\sinh a \sin(2k+1)\pi/2n + j \cosh a \cos(2k+1)\pi/2n, \qquad (3.79)$$
$$k = 0, 1, 2, \ldots, n-1.$$

To avoid the necessity of computing these associated Hurwitz polynomials $p(y)$, they have been tabulated. In Appendix B, the coefficients of the polynomials $p(y)$ are given for various ripple factors and for $n = 1, 2, \ldots, 10$.

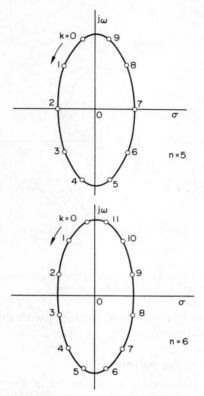

FIG. 3.13. The pole locations of the fifth-order and sixth-order Chebyshev responses.

Comparing (3.79) with (3.13b) shows that the real part of a Chebyshev pole is $\sinh a$ times the real part of the corresponding Butterworth pole and the imaginary part of a Chebyshev pole is $\cosh a$ times the imaginary part of the corresponding Butterworth pole. Thus, by normalizing the major semi-axis to unity, which is equivalent to dividing y_k by $\cosh a$, the Chebyshev and Butterworth poles have the same imaginary part while their real parts are related by the factor $\tanh a$. To obtain the real part of the Chebyshev pole, we simply shift the real part of the corresponding Butterworth pole

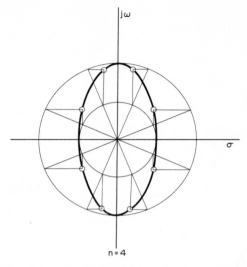

Fig. 3.14. The relations among the pole locations of the Butterworth and Chebyshev responses.

horizontally from a unit circle to an ellipse with a major semi-axis of unity. This is illustrated in Fig. 3.14 for $n = 4$.

2.4. Coefficients of the polynomial $p(y)$

The Hurwitz polynomial $p(y)$ defined in the preceding section cannot be called the Chebyshev polynomial, since the latter is formally defined in (3.45) or more generally in (3.68), while the polynomial $q(y)$ of (3.12) in the Butterworth case is called the Butterworth polynomial. As one might be expected, the recurrence formula for the coefficients of $p(y)$ is much more complicated, both in its derivation and appearance, than the Butterworth counterpart. Starting from the lower orders and carrying out the calculations in detail, compact general expressions for some of these coefficients have been worked out by Green (1954). They are listed below:

$$b_{n-1} = \frac{\sinh a}{\sin \gamma_1}, \tag{3.80a}$$

$$b_{n-2} = \frac{n}{4} + \frac{\sinh^2 a \cos \gamma_1}{\sin \gamma_1 \sin 2\gamma_1}, \tag{3.80b}$$

$$b_{n-3} = \frac{\sinh a}{\sin \gamma_1} \left(\frac{n}{4} - \frac{2 \cos^3 \gamma_1 \sin^2 \gamma_1}{\sin 2\gamma_1 \sin 3\gamma_1} \right)$$
$$+ \frac{\sinh^3 a \cos \gamma_1 \cos 2\gamma_1}{\sin \gamma_1 \sin 2\gamma_1 \sin 3\gamma_1}, \tag{3.80c}$$

where $\gamma_1 = \pi/2n$, as defined in (3.34c), and (Problems 3.13 and 3.86)

$$\begin{aligned} b_0 &= 2^{1-n} \sinh na, & n \text{ odd} \\ &= 2^{1-n} \cosh na, & n \text{ even}. \end{aligned} \tag{3.80d}$$

Some of these results are needed in the theory of broadband matching, to be presented in the next two chapters. They are explicitly given here for ease in reference.

EXAMPLE 3.5. Consider the problem given in Example 3.4, and we wish to determine the polynomial $p(y)$ of the Chebyshev response of order 4 having passband tolerance of 1 dB.

From (3.74) and (3.80), the coefficients of the polynomial $p(y)$ are obtained as follows:

$$a = \frac{1}{4} \sinh^{-1} \frac{1}{0.509} = 1.428/4 = 0.357, \tag{3.81}$$

$$b_3 = \frac{\sinh 0.357}{\sin 22.5°} = \frac{0.3646}{0.38268} = 0.9528, \tag{3.81a}$$

$$b_2 = 1 + \frac{\sinh^2 0.357 \cos 22.5°}{\sin 22.5° \sin 45°} = 1.4539, \tag{3.81b}$$

$$b_1 = 0.9528 \left(1 - \frac{2 \cos^3 22.5° \sin^2 22.5°}{\sin 45° \sin 67.5°} \right)$$
$$+ \frac{\sinh^3 0.357 \cos 22.5° \cos 45°}{\sin 22.5° \sin 45° \sin 67.5°}$$
$$= 0.7426, \tag{3.81c}$$

$$b_0 = 2^{-3} \cosh 1.428 = 2.2051/8 = 0.2756. \tag{3.81d}$$

Hence the desired polynomial is given by

$$p(y) = y^4 + 0.9528y^3 + 1.4539y^2 + 0.7426y + 0.2756, \quad (3.82)$$

coinciding with the values listed in Table B.2.2 of Appendix B.

2.5. Chebyshev networks

We shall consider the same problem as in §1.3 for the Butterworth networks except now that we wish to design a lossless two-port network having a preassigned Chebyshev transducer power-gain characteristic.

Referring again to Fig. 3.4, let $S(s)$ of (3.22) be the scattering matrix of N normalizing to R_1 and R_2. The network possesses an nth-order Chebyshev transducer power-gain characteristic

$$|S_{21}(j\omega)|^2 = G(\omega^2) = \frac{K_n}{1 + \epsilon^2 C_n^2(\omega/\omega_c)}, \quad (3.83)$$

K_n as in (3.24) being bounded between 0 and 1. Following (3.25) and (3.26), the input reflection coefficient $S_{11}(s)$ of the two-port N is given by

$$S_{11}(s)S_{11}(-s) = (1 - K_n)\frac{1 + \hat{\epsilon}^2 C_n^2(-jy)}{1 + \epsilon^2 C_n^2(-jy)}, \quad (3.84a)$$

where

$$\hat{\epsilon} = \epsilon(1 - K_n)^{-1/2}. \quad (3.84b)$$

As in the Butterworth response, the denominator of (3.84a) can be uniquely decomposed as in (3.77). However, the zeros of $S_{11}(s)$ may lie in the RHS and there are no known reasons why they cannot, the only restriction being that the zeros of a complex-conjugate pair must be assigned together. For ease in the subsequent analysis to be presented in Chapters 4 and 5 on the theory of broadband matching, we choose only the LHS zeros. With this restriction, we obtain a minimum-phase decomposition of (3.84a), which yields a bounded-real reflection coefficient

$$S_{11}(s) = \pm \frac{\hat{p}(y)}{p(y)}, \quad (3.85a)$$

where

$$\hat{p}(y) = \hat{b}_0 + \hat{b}_1 y + \cdots + \hat{b}_{n-1} y^{n-1} + \hat{b}_n y^n = \sum_{m=0}^{n} \hat{b}_m y^m \quad (3.85b)$$

with $\hat{b}_n = 1$ is the Hurwitz polynomial formed by the LHS roots of the polynomial $1 + \hat{\epsilon}^2 C_n^2(-jy) = 0$. We remark that unlike (3.27) the leading coefficient in (3.85a) is unity. Thus from (2.112), the input impedance $Z_{11}(s)$ of N with the output port terminating in R_2 is positive-real and is given by

$$Z_{11}(s) = R_1 \frac{p(y) \pm \hat{p}(y)}{p(y) \mp \hat{p}(y)}. \quad (3.86)$$

The quantities R_1, R_2 and K_n are related, and cannot be chosen arbitrarily for all three. In fact, if we substitute $y = 0$ in (3.86) in conjunction with (3.80d) and (3.84b), we obtain the desired relationships for an LC two-port network with $K_n \neq 0$:

$$\frac{R_2}{R_1} = \left[\frac{b_0 + \hat{b}_0}{b_0 - \hat{b}_0}\right]^{\pm 1} \quad (3.87a)$$

$$= \left[\frac{1 + (1 - K_n)^{1/2}}{1 - (1 - K_n)^{1/2}}\right]^{\pm 1} \quad \text{for } n \text{ odd} \quad (3.87b)$$

$$= \left[\frac{(1 + \epsilon^2)^{1/2} + (1 + \epsilon^2 - K_n)^{1/2}}{(1 + \epsilon^2)^{1/2} - (1 + \epsilon^2 - K_n)^{1/2}}\right]^{\pm 1} \quad \text{for } n \text{ even}, \quad (3.87c)$$

the $\pm$ signs being determined respectively according to $R_2 \geqq R_1$ and $R_2 \leqq R_1$. Thus, if n is odd and the dc gain is specified, the ratio of the terminating resistances is determined by (3.87b). On the other hand, if n is even and the minimum passband gain is specified, the ratio of the resistances is given by (3.87c). For design applications where R_1 and R_2 are both specified in advance, then if n is odd, the dc gain is determined from (3.87b) and the passband tolerance may be chosen arbitrarily; and if n is even, the minimum passband gain is fixed in accordance with (3.57) and the passband tolerance may be specified arbitrarily.

For example, let n be odd and $K_n = 1$, the maximum permissible dc gain for a passive two-port. Then from (3.87b) we have $R_1 = R_2$.

On the other hand, if we specify that $R_2/R_1 = 2$ as in Example 3.3, $K_n = 8/9$ is determined outright from (3.87b).

2.6. Chebyshev LC ladder networks

Like the Butterworth LC ladder networks discussed in § 1.4, in the present section we consider in detail the LC ladder networks that possess the Chebyshev transducer power-gain characteristic.

Depending upon the choice of the signs in (3.85a), two cases are treated, each being presented in a separate section.

A. $S_{11}(0) \geqq 0$

With the choice of the plus sign in (3.85a), the input impedance becomes

$$Z_{11}(s) = R_1 \frac{\sum_{m=0}^{n} (b_m + \hat{b}_m) y^m}{\sum_{m=0}^{n} (b_m - \hat{b}_m) y^m}, \qquad (3.88)$$

whose continued fraction expansion is shown in (3.32). Depending upon whether n is odd or even, the corresponding LC ladder network has the configuration as indicated in Fig. 3.5. The first element L_1 can easily be determined from (3.88) in conjunction with (3.80a), and is given by

$$L_1 = 2R_1/(b_{n-1} - \hat{b}_{n-1})\omega_c = \frac{2R_1 \sin \gamma_1}{\omega_c (\sinh a - \sinh \hat{a})}, \qquad (3.89)$$

where $\gamma_1 = \pi/2n$ as in (3.34c) and

$$\hat{a} = \frac{1}{n} \sinh^{-1} \frac{(1 - K_n)^{1/2}}{\epsilon}. \qquad (3.90)$$

Moreover, it can be shown that the values of other elements can be computed by the recurrence formulas (Problem 3.24)

$$L_{2m-1} C_{2m} = \frac{16 \sin \gamma_{4m-3} \sin \gamma_{4m-1}}{\omega_c^2 f_{2m-1}(\sinh a, \sinh \hat{a})}, \qquad (3.91a)$$

$$L_{2m+1} C_{2m} = \frac{16 \sin \gamma_{4m-1} \sin \gamma_{4m+1}}{\omega_c^2 f_{2m}(\sinh a, \sinh \hat{a})} \qquad (3.91b)$$

for $m = 1, 2, \ldots, [\tfrac{1}{2} n]$, terminating in L_n as shown in Fig. 3.5(a) if n

APPROXIMATION AND LADDER REALIZATION

is odd or in C_n as in Fig. 3.5(b) if n is even, where

$$f_m(\sinh a, \sinh \hat{a}) = 4(\sinh^2 a + \sinh^2 \hat{a} + \sin^2 \gamma_{2m}$$
$$- 2 \sinh a \sinh \hat{a} \cos \gamma_{2m}), \quad (3.91c)$$

and $\gamma_m = m\pi/2n$ as given in (3.34c). In addition, the values of the last elements are related to R_2 by

$$L_n = \frac{2R_2 \sin \gamma_1}{\omega_c(\sinh a + \sinh \hat{a})} \quad (3.92a)$$

for n odd, and

$$C_n = \frac{2 \sin \gamma_1}{R_2 \omega_c(\sinh a + \sinh \hat{a})} \quad (3.92b)$$

for n even. Again the above formulas can be derived deductively by carrying out the calculations in detail for the cases of low order and then guessing the final result. A formal elegant proof was first given by Takahasi (1951). Hence we can calculate the element values starting from either the first or the last element.

We shall illustrate the formulas by the following example.

EXAMPLE 3.6. Suppose that we wish to design a low-pass filter that has an equiripple transducer power-gain characteristic. The filter is to be operated between a resistive generator of internal resistance $100 \, \Omega$ and a $200 \text{-} \Omega$ load and such that it gives at least 60-dB attenuation in gain at the frequency five times the radian cutoff frequency $\omega_c = 10^4$ rad/s and beyond. The peak-to-peak ripple in the passband must not exceed 1 dB.

From specifications and Example 3.4, we have

$$\begin{aligned} R_1 &= 100 \, \Omega, & R_2 &= 200 \, \Omega, \\ \omega_c &= 10^4 \text{ rad/s}, & n &= 4. \end{aligned} \quad (3.93)$$

Since R_1 and R_2 are both specified, the minimum passband gain $G_{\min}$ is fixed by (3.87c) in accordance with (3.57). For $R_2 > R_1$, we must select the plus sign in (3.85a) and (3.87c). From (3.87c), we obtain

$$\frac{200}{100} = \frac{1 + (1 - G_{\min})^{1/2}}{1 - (1 - G_{\min})^{1/2}}. \quad (3.94)$$

Solving for G_{min} yields

$$G_{min} = K_4/(1 + \epsilon^2) = 8/9. \quad (3.95)$$

According to (3.62), the 1-dB ripple in the passband corresponds to a ripple factor $\epsilon = 0.509$. If we use this value of ϵ in (3.95), we obtain $K_4 = 1.119$, which is too large for the network to be physically realizable. Thus, let $K_4 = 1$, the maximum permissible value. This according to (3.95) corresponds to $\epsilon = 0.354$ or 0.51-dB ripple, well within the 1-dB specification. With this selection of parameters, we have

$$\hat{a} = 0, \quad (3.96a)$$

$$a = \frac{1}{4} \sinh^{-1} \frac{1}{0.354} = 0.441, \quad (3.96b)$$

$$f_m(\sinh 0.441, 0) = 0.828 + 4 \sin^2 \gamma_{2m}. \quad (3.96c)$$

From (3.89), the value of the first inductance is given by

$$L_1 = \frac{200 \sin 22.5°}{10^4 \sinh 0.441} = 16.82 \text{ mH}. \quad (3.97a)$$

Using (3.91a) with $m = 1$ in conjunction with (3.97a) yields

$$C_2 = \frac{16 \sin 22.5° \sin 67.5°}{1.682 \times 10^6 \times (0.828 + 4 \sin^2 45°)} = 1.1892 \, \mu\text{F}. \quad (3.97b)$$

Using (3.91b) with $m = 1$ and (3.97b) gives

$$L_3 = \frac{16 \sin 67.5° \sin 112.5°}{1.1892 \times 10^2 \times (0.828 + 4 \sin^2 90°)} = 23.786 \text{ mH}. \quad (3.97c)$$

Repeat the above process for $m = 2$ in (3.91a), and we obtain

$$C_4 = \frac{16 \sin 112.5° \sin 157.5°}{2.3786 \times 10^6 \times (0.828 + 4 \sin^2 135°)} = 0.841 \, \mu\text{F}. \quad (3.97d)$$

The last capacitance can also be computed directly from (3.92b),

APPROXIMATION AND LADDER REALIZATION

yielding

$$C_4 = \frac{2 \sin 22.5°}{200 \times 10^4 \sinh 0.441} = 0.841 \ \mu\text{F}, \quad (3.97\text{e})$$

coinciding with (3.97d). The LC ladder network together with its terminations is presented in Fig. 3.15.

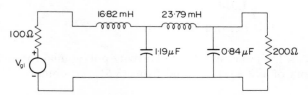

FIG. 3.15. A low-pass filter possessing the fourth-order Chebyshev transducer power-gain characteristic.

For illustrative purposes, we compute the element values by expanding the input impedance $Z_{11}(s)$ as a continued fraction as in (3.32). To this end, we first determine the poles of the Chebyshev response by means of (3.79), yielding

$$\begin{aligned} y_2, y_3 &= -\sinh 0.441 \sin 67.5° \pm j \cosh 0.441 \cos 67.5° \\ &= -0.4204 \pm j0.4204, \end{aligned} \quad (3.98\text{a})$$

$$\begin{aligned} y_1, y_4 &= -\sinh 0.441 \sin 22.5° \pm j \cosh 0.441 \cos 22.5° \\ &= -0.1741 \pm j1.015. \end{aligned} \quad (3.98\text{b})$$

The associated Hurwitz polynomial is given by

$$\begin{aligned} p(y) &= (y - y_1)(y - y_2)(y - y_3)(y - y_4) \\ &= y^4 + 1.189y^3 + 1.7068y^2 + 1.0147y + 0.375, \end{aligned} \quad (3.99)$$

whose coefficients can also be computed directly from (3.80), giving

$$b_3 = \frac{\sinh 0.441}{\sin 22.5°} = 1.189, \quad (3.100\text{a})$$

$$b_2 = 1 + \frac{\sinh^2 0.441 \cos 22.5°}{\sin 22.5° \sin 45°} = 1.7069, \qquad (3.100b)$$

$$b_1 = \frac{\sinh 0.441}{\sin 22.5°}\left(1 - \frac{2\cos^3 22.5° \sin^2 22.5°}{\sin 45° \sin 67.5°}\right)$$
$$+ \frac{\sinh^3 0.441 \cos 22.5° \cos 45°}{\sin 22.5° \sin 45° \sin 67.5°}$$
$$= 1.0147, \qquad (3.100c)$$

$$b_0 = \tfrac{1}{8} \cosh 1.764 = 0.375. \qquad (3.100d)$$

For $K_4 = 1$, $\hat{p}(y)$ is simply the Hurwitz polynomial formed by the LHS roots of $C_4^2(-jy) = 0$, and from (3.51) we obtain

$$\hat{p}(y) = y^4 + y^2 + 0.125. \qquad (3.101)$$

Substituting (3.101) and (3.99) in (3.86) yields

$$Z_{11}(s) = 100\frac{2y^4 + 1.189y^3 + 2.7068y^2 + 1.0147y + 0.5}{1.189y^3 + 0.7068y^2 + 1.0147y + 0.25}$$
$$= 16.82 \times 10^{-3}s + \cfrac{1}{1.189 \times 10^{-6}s + \cfrac{1}{23.786 \times 10^{-3}s + \cfrac{1}{0.841 \times 10^{-6} + \cfrac{1}{200}}}}$$
$$(3.102)$$

which can be identified as an *LC* ladder terminating in a 200-Ω resistor, as required, whose element values are given in (3.97).

B. $S_{11}(0) < 0$

With the choice of the minus sign in (3.85a), the input impedance, aside from the constant R_1, becomes the reciprocal of that in (3.88):

$$Z_{11}(s) = R_1 \frac{\sum_{m=0}^{n}(b_m - \hat{b}_m)y^m}{\sum_{m=0}^{n}(b_m + \hat{b}_m)y^m}, \qquad (3.103)$$

whose continued fraction expansion is shown in (3.41). Depending upon whether n is even or odd, the LC ladder network has the configuration as shown in Fig. 3.8. Formulas for the element values are the same as those given in (3.89)–(3.92) except that the roles of C's and L's are interchanged and that R_1 and R_2 are replaced by their reciprocals (also see Problem 3.25):

$$C_1 = \frac{2 \sin \gamma_1}{\omega_c R_1 (\sinh a - \sinh \hat{a})}, \qquad (3.104a)$$

$$C_{2m-1} L_{2m} = \frac{16 \sin \gamma_{4m-3} \sin \gamma_{4m-1}}{\omega_c^2 f_{2m-1}(\sinh a, \sinh \hat{a})}, \qquad (3.104b)$$

$$C_{2m+1} L_{2m} = \frac{16 \sin \gamma_{4m-1} \sin \gamma_{4m+1}}{\omega_c^2 f_{2m}(\sinh a, \sinh \hat{a})} \qquad (3.104c)$$

for $m = 1, 2, \ldots, [\tfrac{1}{2}n]$, and

$$C_n = \frac{2 \sin \gamma_1}{\omega_c R_2 (\sinh a + \sinh \hat{a})} \qquad (3.104d)$$

for n odd, and

$$L_n = \frac{2 R_2 \sin \gamma_1}{\omega_c (\sinh a + \sinh \hat{a})} \qquad (3.104e)$$

for n even; where γ_m and $f_m(\sinh a, \sinh \hat{a})$ are defined in (3.34c) and (3.91c).

As an illustration, in Example 3.6 if we choose the minus sign in (3.85a) instead of the plus sign, the resulting LC ladder network together with its terminations is presented in Fig. 3.16, whose normalized input impedance $Z_{11}(s)/R_1$ is the reciprocal of that obtained for the corresponding ladder of Fig. 3.15. We recognize that in selecting the minus sign in (3.85a) we must choose the minus sign in (3.87c). Thus, for a specified minimum passband gain and a fixed generator resistance R_1, the terminating resistance R_2 is determined by (3.87c). In the present situation, we have $R_1 = 100\ \Omega$, $K_4 = 1$ and $\epsilon = 0.354$ (or 0.51-dB ripple in the passband). Substituting these in (3.87c) gives $R_2 = 50\ \Omega$, as indicated in Fig. 3.16.

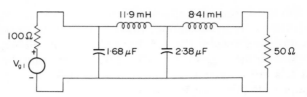

FIG. 3.16. A low-pass filter possessing the fourth-order Chebyshev transducer power-gain characteristic.

3. Elliptic functions

In the foregoing, we have presented two popular schemes that approximate the ideal low-pass characteristic: the Butterworth response and the Chebyshev response. The Butterworth response is maximally flat in both the passband and stopband, while the Chebyshev response yields an equiripple characteristic in the passband and maximally flat in the stopband. Considering the advantages of the Chebyshev response over the Butterworth, one naturally is led to ask the possibility of designing a filter which has equiripple characteristics in both the passband and stopband. This may be achieved by the use of the elliptic functions and is called an *elliptic filter*. Since many readers are not familiar with the subject of elliptic functions, in the present section we shall discuss some of their fundamental properties that are needed in the subsequent analysis.

3.1. Jacobian elliptic functions

Consider the integral

$$u \equiv F(k, \phi) = \int_0^x \frac{dx}{(1-x^2)^{1/2}(1-k^2x^2)^{1/2}} = \int_0^\phi \frac{d\phi}{(1-k^2 \sin^2 \phi)^{1/2}},$$
(3.105a)

where $0 \leq k \leq 1$ and

$$x \equiv \sin \phi.$$
(3.105b)

The function $F(k, \phi)$ is called the *Legendre standard form of the elliptic integral of the first kind of modulus k*, whose values are known and have been tabulated in most mathematical tables for various values of k and ϕ (see, for example, Abramowitz and Stegun, 1965, and Jahnke and Emde, 1945).

It was suggested by Abel in 1823 that we can consider the upper limit x as a function of $u = F(k, \phi)$ and k. Equivalently, we can study the upper limit ϕ as a function of u and k. To this end, Jacobi introduced the notation

$$\phi = \text{am}(u, k), \tag{3.106}$$

which reads that ϕ is the *amplitude* of u of modulus k. Thus, we have

$$x = \sin \phi = \sin \text{am}(u, k). \tag{3.107}$$

To simplify this expression, Gudermann proposed the abbreviated notation

$$x \equiv \text{sn}(u, k), \tag{3.108}$$

which is called the *Jacobian elliptic sine function of modulus k*. For convenience, define

$$\text{cn}(u, k) \equiv \cos \phi, \tag{3.109a}$$

$$\text{dn}(u, k) \equiv (1 - k^2 x^2)^{1/2}. \tag{3.109b}$$

It follows at once that (Problems 3.31 and 3.32)

$$\text{sn}^2(u, k) + \text{cn}^2(u, k) = 1, \tag{3.110a}$$

$$\text{dn}^2(u, k) + k^2 \text{sn}^2(u, k) = 1, \tag{3.110b}$$

$$\frac{d}{du} \text{sn}(u, k) = \text{cn}(u, k) \text{dn}(u, k), \tag{3.111a}$$

$$\frac{d}{du} \text{cn}(u, k) = -\text{sn}(u, k) \text{dn}(u, k), \tag{3.111b}$$

$$\frac{d}{du} \text{dn}(u, k) = -k^2 \text{sn}(u, k) \text{cn}(u, k). \tag{3.111c}$$

Furthermore, if $u = 0$ then the upper limit $\phi = 0$ in (3.105), giving (Problem 3.34)

$$\operatorname{sn}(0, k) = 0, \qquad (3.112a)$$
$$\operatorname{cn}(0, k) = 1, \qquad (3.112b)$$
$$\operatorname{dn}(0, k) = 1. \qquad (3.112c)$$

Since $\operatorname{am}(-u, k) = -\operatorname{am}(u, k)$, we have

$$\operatorname{sn}(-u, k) = -\operatorname{sn}(u, k), \qquad (3.113a)$$
$$\operatorname{cn}(-u, k) = \operatorname{cn}(u, k), \qquad (3.113b)$$
$$\operatorname{dn}(-u, k) = \operatorname{dn}(u, k), \qquad (3.113c)$$

indicating that $\operatorname{sn}(u, k)$ is an odd function while $\operatorname{cn}(u, k)$ and $\operatorname{dn}(u, k)$ are even.

3.2. Jacobi's imaginary transformations

In this section, we show that the function $\operatorname{sn}(ju, k)$ is imaginary while $\operatorname{cn}(ju, k)$ and $\operatorname{dn}(ju, k)$ are real. Let

$$\sin\theta = j\tan\phi. \qquad (3.114)$$

Then

$$\cos\theta = (1 - \sin^2\theta)^{1/2} = (1 + \tan^2\phi)^{1/2} = \sec\phi, \qquad (3.115a)$$
$$\sin\phi = \cos\phi\tan\phi = -j\tan\theta, \qquad (3.115b)$$
$$d\theta = j\sec^2\phi\sec\theta\,d\phi = j\sec\phi\,d\phi. \qquad (3.115c)$$

Substituting these in (3.105) gives

$$F(k, \theta) = \int_0^\theta \frac{d\theta}{(1 - k^2\sin^2\theta)^{1/2}} = \int_0^\phi \frac{j\sec\phi\,d\phi}{(1 + k^2\tan^2\phi)^{1/2}}$$
$$= j\int_0^\phi \frac{d\phi}{(1 - k'^2\sin^2\phi)^{1/2}} = jF(k', \phi) = ju, \qquad (3.116a)$$

where

$$k' = (1 - k^2)^{1/2} \qquad (3.116b)$$

is called the *complementary modulus* of k. Using the notation (3.106) yields

$$\phi = \text{am}(u, k'), \tag{3.117a}$$

$$\theta = \text{am}(ju, k). \tag{3.117b}$$

From (3.114), we have

$$\sin \theta = j \frac{\sin \phi}{\cos \phi}, \tag{3.118}$$

which according to (3.107), (3.108) and (3.109a) is equivalent to

$$\text{sn}(ju, k) = j \frac{\text{sn}(u, k')}{\text{cn}(u, k')} = j \text{ tn}(u, k'), \tag{3.119}$$

where

$$\text{tn}(u, k) \equiv \frac{\text{sn}(u, k)}{\text{cn}(u, k)}. \tag{3.120}$$

Similarly, we can show that (Problem 3.38)

$$\text{cn}(ju, k) = 1/\text{cn}(u, k'), \tag{3.121a}$$

$$\text{dn}(ju, k) = \frac{\text{dn}(u, k')}{\text{cn}(u, k')}. \tag{3.121b}$$

Thus, we have shown that an elliptic function with an imaginary argument can be expressed through elliptic functions with a real argument, whose modulus is the complement of the original modulus.

3.3. Periods of elliptic functions

Unlike the trigonometric functions, here we demonstrate that the elliptic functions have two periods: a real period and an imaginary period. Thus an elliptic function is doubly periodic with respect to both the real and imaginary axes. For this purpose, we first consider the graphs of the elliptic integral $F(k, \phi)$.

Let

$$w = (1 - k^2 \sin^2 \phi)^{-1/2}. \tag{3.122}$$

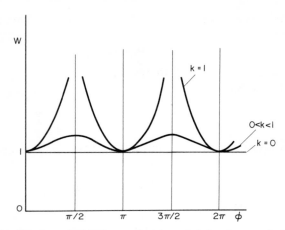

FIG. 3.17. Sketches of (3.122) as a function of ϕ for three typical values of k.

A plot of w as a function of ϕ for three typical values of k is presented in Fig. 3.17. It is clear from (3.105a) that the areas under these curves represent the integral $F(k, \phi)$. For a given value of k, $F(k, \phi)$ increases continuously as ϕ is increased. Furthermore, for a given value of ϕ, the integral $F(k, \phi)$ increases as k is increased. At $\phi = \frac{1}{2}\pi$, the integral $F(k, \frac{1}{2}\pi)$ written as

$$K(k) = F(k, \tfrac{1}{2}\pi) \qquad (3.123)$$

is called the *complete elliptic integral of the first kind of modulus k*. In practice, it is convenient to express k as

$$k = \sin \theta. \qquad (3.124)$$

A plot of $F(\theta, \phi)$ as a function of ϕ for various values of θ is shown in Fig. 3.18.

Since the curves in Fig. 3.17 are symmetric about the line $\phi = \frac{1}{2}\pi$, it is evident that for a fixed value of k, it is sufficient to know the values ϕ from 0 to $\frac{1}{2}\pi$. For $\phi = \pi$, we have $F(k, \pi) = 2K(k)$, and for any value of ϕ we have

$$F(k, \phi) = F(k, \pi) - F(k, \pi - \phi) \qquad (3.125a)$$

APPROXIMATION AND LADDER REALIZATION 157

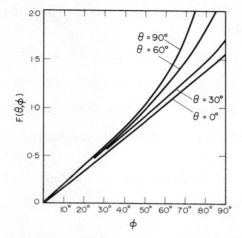

FIG. 3.18. The elliptic integral $F(\theta, \phi)$ of the first kind of modulus $k = \sin \theta$.

or
$$F(k, \phi) = 2K(k) - F(k, \pi - \phi), \tag{3.125b}$$

showing that the values of $F(k, \phi)$ for ϕ from $\frac{1}{2}\pi$ to π are related to those for ϕ from 0 to $\frac{1}{2}\pi$. Since $F(k, -\phi) = -F(k, \phi)$, (3.125b) can also be written as

$$F(k, \phi) = 2K(k) + F(k, \phi - \pi). \tag{3.125c}$$

In general, we have the formula

$$F(k, m\pi \pm \phi) = 2mK(k) \pm F(k, \phi), \tag{3.126}$$

m being any integer.

3.3.1. *The real periods*

We now proceed to derive the real periods of the Jacobian elliptic functions. To simplify our notation, we write

$$K \equiv K(k) \tag{3.127}$$

if the modulus k is clearly understood. From (3.105) and (3.106), formula (3.126) gives

$$m\pi \pm \phi = \text{am}\,(2mK \pm u, k) \qquad (3.128)$$

with $u = F(k, \phi)$ or in terms of the sine function of the angle

$$\sin(m\pi \pm \phi) = \sin\text{am}\,(2mK \pm u, k). \qquad (3.129)$$

Expanding both sides of (3.129) and applying (3.108) and (3.109) yield

$$\sin m\pi \,\text{cn}\,(u, k) \pm \cos m\pi \,\text{sn}\,(u, k) = \text{sn}\,(2mK \pm u, k), \qquad (3.130)$$

which shows that

$$\text{sn}\,(u \pm 2K, k) = -\text{sn}\,(u, k), \qquad (3.131\text{a})$$

$$\text{sn}\,(u \pm 4K, k) = \text{sn}\,(u, k). \qquad (3.131\text{b})$$

Similarly, we can show that (Problem 3.42)

$$\text{cn}\,(u \pm 2K, k) = -\text{cn}\,(u, k), \qquad (3.132\text{a})$$

$$\text{cn}\,(u \pm 4K, k) = \text{cn}\,(u, k), \qquad (3.132\text{b})$$

$$\text{dn}\,(u \pm 2K, k) = \text{dn}\,(u, k), \qquad (3.132\text{c})$$

$$\text{dn}\,(u \pm 4K, k) = \text{dn}\,(u, k). \qquad (3.132\text{d})$$

Thus, $4K$ is a period of $\text{sn}\,(u, k)$, $\text{cn}\,(u, k)$ and $\text{dn}\,(u, k)$. In fact, $2K$ is also a period of $\text{dn}\,(u, k)$. Finally, setting $u = 0$ and using $\text{am}\,(K, k) = \tfrac{1}{2}\pi$ give (Problem 3.45)

$$\text{sn}\,(K, k) = 1, \qquad \text{sn}\,(2K, k) = 0, \qquad \text{sn}\,(4K, k) = 0, \qquad (3.133\text{a})$$

$$\text{cn}\,(K, k) = 0, \qquad \text{cn}\,(2K, k) = -1, \qquad \text{cn}\,(4K, k) = 1, \qquad (3.133\text{b})$$

$$\text{dn}\,(K, k) = k', \qquad \text{dn}\,(2K, k) = 1, \qquad \text{dn}\,(4K, k) = 1. \qquad (3.133\text{c})$$

3.3.2. *The imaginary periods*

In addition to the real periods, we show that the Jacobian elliptic functions are also periodic with respect to the imaginary axis. From

APPROXIMATION AND LADDER REALIZATION   159

(3.119), (3.131a) and (3.132a), it follows at once that

$$\operatorname{sn}[j(u \pm 2K'), k] = j \operatorname{tn}(u \pm 2K', k') = j \operatorname{tn}(u, k')$$
$$= \operatorname{sn}(ju, k), \quad (3.134)$$

where
$$K' = K(k') \quad (3.135)$$

if k' is clearly understood. Changing ju to u gives

$$\operatorname{sn}(u \pm j2K', k) = \operatorname{sn}(u, k), \quad (3.136)$$

indicating that $j2K'$ is a period of $\operatorname{sn}(u, k)$. In a similar way, we can show that (Problems 3.48–3.50)

$$\operatorname{sn}(u \pm j4K', k) = \operatorname{sn}(u, k), \quad (3.137)$$
$$\operatorname{cn}(u \pm j2K', k) = -\operatorname{cn}(u, k), \quad (3.138a)$$
$$\operatorname{cn}(u \pm j4K', k) = \operatorname{cn}(u, k), \quad (3.138b)$$
$$\operatorname{dn}(u \pm j2K', k) = -\operatorname{dn}(u, k), \quad (3.139a)$$
$$\operatorname{dn}(u \pm j4K', k) = \operatorname{dn}(u, k), \quad (3.139b)$$

yielding

$$\operatorname{sn}(j2K', k) = 0, \quad \operatorname{cn}(j2K', k) = -1, \quad \operatorname{dn}(j2K', k) = -1, \quad (3.140a)$$
$$\operatorname{sn}(j4K', k) = 0, \quad \operatorname{cn}(j4K', k) = 1, \quad \operatorname{dn}(j4K', k) = 1. \quad (3.140b)$$

From these, we conclude that $2K + j2K'$ is a period of $\operatorname{cn}(u, k)$ and that $j4K'$ is a period of $\operatorname{dn}(u, k)$.

3.4. Poles and zeros of the Jacobian elliptic functions

Pole locations of the Jacobian elliptic sine functions are needed in the subsequent design of the elliptic filters. For this we derive formulas for the determination of the locations of these singularities.

Let

$$\sin \phi = \frac{\cos \theta}{(1 - k^2 \sin^2 \theta)^{1/2}}. \quad (3.141)$$

Then from (3.105a) we have

$$\int_0^\phi \frac{d\phi}{(1 - k^2 \sin^2 \phi)^{1/2}} = -\int_{\pi/2}^\theta \frac{d\theta}{(1 - k^2 \sin^2 \theta)^{1/2}}$$

$$= \int_0^{\pi/2} \frac{d\theta}{(1 - k^2 \sin^2 \theta)^{1/2}} - \int_0^\theta \frac{d\theta}{(1 - k^2 \sin^2 \theta)^{1/2}}$$

$$= K - u, \quad (3.142)$$

which in conjunction with (3.109), (3.110b) and (3.141) shows that

$$\operatorname{sn}(K - u, k) = \sin \phi = \frac{\operatorname{cn}(u, k)}{\operatorname{dn}(u, k)} \quad (3.143a)$$

or

$$\operatorname{sn}(u \pm K, k) = \pm \frac{\operatorname{cn}(u, k)}{\operatorname{dn}(u, k)}. \quad (3.143b)$$

Similarly, we obtain (Problem 3.54)

$$\operatorname{cn}(u \pm K, k) = \mp \frac{k' \operatorname{sn}(u, k)}{\operatorname{dn}(u, k)}, \quad (3.144a)$$

$$\operatorname{dn}(u \pm K, k) = \frac{k'}{\operatorname{dn}(u, k)}. \quad (3.144b)$$

Using (3.119), (3.143) and (3.144) gives

$$\operatorname{sn}(ju \pm jK', k) = -j \frac{\operatorname{cn}(u, k')}{k \operatorname{sn}(u, k')}. \quad (3.145)$$

Changing ju to u results in

$$\operatorname{sn}(u \pm jK', k) = \frac{1}{k \operatorname{sn}(u, k)}. \quad (3.146)$$

APPROXIMATION AND LADDER REALIZATION

Thus, the poles of $\operatorname{sn}(u \pm jK', k)$ occur at the zeros of $\operatorname{sn}(u, k)$, which according to (3.112a) vanishes at $u = 0$. This implies that jK' is a pole of $\operatorname{sn}(u, k)$. Since $\operatorname{sn}(u, k)$ is doubly periodic with respect to $4K$ and $j2K'$, the poles of $\operatorname{sn}(u, k)$ clearly are located at the points

$$2mK + j(2n + 1)K', \tag{3.147}$$

m and n being any integers. In a similar manner, we can show that (Problem 3.56)

$$\operatorname{cn}(u \pm jK', k) = \mp \frac{j \operatorname{dn}(u, k)}{k \operatorname{sn}(u, k)}, \tag{3.148a}$$

$$\operatorname{dn}(u \pm jK', k) = \mp j \cot \operatorname{am}(u, k), \tag{3.148b}$$

whose pole locations are the same as those given in (3.147).

If in these formulas $u = 0$, then

$$\operatorname{sn}(\pm K, k) = \pm 1, \quad \operatorname{cn}(K, k) = 0, \quad \operatorname{dn}(K, k) = k', \tag{3.149}$$

and if $u = K$,

$$\operatorname{sn}(K \pm jK', k) = \frac{1}{k}, \tag{3.150a}$$

$$\operatorname{cn}(K \pm jK', k) = \mp j \frac{k'}{k}, \tag{3.150b}$$

$$\operatorname{dn}(K \pm jK', k) = 0. \tag{3.150c}$$

Finally, appealing again to the doubly periodic property of the Jacobian elliptic functions and using (3.112a), (3.133a), (3.133b) and (3.150c) yield the locations of the zeros of these functions:

$$2mK + j2nK' \quad \text{for } \operatorname{sn}(u, k), \tag{3.151a}$$

$$(2m + 1)K + j2nK' \quad \text{for } \operatorname{cn}(u, k), \tag{3.151b}$$

$$(2m + 1)K + j(2n + 1)K' \quad \text{for } \operatorname{dn}(u, k). \tag{3.151c}$$

We now summarize the above results for the Jacobian elliptic sine function in Fig. 3.19, in which the periods are represented by a

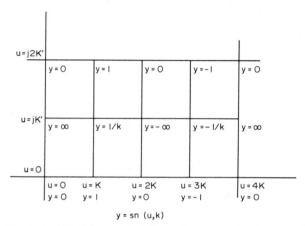

FIG. 3.19. The representation of the Jacobian elliptic sine function in terms of the periodic parallelograms.

parallelogram. On the boundaries of the periodic parallelogram, there are six points at which the function vanishes; but if the adjacent periodic parallelograms are constructed, it will be seen that only two zeros belong to each parallelogram. Similarly, there are two poles belonging to each parallelogram. The images of these poles and zeros appear in congruent rectangles throughout the entire plane. In fact, in each periodic parallelogram there are *two* values of u which gives a prescribed value of the function. The other important quantities are outlined in Table 3.1.

TABLE 3.1.

Functions	Zeros	Poles	Periods
$\text{sn}(u, k)$	$2mK + j2nK'$	$2mK + j(2n+1)K'$	$4K, j2K'$
$\text{cn}(u, k)$	$(2m+1)K + j2nK'$	$2mK + j(2n+1)K'$	$4K, 2K + j2K'$
$\text{dn}(u, k)$	$(2m+1)K + j(2n+1)K'$	$2mK + j(2n+1)K'$	$2K, j4K'$

m and n are any integers.

3.5. Addition theorems and complex arguments

In order to evaluate Jacobian elliptic functions of a complex argument, it is necessary that we express them in terms of functions

APPROXIMATION AND LADDER REALIZATION 163

of a real argument. For this, we first state the well-known addition theorems, which relate the functions of $u + v$ to those of u and those of v:

$$\operatorname{sn}(u \pm v, k)$$
$$= \frac{\operatorname{sn}(u, k) \operatorname{cn}(v, k) \operatorname{dn}(v, k) \pm \operatorname{sn}(v, k) \operatorname{cn}(u, k) \operatorname{dn}(u, k)}{1 - k^2 \operatorname{sn}^2(u, k) \operatorname{sn}^2(v, k)}, \quad (3.152\text{a})$$

$$\operatorname{cn}(u \pm v, k)$$
$$= \frac{\operatorname{cn}(u, k) \operatorname{cn}(v, k) \mp \operatorname{sn}(u, k) \operatorname{sn}(v, k) \operatorname{dn}(u, k) \operatorname{dn}(v, k)}{1 - k^2 \operatorname{sn}^2(u, k) \operatorname{sn}^2(v, k)}, \quad (3.152\text{b})$$

$$\operatorname{dn}(u \pm v, k)$$
$$= \frac{\operatorname{dn}(u, k) \operatorname{dn}(v, k) \mp k^2 \operatorname{sn}(u, k) \operatorname{sn}(v, k) \operatorname{cn}(u, k) \operatorname{cn}(v, k)}{1 - k^2 \operatorname{sn}^2(u, k) \operatorname{sn}^2(v, k)}. \quad (3.152\text{c})$$

Applying Jacobi's imaginary transformations of (3.119) and (3.121) to (3.152) yields the desired formulas:

$$\operatorname{sn}(u \pm jv, k)$$
$$= \frac{\operatorname{sn}(u, k) \operatorname{dn}(v, k') \pm j \operatorname{cn}(u, k) \operatorname{dn}(u, k) \operatorname{sn}(v, k') \operatorname{cn}(v, k')}{\operatorname{cn}^2(v, k') + k^2 \operatorname{sn}^2(u, k) \operatorname{sn}^2(v, k')}, \quad (3.153\text{a})$$

$$\operatorname{cn}(u \pm jv, k)$$
$$= \frac{\operatorname{cn}(u, k) \operatorname{cn}(v, k') \mp j \operatorname{sn}(u, k) \operatorname{dn}(u, k) \operatorname{sn}(v, k') \operatorname{dn}(v, k')}{\operatorname{cn}^2(v, k') + k^2 \operatorname{sn}^2(u, k) \operatorname{sn}^2(v, k')}, \quad (3.153\text{b})$$

$$\operatorname{dn}(u \pm jv, k)$$
$$= \frac{\operatorname{dn}(u, k) \operatorname{cn}(v, k') \operatorname{dn}(v, k') \mp jk^2 \operatorname{sn}(u, k) \operatorname{cn}(u, k) \operatorname{sn}(v, k')}{\operatorname{cn}^2(v, k') + k^2 \operatorname{sn}^2(u, k) \operatorname{sn}^2(v, k')}. \quad (3.153\text{c})$$

We illustrate the above formulas by the following example.

EXAMPLE 3.7. We wish to evaluate the Jacobian elliptic sine function

$$\operatorname{sn}(1 + j1.5, 0.5),$$

which gives the following set of parameters:

$$k = 0.5, \qquad k' = (1 - 0.5^2)^{1/2} = 0.866,$$
$$\theta = \sin^{-1} k = 30°, \qquad \theta' = \sin^{-1} k' = 60°,$$
$$u = 1, \qquad v = 1.5.$$

From tables of elliptic integrals, we obtain

$$F(0.5, \phi) = 1 \quad \text{giving} \quad \phi = 55°21',$$
$$F(0.866, \phi) = 1.5 \quad \text{giving} \quad \phi = 70°11',$$

which were computed by interpolation. Thus, we have

$$\text{sn}\,(1, 0.5) = \sin 55°21' = 0.82264,$$
$$\text{sn}\,(1.5, 0.866) = \sin 70°11' = 0.94078,$$
$$\text{cn}\,(1, 0.5) = \cos 55°21' = 0.56856,$$
$$\text{cn}\,(1.5, 0.866) = \cos 70°11' = 0.33901.$$

From (3.110b) we can compute

$$\text{dn}\,(1.5, 0.866) = 0.57986,$$
$$\text{dn}\,(1, 0.5) = 0.91149.$$

Substituting these in (3.153a) gives

$$\text{sn}\,(1 + j1.5, 0.5)$$
$$= \frac{0.82264 \times 0.57986 + j0.56856 \times 0.91149 \times 0.94078 \times 0.33901}{(0.33901)^2 + 0.25 \times (0.82264)^2 (0.94078)^2}$$
$$= 1.80233 + j0.62450. \qquad (3.154)$$

For illustrative purposes, we also compute $\text{cn}\,(1 + j1.5, 0.5)$ directly from (3.153b).

APPROXIMATION AND LADDER REALIZATION

$$\operatorname{cn}(1+j1.5, 0.5)$$
$$= \frac{0.56856 \times 0.33901 - j0.82264 \times 0.91149 \times 0.94078 \times 0.57986}{(0.33901)^2 + 0.25 \times (0.82264)^2 (0.94078)^2}$$
$$= 0.72826 - j1.54552. \tag{3.155}$$

Combining (3.154) and (3.155) according to (3.110a) gives

$$\operatorname{sn}^2(1+j1.5, 0.5) + \operatorname{cn}^2(1+j1.5, 0.5)$$
$$= (2.85832 + j2.25109) + (-1.85832 - j2.25109) = 1, \tag{3.156}$$

as expected. From (3.153c), we obtain

$$\operatorname{dn}(1+j1.5, 0.5)$$
$$= \frac{0.91149 \times 0.33901 \times 0.57986 - j0.25 \times 0.82264 \times 0.56856 \times 0.94078}{(0.33901)^2 + 0.25 \times (0.82264)^2 (0.94078)^2}$$
$$= 0.67699 - j0.41564. \tag{3.157}$$

Alternatively, $\operatorname{dn}(1+j1.5, 0.5)$ can be computed directly from (3.110b), giving

$$\operatorname{dn}^2(1+j1.5, 0.5) = 1 - 0.25 \operatorname{sn}^2(1+j1.5, 0.5)$$
$$= 0.28541 - j0.56277.$$

or $\operatorname{dn}(1+j1.5, 0.5)$ as given in (3.157).

4. The elliptic response

After this digression into a discussion of the theory of Jacobian elliptic functions, we now return to our problem of designing a low-pass filter that yields a transducer power-gain characteristic having equiripple properties in both the passband and stopband.

Consider the nth-order low-pass elliptic transducer power-gain characteristic†

$$G(\omega^2) = \frac{H_n}{1 + \epsilon^2 F_n^2(\omega/\omega_c)}, \quad H_n \geq 0, \tag{3.158}$$

†Since the symbol K denotes the complete elliptic integral of the first kind, to avoid possible confusion we shall use the symbol H_n instead of K_n as used in the previous two responses.

where
$$F_n(\omega/\omega_c) = \text{sn}\left[\frac{nK_1}{K}\text{sn}^{-1}(\omega/\omega_c, k), k_1\right] \quad (3.159a)$$
for n odd, and
$$F_n(\omega/\omega_c) = \text{sn}\left[K_1 + \frac{nK_1}{K}\text{sn}^{-1}(\omega/\omega_c, k), k_1\right] \quad (3.159b)$$
for n even; where
$$K_1 = K(k_1), \quad (3.160a)$$
$$K = K(k) \quad (3.160b)$$

are the complete elliptic integrals of moduli k_1 and k, respectively, and the notation $\text{sn}^{-1}(u, k)$ denotes an inverse elliptic function which is defined as if $y = \text{sn}(u, k)$ then $u = \text{sn}^{-1}(y, k)$. In the remainder of this section, we show that this characteristic approximates the ideal low-pass response of Fig. 3.1 with equiripple properties in both the passband and stopband. The constants H_n and ω_c have the same interpretation as in the Chebyshev response, and the real parameters ϵ, k and k_1, all bounded between 0 and 1, are to be determined from the specifications. However, as will be shown later, the three quantities n, k and k_1 are not completely independent and are related by the equation

$$\frac{K_1'}{K'} = \frac{nK_1}{K}, \quad (3.161)$$

where
$$K_1' = K(k_1'), \quad (3.162a)$$
$$K' = K(k'), \quad (3.162b)$$
and
$$k_1' = (1 - k_1^2)^{1/2}, \quad (3.163a)$$
$$k' = (1 - k^2)^{1/2} \quad (3.163b)$$

are the complementary moduli of k_1 and k, respectively. This restriction is dictated by the desire to obtain design formulas that are simple and that will give the equiripple characteristic in both the

APPROXIMATION AND LADDER REALIZATION 167

passband and stopband; its justification will be apparent after we present the formulas.

We remark that since the application of the Jacobian elliptic functions to the design of electrical filters was first suggested by Cauer (1931), they are also referred to as the *Cauer-parameter filters*.

4.1. The characteristic function $F_n(\omega)$

We first show that the choice of the characteristic function $F_n(\omega)$ as given in (3.159) will result in a rational function, i.e. the ratio of two polynomials in ω, all of whose zeros lie within the passband and all of whose poles lie in the stopband. As in (3.159), two cases are considered, each being presented in a separate section.

To simplify our notation and without loss of generality, let $\omega_c = 1$. This is equivalent to having frequency-scaled by a factor of ω_c.

A. n odd

In this case, the normalized characteristic function becomes

$$F_n(\omega) = \operatorname{sn}\left[\frac{nK_1}{K}\operatorname{sn}^{-1}(\omega, k), k_1\right]. \tag{3.164}$$

According to (3.151a), the zeros of this function are defined by the equation

$$\frac{nK_1}{K}\operatorname{sn}^{-1}(\omega, k) = 2mK_1 + j2m_1 K'_1, \tag{3.165}$$

where m and m_1 are any integers. Solving for ω and appealing to or defining (3.161) yield the desired locations for all the distinct zeros of $F_n(\omega)$ as

$$\begin{aligned}\omega_{zm} &= \operatorname{sn}(2mK/n + j2m_1 K', k) \\ &= \operatorname{sn}(2mK/n, k), \quad m = 0, \pm 1, \ldots, \pm \tfrac{1}{2}(n-1).\end{aligned} \tag{3.166}$$

The second line follows from (3.136) and the fact that $j2K'$ is a period of $\operatorname{sn}(u, k)$. Since $\omega_{zm} \le 1$, all of the zeros lie within the passband.

Similarly, from (3.147) the poles of $F_n(\omega)$ are determined by the equation

$$\frac{nK_1}{K}\operatorname{sn}^{-1}(\omega, k) = 2mK_1 + j(2m_1 + 1)K_1'. \quad (3.167)$$

Solving for ω and applying (3.161) give all the distinct pole locations of $F_n(\omega)$ as

$$\begin{aligned}\omega_{pm} &= \operatorname{sn}[2mK/n + j(2m_1 + 1)K', k] \\ &= \operatorname{sn}(2mK/n + jK', k) \\ &= \frac{1}{k\operatorname{sn}(2mK/n, k)}, \quad m = 0, \pm 1, \ldots, \pm\tfrac{1}{2}(n-1). \quad (3.168)\end{aligned}$$

The third line follows directly from (3.146). For $m = 0$, the right-hand side of (3.168) is infinite, indicating that the function $F_n(\omega)$ has a pole at infinity. Since $\operatorname{sn}(2mK/n, k) \leq 1$, all of the poles lie in the stopband.

Observe that in (3.166) and (3.168) we have the relations

$$\omega_{zm} = -\omega_{z(-m)}, \quad m \neq 0, \quad (3.169a)$$

$$\omega_{pm} = -\omega_{p(-m)}, \quad m \neq 0. \quad (3.169b)$$

Using these, the characteristic function can now be rewritten as

$$F_n(\omega) = H_o\frac{\omega(\omega_1^2 - \omega^2)(\omega_2^2 - \omega^2)\cdots(\omega_q^2 - \omega^2)}{(1 - k^2\omega_1^2\omega^2)(1 - k^2\omega_2^2\omega^2)\cdots(1 - k^2\omega_q^2\omega^2)}, \quad (3.170)$$

where $F_1(\omega) = H_o\omega$, and for $n > 1$,

$$\omega_m = \operatorname{sn}(2mK/n, k), \quad m = 1, 2, \ldots, q, \quad q = \tfrac{1}{2}(n-1), \quad (3.171)$$

and H_o is a real constant to be determined from (3.164). At $\omega = 1$, we have from (3.133a), $\operatorname{sn}^{-1}(1, k) = K$ and (3.164) becomes

$$F_n(1) = \operatorname{sn}(nK_1, k_1) = (-1)^q, \quad (3.172)$$

which follows from (3.131) and (3.133a) and is valid only for n odd.

APPROXIMATION AND LADDER REALIZATION

Using this in (3.170) gives

$$H_o = \frac{(1-k^2\omega_1^2)(1-k^2\omega_2^2)\cdots(1-k^2\omega_q^2)}{(1-\omega_1^2)(1-\omega_2^2)\cdots(1-\omega_q^2)}. \quad (3.173)$$

At $\omega = 1/k$, we have from (3.150a), $\mathrm{sn}^{-1}(1/k, k) = K \pm jK'$ and (3.164) becomes

$$\begin{aligned} F_n(1/k) &= \mathrm{sn}\,(nK_1 \pm jnK_1K'/K, k_1) \\ &= \mathrm{sn}\,(nK_1 \pm jK_1', k_1) = (-1)^q/k_1. \end{aligned} \quad (3.174)$$

The second line is obtained from (3.161), (3.146) and (3.172). Substituting $\omega = 1/k$ in (3.170) in conjunction with (3.173) and (3.174) yields

$$H_o = \left(\frac{k^n}{k_1}\right)^{1/2}. \quad (3.175)$$

Thus, k_1 is determined by the formula (Problem 3.53)

$$k_1 = k^n \left[\frac{(1-\omega_1^2)(1-\omega_2^2)\cdots(1-\omega_q^2)}{(1-k^2\omega_1^2)(1-k^2\omega_2^2)\cdots(1-k^2\omega_q^2)} \right]^2, \quad (3.176)$$

once k and n are known.

EXAMPLE 3.8. Let

$n = 3$,

$k = 1/1.4 = 0.71429$.

Determine the characteristic function $F_n(\omega)$.

From tables of complete elliptic integrals of the first kind, we have

$$K = K(k) = K(0.71429) = 1.86282,$$

where $\theta = \sin^{-1} k = 45°35'$. Using (3.171) yields

$$\begin{aligned} \omega_1 &= \mathrm{sn}\,(2K/3, k) = \mathrm{sn}\,(1.24188, 0.71429) \\ &= \sin 64°18' = 0.90114. \end{aligned} \quad (3.177)$$

From (3.176), we compute the constant

$$k_1 = k^3 \left[\frac{1-\omega_1^2}{1-k^2\omega_1^2}\right]^2$$

$$= 0.36444 \left[\frac{1-0.81206}{1-0.51021 \times 0.81206}\right]^2 = 0.037527. \quad (3.178)$$

Substituting these in (3.170) yields the characteristic function

$$F_3(\omega) = \frac{k^{3/2}}{k_1^{1/2}} \cdot \frac{\omega(\omega_1^2 - \omega^2)}{1-k^2\omega_1^2\omega^2}$$

$$= 3.11629 \frac{\omega(0.81206 - \omega^2)}{1 - 0.41432\omega^2}. \quad (3.179)$$

For illustrative purposes, we also compute

$$F_3(1) = 3.11629 \frac{0.81206 - 1}{1 - 0.41432} = -1,$$

$$F_3(1/k) = F_3(1.4) = 3.11629 \frac{1.4(0.81206 - 1.96)}{1 - 0.41432 \times 1.96}$$

$$= -26.64748 = -1/0.037527 = -1/k_1,$$

as expected from (3.172) and (3.174). A plot of the characteristic function $F_3(\omega)$ as a function of ω is presented in Fig. 3.20.

B. n **even**

In this case, the normalized characteristic function becomes

$$F_n(\omega) = \operatorname{sn}\left[K_1 + \frac{nK_1}{K}\operatorname{sn}^{-1}(\omega, k), k_1\right]. \quad (3.180)$$

Proceeding as in (3.165) and (3.166) gives the zeros of $F_n(\omega)$ as

$$\omega_{zm} = \operatorname{sn}\left[(2m-1)K/n + j2m_1K', k\right]$$
$$= \operatorname{sn}\left[(2m-1)K/n, k\right], \quad m = -\tfrac{1}{2}n+1, -\tfrac{1}{2}n+2, \ldots, \tfrac{1}{2}n. \quad (3.181)$$

APPROXIMATION AND LADDER REALIZATION 171

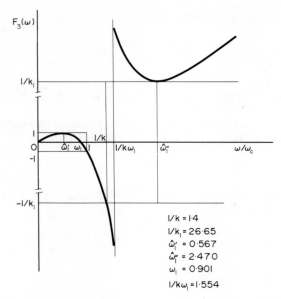

FIG. 3.20. A plot of the characteristic function $F_3(\omega)$ as a function of ω.

Similarly, following (3.167) and (3.168) yields the pole locations of $F_n(\omega)$ as

$$\omega_{pm} = \text{sn}\,[(2m-1)K/n + jK', k]$$

$$= \frac{1}{k\,\text{sn}\,[(2m-1)K/n, k]},$$

$$m = -\tfrac{1}{2}n + 1, -\tfrac{1}{2}n + 2, \ldots, \tfrac{1}{2}n - 1, \tfrac{1}{2}n. \quad (3.182)$$

Observe that

$$\omega_{z[-(1/2)n+v]} = \text{sn}\,[(-n + 2v - 1)K/n, k]$$

$$= -\text{sn}\,[(n - 2v + 1)K/n, k]$$

$$= -\omega_{z[(1/2)n-v+1]}, \quad v = 1, 2, \ldots, \tfrac{1}{2}n, \quad (3.183a)$$

$$\omega_{p[-(1/2)n+v]} = -\omega_{p[(1/2)n-v+1]}, \quad v = 1, 2, \ldots, \tfrac{1}{2}n, \quad (3.183b)$$

meaning that $F_n(\omega)$ can be written as in (3.170). We obtain

$$F_n(\omega) = H_e \frac{(\omega_1^2 - \omega^2)(\omega_2^2 - \omega^2)\cdots(\omega_{n/2}^2 - \omega^2)}{(1 - k^2\omega_1^2\omega^2)(1 - k^2\omega_2^2\omega^2)\cdots(1 - k^2\omega_{n/2}^2\omega^2)}, \quad (3.184)$$

where

$$\omega_m = \text{sn}\,[(2m - 1)K/n, k], \qquad m = 1, 2, \ldots, \tfrac{1}{2}n, \quad (3.185)$$

and H_e as in the case for n odd is a real constant to be determined from (3.180). As before, at $\omega = 1$, (3.180) becomes

$$F_n(1) = \text{sn}\,[(n + 1)K_1, k_1] = (-1)^{n/2} \quad (3.186)$$

which is valid only for n even. Substituting $\omega = 1$ in (3.184) in conjunction with (3.186) gives

$$H_e = \frac{(1 - k^2\omega_1^2)(1 - k^2\omega_2^2)\cdots(1 - k^2\omega_{n/2}^2)}{(1 - \omega_1^2)(1 - \omega_2^2)\cdots(1 - \omega_{n/2}^2)}. \quad (3.187)$$

At $\omega = 1/k$, (3.180) as in (3.174) becomes

$$F_n(1/k) = \text{sn}\,[(n + 1)K_1 \pm jK_1', k_1] = (-1)^{n/2}/k_1. \quad (3.188)$$

Substituting $\omega = 1/k$ in (3.184) and using (3.187) and (3.188) results in

$$H_e = \left(\frac{k^n}{k_1}\right)^{1/2}, \quad (3.189)$$

being the same as in (3.175). Thus, k_1 is determined by the formula (Problem 3.53)

$$k_1 = k^n \left[\frac{(1 - \omega_1^2)(1 - \omega_2^2)\cdots(1 - k^2\omega_{n/2}^2)}{(1 - k^2\omega_1^2)(1 - k^2\omega_2^2)\cdots(1 - k^2\omega_{n/2}^2)}\right]^2, \quad (3.190)$$

once k and n are known. Also it is easy to show that (Problem 3.76)

$$F_n(\infty) = 1/k_1, \quad (3.191)$$

indicating that the function has neither a pole nor a zero at the infinity.

EXAMPLE 3.9. Let
$$n = 4,$$
$$k = 1/1.4 = 0.71429.$$

Determine the characteristic function $F_n(\omega)$.
From (3.185), we obtain

$$\omega_1 = \text{sn}(K/4, k) = \text{sn}(0.46571, 0.71429)$$
$$= \sin 26°13' = 0.44173, \quad (3.192a)$$
$$\omega_2 = \text{sn}(3K/4, k) = \text{sn}(1.39712, 0.71429)$$
$$= \sin 70°59' = 0.94545, \quad (3.192b)$$

where K was computed in Example 3.8. From (3.190), we can determine the constant

$$k_1 = k^4 \left[\frac{(1-\omega_1^2)(1-\omega_2^2)}{(1-k^2\omega_1^2)(1-k^2\omega_2^2)} \right]^2$$
$$= 0.26031 \left[\frac{(1-0.19512)(1-0.89387)}{(1-0.09955)(1-0.45606)} \right]^2$$
$$= 0.0079183. \quad (3.192c)$$

Substituting these in (3.184) yields the characteristic function

$$F_4(\omega) = \frac{k^2}{k_1^{1/2}} \cdot \frac{(\omega_1^2 - \omega^2)(\omega_2^2 - \omega^2)}{(1-k^2\omega_1^2\omega^2)(1-k^2\omega_2^2\omega^2)}$$
$$= 5.73367 \frac{(0.19512 - \omega^2)(0.89387 - \omega^2)}{(1-0.09955\omega^2)(1-0.45606\omega^2)}, \quad (3.193)$$

which gives $F_4(1) = 1$,

$$F_4(1/k) = F_4(1.4) = 5.73367 \times \frac{1.76488 \times 1.06613}{0.80488 \times 0.10612} = 126.29 = 1/k_1,$$

$$F_4(\infty) = 5.73367 \frac{1}{0.09955 \times 0.45606} = 126.29 = 1/k_1,$$

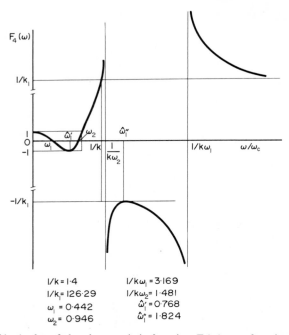

Fig. 3.21. A plot of the characteristic function $F_4(\omega)$ as a function of ω.

as expected. A plot of the characteristic function $F_4(\omega)$ as a function of ω is presented in Fig. 3.21.

4.2. Equiripple characteristic in passband and stopband

We now examine the manner in which the characteristic function $F_n(\omega)$ as used in (3.158) approximates the ideal response of Fig. 3.1. First, we denormalize $F_n(\omega)$ to the radian cutoff frequency ω_c, which is equivalent to replacing ω by ω/ω_c. Then examine the gain equation (3.158) which directs that, apart from the constant H_n, we square $F_n(\omega/\omega_c)$, multiply it by ϵ^2, not greater than unity, add unity, and form the reciprocal. If we carry out all these steps in conjunction with the plots of $F_3(\omega)$ and $F_4(\omega)$ as shown in Figs. 3.20 and 3.21, the response that results has equal maxima and equal

APPROXIMATION AND LADDER REALIZATION

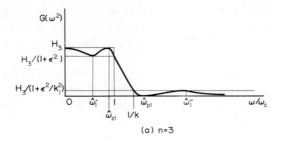

(a) n=3

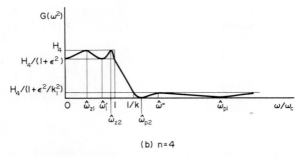

(b) n=4

FIG. 3.22. The third-order and fourth-order elliptic responses.

minima in both the passband and stopband. Two typical plots corresponding to $n = 3$ and $n = 4$ are shown in Fig. 3.22. From these plots, it is clear that the total number of peaks and troughs for positive ω is equal to $2n$, half of them lying within the passband between $\omega = 0$ and $\omega = \omega_c$ and the other half lying in the stopband which extends from $\omega = \omega_s$ to $\omega = \infty$. Between the passband and the stopband, we have the transitional frequency range from $\omega = \omega_c$ to $\omega = \omega_s$, which will be defined and elaborated shortly.

Inspection of the characteristic function $F_n(\omega)$ of (3.170) or (3.184), we recognize that its poles and zeros are inversely proportional, the constant of proportionality being $1/k$ or ω_c^2/k after denormalization. Because of this property, it is easy to check that

$$F_n(\omega_c/k\omega) = \frac{1}{k_1 F_n(\omega/\omega_c)}, \qquad (3.194)$$

176   BROADBAND MATCHING NETWORKS

which means that if the value of the function $k_1 F_n(\omega)$ is known at ω/ω_c, the value of $F_n(\omega)$ at the reciprocal of $k\omega/\omega_c$ is determined and is equal to the reciprocal of $k_1 F_n(\omega/\omega_c)$. Using this fact in (3.158) shows that if the gain characteristic has equal ripple property in the passband, it will automatically have equal ripple property in the stopband. To be more specific, we consider the design parameters associated with the three frequency ranges (passband, transitional band, and stopband) as follows.

A. *Maxima and minima in the passband*

Like the Chebyshev response, the maximum value of gain occurs at the points of ω where $F_n(\omega)$ vanishes. From (3.166) and (3.181) we have

$$G(\omega^2)_{\max} = G(\omega_{zm}^2) = H_n, \qquad (3.195)$$

where

$$\omega_{zm} = \omega_c \, \text{sn}\,(2mK/n, k), \qquad m = 0, 1, 2, \ldots, \tfrac{1}{2}(n-1) \qquad (3.196a)$$

for n odd, and

$$\omega_{zm} = \omega_c \, \text{sn}\,[(2m-1)K/n, k], \qquad m = 1, 2, \ldots, \tfrac{1}{2}n \qquad (3.196b)$$

for n even. The minimum value of gain in the passband is given by

$$G(\omega^2)_{\min} = \frac{H_n}{1+\epsilon^2}, \qquad \omega < \omega_c, \qquad (3.197)$$

occurring at the points of ω where $F_n(\omega) = \pm 1$. Since ω is real, according to (3.131a) and (3.133a) in conjunction with (3.159), these points are defined by the equations

$$\frac{nK_1}{K}\,\text{sn}^{-1}(\omega'_m/\omega_c, k) = (2m+1)K_1 + j2m_1 K'_1 \qquad (3.198a)$$

for n odd and

$$K_1 + \frac{nK_1}{K}\,\text{sn}^{-1}(\omega'_m/\omega_c, k) = (2m+1)K_1 + j2m_1 K'_1 \qquad (3.198b)$$

for n even, where m and m_1 as in (3.165) are any integers. Appealing to (3.161) and the fact that sn (u, k) is of periods $4K$ and $j2K'$ and solving for ω'_m yield (Problem 3.77)

$$\omega'_m = \omega_c \, \text{sn}\,[(2m-1)K/n, k], \qquad m = 1, 2, \ldots, \tfrac{1}{2}(n-1) \qquad (3.199a)$$

for n odd and $n > 1$ (Problem 3.57), and

$$\omega'_m = \omega_c \, \text{sn}\,(2mK/n, k), \qquad m = 0, 1, 2, \ldots, \tfrac{1}{2}n - 1 \qquad (3.199b)$$

for n even. Thus, we see that the dc gain is given by

$$G(0) = H_n, \qquad n \text{ odd} \qquad (3.200a)$$

$$= \frac{H_n}{1 + \epsilon^2}, \qquad n \text{ even}. \qquad (3.200b)$$

At the edge of the passband $\omega = \omega_c$, the gain goes through a minimum point, as depicted in Fig. 3.22, and is given by

$$G(\omega_c) = \frac{H_n}{1 + \epsilon^2} \qquad (3.201)$$

for all n, which follows directly from (3.172), (3.186) and (3.197) or equivalently by letting $m = \tfrac{1}{2}(n+1)$ in (3.199a) or $m = \tfrac{1}{2}n$ in (3.199b). Observe that for a fixed H_n the quantity ϵ, again called the *ripple factor*, controls the peak-to-peak ripple in the passband. In terms of decibels, the passband tolerance in dB is related to the ripple factor by

$$\alpha = 10 \log (1 + \epsilon^2) \text{ dB}. \qquad (3.202)$$

Thus, for a specified passband tolerance, the ripple factor is determined.

B. *Maxima and minima in the stopband*

Evidently, the minimum value of gain in the stopband is zero which occurs at the poles of $F_n(\omega)$. From (3.168) and (3.182), they are given by

$$\omega_{pm} = \frac{\omega_c^2}{k\omega_{zm}}, \qquad (3.203)$$

ω_{zm} being defined in (3.196). From the relation (3.194), it is clear that if ω'_m are the frequencies at which the passband gain is minimum, meaning that $F_n^2(\omega'_m)$ is maximum, the corresponding frequencies $\omega_c^2/k\omega'_m$ at which $F_n(\omega)$ will be minimum in the stopband are the points in the stopband that yield the maximum value of gain. Thus, the maximum value of gain in the stopband is given by

$$G(\omega^2)_{\max} = \frac{H_n}{1 + \epsilon^2/k_1^2}, \qquad \omega > \omega_s, \qquad (3.204)$$

occurring at the frequencies

$$\omega''_m = \frac{\omega_c^2}{k\omega'_m}, \qquad (3.205)$$

ω'_m being given in (3.199), since $F_n^2(\omega'_m) = 1$. In terms of decibels, the attenuation in the stopband is related to k_1 and the ripple factor by

$$\alpha = 10 \log (1 + \epsilon^2/k_1^2) \qquad (3.206)$$

Thus, with a specified passband tolerance and a stopband attenuation, the constant k_1 is determined. Recall that k_1 is also related to k and n through (3.176) or (3.190), meaning that if k and n are specified, the stopband attenuation cannot be chosen arbitrarily; it is determined by the formula (3.206).

C. *Transitional band*

As depicted in Fig. 3.22, between the passband and the stopband, the gain characteristic is attenuated monotonically from the edge of the passband to the edge of the stopband. It is convenient that we define the *edge of the stopband* to be the frequency ω_s at which the gain is dropped to the value of the maxima in the stopband. This is equivalent to requiring that

$$G(\omega_s^2) = \frac{H_n}{1 + \epsilon^2 F_n^2(\omega_s/\omega_c)} = \frac{H_n}{1 + \epsilon^2/k_1^2}. \qquad (3.207)$$

APPROXIMATION AND LADDER REALIZATION   179

Solving for $F_n^2(\omega_s/\omega_c)$ gives

$$F_n(\omega_s/\omega_c) = \pm 1/k_1. \tag{3.208}$$

From (3.174) and (3.188) we obtain

$$k = \frac{\omega_c}{\omega_s}, \tag{3.209}$$

indicating that the parameter k is a measure of the steepness of the gain characteristic in the transitional band, and is referred to as the *selectivity factor*. For our purposes, the reciprocal of the selectivity factor is called the *steepness*.

We remark that in many of the derivations presented in the foregoing, we have employed the identity (3.161) in order to simplify and derive the required formulas. This restriction imposed among the parameters k, k_1 and n can also be used to estimate the order of the elliptic response for given passband and stopband attenuations and a required steepness of the transitional band. Once the required order of the elliptic response is determined, we must recompute k or k_1. In fact, knowing any two of the three parameters k, k_1 and n would determine the third one. Hence, we assume that the quantities k, k_1, n and ϵ are all known in accordance with the desired specifications.

We illustrate the above results by the following examples.

EXAMPLE 3.10.   Suppose that we wish to design a low-pass filter having an elliptic transducer power-gain characteristic. The peak-to-peak ripple within the passband, which extends from 0 to $50/\pi$ MHz, must not exceed 0.5 dB, and at $70/\pi$ MHz the gain must be attenuated by at least 18 dB. Determine the transducer power-gain characteristic.

From (3.209), we obtain

$$k = \frac{\omega_c}{\omega_s} = \frac{2\pi(50/\pi) \times 10^6}{2\pi(70/\pi) \times 10^6} = 0.71429, \tag{3.210a}$$

giving

$$K = K(0.71429) = 1.86282, \tag{3.210b}$$

$$K' = K(0.69985) = 1.84553, \tag{3.210c}$$

where $k' = 0.69985$. They were computed by interpolation. The passband tolerance is 0.5 dB, and according to (3.202) we have

$$0.5 = 10 \log (1 + \epsilon^2),$$

yielding a value for the ripple factor

$$\epsilon = 0.34931. \qquad (3.210d)$$

The constant k_1 is determined from (3.206) by

$$18 = 10 \log (1 + \epsilon^2/k_1^2).$$

Solving for k_1 gives

$$k_1 = 0.044323, \qquad (3.210e)$$

yielding

$$K_1 = K(0.044323) = 1.57157, \qquad (3.210f)$$

$$K_1' = K(0.99902) = 4.49461, \qquad (3.210g)$$

where $k_1' = 0.99902$, $\sin^{-1} k_1 = 2°32'$, and $\sin^{-1} k_1' = 87°28'$. Finally, we compute n from (3.161), giving

$$n = \frac{KK_1'}{K'K_1} = \frac{1.86282 \times 4.49461}{1.84553 \times 1.57157} = 2.88674. \qquad (3.211a)$$

Since n must be an integer, we choose $n = 3$. For $n = 3$ and $k = 0.71429$, we must recompute k_1 by means of (3.176), which was done in Example 3.8 giving $k_1 = 0.037527$. Using this k_1, the actual stopband attenuation is given by

$$\alpha = 10 \log [1 + (0.34931/0.037527)^2] = 19.427 \text{ dB}, \qquad (3.211b)$$

which is above the minimum requirement of 18 dB. Substituting (3.210d) and (3.179) in (3.158) yields the desired transducer power-gain characteristic

$$G(\omega^2) = H_3 \frac{0.17166\hat{\omega}^4 - 0.82864\hat{\omega}^2 + 1}{1.18496\hat{\omega}^6 - 1.75283\hat{\omega}^4 - 0.04725\hat{\omega}^2 + 1}, \qquad (3.212)$$

where $\hat{\omega} = \omega/10^8$. According to (3.196a), the maximum value of gain in the passband, which is H_3, occurs at the frequencies $\omega_{z0} = 0$ and

$$\omega_{z1} = 10^8 \text{ sn } (2K/3, k) = 10^8 \text{ sn } (1.24188, 0.71429)$$
$$= 0.90114 \times 10^8 \text{ rad/s.} \quad (3.213)$$

The minimum value of gain in the passband is given by

$$G(\omega^2)_{min} = \frac{H_3}{1 + (0.34931)^2} = 0.89125 H_3, \quad (3.214a)$$

or 0.5 dB down from the maximum gain, which from (3.199a) occurs at the frequency

$$\omega_1' = 10^8 \text{ sn } (K/3, k) = 10^8 \text{ sn } (0.62094, 0.71429)$$
$$= 0.56689 \times 10^8 \text{ rad/s.} \quad (3.214b)$$

As a check, we substitute (3.214b) in (3.212), yielding $G(\omega_1'^2) = 0.89125 H_3$. Also we have $G(\omega_c^2) = 0.89125 H_3$.

From (3.204), the maximum value of gain in the stopband is given by

$$G(\omega^2)_{max} = \frac{H_3}{1 + (0.34931/0.037527)^2} = 0.01141 H_3, \quad (3.215a)$$

which according to (3.205) occurs at the frequency

$$\omega_1'' = \frac{\omega_c^2}{k\omega_1'} = \frac{10^{16}}{0.71429 \times 0.56689 \times 10^8} = 2.4696 \times 10^8 \text{ rad/s.} \quad (3.215b)$$

Substituting it in (3.212) gives $G(\omega_1''^2) = 0.01141 H_3$. The minimum value of gain in the stopband is of course zero, and according to (3.203) occurs at the frequencies $\omega_{p0} = \infty$ and

$$\omega_{p1} = \frac{\omega_c^2}{k\omega_{z1}} = \frac{10^{16}}{0.71429 \times 0.90114 \times 10^8} = 1.55358 \times 10^8 \text{ rad/s,} \quad (3.216)$$

which is a pole of $F_3(\omega/\omega_c)$, as it must. These are indicated in Fig. 3.22(a).

EXAMPLE 3.11. Consider the same problem as in Example 3.10 except that now we wish to achieve the fourth-order elliptic response.

For $n = 4$ and $k = 0.71429$, we first compute k_1 which was done in Example 3.9, giving $k_1 = 0.0079183$. From (3.206), the stopband attenuation becomes

$$\alpha = 10 \log [1 + (0.34931/0.0079183)^2] = 32.89 \text{ dB}. \quad (3.217)$$

Substituting (3.210d) and (3.193) in (3.158) yields the desired transducer power-gain characteristic

$$G(\omega^2)$$
$$= H_4 \frac{0.00206\hat{\omega}^8 - 0.05045\hat{\omega}^6 + 0.39950\hat{\omega}^4 - 1.11122\hat{\omega}^2 + 1.00000}{4.01338\hat{\omega}^8 - 8.78702\hat{\omega}^6 + 6.55577\hat{\omega}^4 - 2.63498\hat{\omega}^2 + 1.12202},$$
$$(3.218)$$

where $\hat{\omega} = \omega/10^8$. According to (3.196b), the maximum value of gain in the passband, which is H_4, occurs at the frequencies

$$\omega_{z1} = 10^8 \text{ sn } (K/4, k) = 10^8 \text{ sn } (0.46571, 0.71429)$$
$$= 0.44173 \times 10^8 \text{ rad/s,} \quad (3.219a)$$
$$\omega_{z2} = 10^8 \text{ sn } (3K/4, k) = 10^8 \text{ sn } (1.39712, 0.71429)$$
$$= 0.94545 \times 10^8 \text{ rad/s.} \quad (3.219b)$$

The minimum value of gain in the passband is the same as that given in (3.214a) except that we replace H_3 by H_4 since they have the same ripple factor, and from (3.199b) occurs at the frequencies $\omega_0' = 0$ and

$$\omega_1' = 10^8 \text{ sn } (K/2, k) = 10^8 \text{ sn } (0.93141, 0.71429)$$
$$= 0.76773 \times 10^8 \text{ rad/s.} \quad (3.220)$$

As a check, we substitute it in (3.218), giving $G(\omega_1'^2) = 0.89125 H_4$.

From (3.204), the maximum value of gain in the stopband is given by

$$G(\omega^2)_{max} = \frac{H_4}{1 + (0.34931/0.0079183)^2} = 0.0005135 H_4, \quad (3.221a)$$

occurring at the frequencies $\omega_0'' = \infty$ and

$$\omega_1'' = \frac{\omega_c^2}{k\omega_1'} = \frac{10^8}{0.71429 \times 0.76773} = 1.82356 \times 10^8 \text{ rad/s}. \quad (3.221b)$$

Substituting it in (3.218) yields $G(\omega_1''^2) = 0.0005135 H_4$. The minimum value of gain in the stopband is of course zero, and according to (3.203) occurs at the frequencies

$$\omega_{p1} = \frac{\omega_c^2}{k\omega_{z1}} = \frac{10^8}{0.71429 \times 0.44173} = 3.16936 \times 10^8 \text{ rad/s}, \quad (3.222a)$$

$$\omega_{p2} = \frac{\omega_c^2}{k\omega_{z2}} = \frac{10^8}{0.71429 \times 0.94545} = 1.48078 \times 10^8 \text{ rad/s}, \quad (3.222b)$$

which are poles of $F_4(\omega/\omega_c)$, as expected. These are indicated in Fig. 3.22(b).

4.3. Poles and zeros of elliptic response

Like the Butterworth and Chebyshev responses, we now proceed to determine the poles and zeros of the elliptic response. For this we again appeal to the theory of analytic continuation by replacing ω by $-js$ in (3.158), resulting in

$$G(-s^2) = \frac{H_n}{1 + \epsilon^2 F_n^2(-jy)}, \quad (3.223a)$$

where $y = s/\omega_c$ as in (3.8b),

$$F_n(-jy) = \text{sn}\left[\frac{nK_1}{K} \text{sn}^{-1}(-jy, k), k_1\right] \quad (3.223b)$$

for n odd, and

$$F_n(-jy) = \text{sn}\left[K_1 + \frac{nK_1}{K} \text{sn}^{-1}(-jy, k), k_1\right] \quad (3.223c)$$

for n even. As shown in §4.1, since $F_n(-jy)$ is a rational function, it is clear that the zeros of $G(-s^2)$ are the double poles of $F_n(-jy)$, which from (3.168) and (3.182) are located at

$$s_{zm} = j\frac{\omega_c}{k\operatorname{sn}(2mK/n, k)}, \quad m = 0, \pm 1, \ldots, \pm\tfrac{1}{2}(n-1) \quad (3.224a)$$

for n odd, and

$$s_{zm} = j\frac{\omega_c}{k\operatorname{sn}[(2m-1)K/n, k]}, \quad m = -\tfrac{1}{2}n+1, -\tfrac{1}{2}n+2, \ldots, \tfrac{1}{2}n \quad (3.224b)$$

for n even. Thus, all of its zeros are imaginary and lie on the segments $|\omega/\omega_c| > 1/k$. The poles of $G(-s^2)$ are defined by the zeros of the equation

$$1 + \epsilon^2 F_n^2(-jy) = 0 \quad (3.225a)$$

or

$$F_n(-jy) = \pm j/\epsilon. \quad (3.225b)$$

To facilitate our discussion, two cases are considered.

A. n odd

In this case, we substitute (3.223b) in (3.225b), giving

$$\operatorname{sn}\left[\frac{nK_1}{K}\operatorname{sn}^{-1}(-jy, k), k_1\right] = \pm j/\epsilon. \quad (3.226)$$

Since $\operatorname{sn}(u, k)$ is of periods $4K$ and $j2K'$ and since from (3.131a) $\operatorname{sn}(u \pm 2K, k) = -\operatorname{sn}(u, k)$, it is evident that the general solution of (3.226) must be of the form

$$2mK_1 + j2m_1 K_1' + \frac{nK_1}{K}\operatorname{sn}^{-1}(-jy, k) = \pm\operatorname{sn}^{-1}(j/\epsilon, k_1), \quad (3.227)$$

where m and m_1 as in (3.165) are any integers. Solving for y and appealing once again to (3.161) yield the locations for all the distinct poles of $G(-s^2)$ as

$$y_{pm} = s_{pm}/\omega_c = j \operatorname{sn}(2mK/n + j2m_1K' \pm ja, k)$$
$$= j \operatorname{sn}(2mK/n \pm ja, k), \quad m = 0, \pm 1, \ldots, \pm \tfrac{1}{2}(n-1), \quad (3.228a)$$

where

$$a = -j\frac{K}{nK_1} \operatorname{sn}^{-1}(j/\epsilon, k_1), \quad (3.228b)$$

which according to Jacobi's imaginary transformation (3.119) is real. For each choice of m, we can choose either the plus or the minus sign for ja, giving two distinct poles of $G(-s^2)$. For convenience, we consider y_{pm} to be associated with the plus sign and y'_{pm} with the minus sign. Then as in (3.169) we have

$$y_{pm} = -y'_{p(-m)}, \quad (3.229a)$$
$$y'_{pm} = -y_{p(-m)}, \quad (3.229b)$$

showing that the denominator of $G(-s^2)$, apart from a constant, can be factored as

$$r(y)r(-y) = -(y^2 - y_{p0}^2)(y^2 - y_{p1}^2)(y^2 - y_{p1}'^2)(y^2 - y_{p2}^2)(y^2 - y_{p2}'^2) \cdots$$
$$(y^2 - y_{pq}^2)(y^2 - y_{pq}'^2), \quad q = \tfrac{1}{2}(n-1), \quad (3.230)$$

where

$$r(y) = c_0 + c_1 y + \cdots + c_{n-1} y^{n-1} + c_n y^n = \sum_{m=0}^{n} c_m y^m \quad (3.231)$$

with $c_n = 1$, is the Hurwitz polynomial of degree n formed by the LHS zeros of (3.225), which are given by (Problem 3.60)

$$y_{pm} = j \operatorname{sn}(2mK/n + ja, k), \quad m = 0, \pm 1, \ldots, \pm \tfrac{1}{2}(n-1). \quad (3.232)$$

To avoid the necessity of computing these zeros y_{pm} and their associated polynomials $r(y)$, the coefficients of the polynomials $r(y)$ are tabulated in Appendix C for various values of ϵ and k and for $n = 1, 3, 5, 7$ and 9.

For $n = 3$ and $n = 5$, the zero and pole locations of the gain function $G(-s^2)$ are presented in Fig. 3.23. Like the previous two

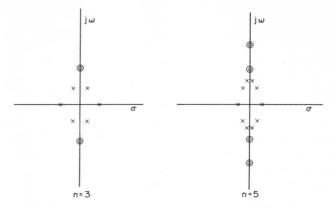

FIG. 3.23. The zero and pole locations of the third-order and fifth-order elliptic responses.

responses, its poles again possess quadrantal symmetry. All of its zeros are imaginary and of multiplicity 2 and lie on the segments $|\omega/\omega_c| > 1/k$ of the $j\omega$-axis, a fact that was mentioned in (3.224). Note that a pair of poles are located on the real axis.

Unlike the Butterworth and Chebyshev responses, explicit formulas for the coefficients of the Hurwitz polynomial $r(y)$ are not available.

EXAMPLE 3.12. Let

$$k = 0.71429, \quad (3.233a)$$

$$n = 3, \quad (3.233b)$$

$$\epsilon = 0.34931 \quad (0.5\text{-dB ripple}), \quad (3.233c)$$

as given in Example 3.10. We wish to determine the Hurwitz polynomial $r(y)$.

From Example 3.8 we have

$$K = 1.86282,$$

$$k_1 = 0.03753,$$

giving
$$K_1 = K(k_1) = 1.57135. \quad (3.234a)$$

Substituting these in (3.228b) yields

$$\begin{aligned}a &= -j\frac{1.86282}{3 \times 1.57135}\,\text{sn}^{-1}\,(j/0.34931, 0.03753) \\ &= 0.39516F(\sin^{-1} 0.94406, 0.99930) \\ &= 0.39516 \times 1.77171 = 0.70012.\end{aligned} \quad (3.234b)$$

The second line follows from the fact that if $ju = \text{sn}^{-1}(j/\epsilon, k_1)$ then from (3.110a) and (3.119) we obtain

$$\text{sn}\,(u, k'_1) = (1 + \epsilon^2)^{-1/2} = \sin \phi \quad (3.234c)$$

or

$$u = F(\phi, k'_1). \quad (3.234d)$$

Thus, from (3.232) the LHS poles of the gain function are given by

$$\begin{aligned}y_{p0} &= j\,\text{sn}\,(j0.70012, 0.71429) \\ &= -\text{tn}\,(0.70012, 0.71429) = -\tan 38°40' \\ &= -0.8002,\end{aligned} \quad (3.235a)$$

$$\begin{aligned}y_{p1}, y_{p(-1)} &= j\,\text{sn}\,(\pm 2K/3 + ja, k) \\ &= j\,\text{sn}\,(\pm 1.24188 + j0.70012, 0.71429) \\ &= -0.20983 \pm j1.05066.\end{aligned} \quad (3.235b)$$

The last line is obtained from (3.153a) and the details are left as an exercise (Problem 3.33).

The desired Hurwitz polynomial $r(y)$ is given by

$$\begin{aligned}r(y) &= (y - y_{p0})(y - y_{p1})(y - \bar{y}_{p1}) \\ &= y^3 + 1.21995y^2 + 1.48376y + 0.91865\end{aligned} \quad (3.236)$$

confirming our results tabulated in Appendix C.

Suppose now that the passband ripple is changed from 0.5 dB to

0.43 dB, everything else being the same. This is equivalent to having $\epsilon = 0.32261$, giving

$$a = -j0.39516 \, \text{sn}^{-1} \, (j/0.32261, 0.03753) = 0.72972, \quad (3.237a)$$

$$y_{p0} = j \, \text{sn} \, (j0.72972, 0.71429) = -0.84473, \quad (3.237b)$$

$$y_{p1}, \bar{y}_{p1} = j \, \text{sn} \, (\pm 1.24188 + j0.72972, 0.71429)$$
$$= -0.21631 \pm j1.06335. \quad (3.237c)$$

The corresponding Hurwitz polynomial is given by

$$r(y) = y^3 + 1.27735y^2 + 1.54296y + 0.99468. \quad (3.238)$$

Observe that by reducing the passband tolerance, the poles of the gain function move away from the imaginary axis.

B. n **even**

In this case, we substitute (3.223c) in (3.225b) and obtain

$$\text{sn} \left[K_1 + \frac{nK_1}{K} \, \text{sn}^{-1} \, (-jy, k), k_1 \right] = \pm j/\epsilon. \quad (3.239)$$

Proceeding as in (3.227) yields the locations for all the distinct poles of the gain function $G(-s^2)$ as

$$y_{pm} = s_{pm}/\omega_c = j \, \text{sn} \, [(2m-1)K/n \pm ja, k], \quad (3.240)$$
$$m = -\tfrac{1}{2}n + 1, -\tfrac{1}{2}n + 2, \ldots, \tfrac{1}{2}n,$$

where a is defined in (3.228b). Like in the odd case, let y_{pm} be the term in (3.240) associated with the plus sign of ja, and y'_{pm} with the minus sign. Then

$$y_{p[-(1/2)n+v]} = j \, \text{sn} \, [(-n+2v-1)K/n + ja, k]$$
$$= -y'_{p[(1/2)n-v+1]}, \quad v = 1, 2, \ldots, n. \quad (3.241)$$

The denominator of $G(-s^2)$, apart from a constant, can be

factored as

$$r(y)r(-y) = \prod_{m=-(1/2)n+1}^{n/2} (y^2 - y_{pm}^2), \qquad (3.242)$$

where the Hurwitz polynominal $r(y)$ is formed by the LHS zeros of (3.225), which are given by (Problem 3.60)

$$y_{pm} = j \operatorname{sn}[(2m-1)K/n + ja, k], \qquad m = -\tfrac{1}{2}n + 1, -\tfrac{1}{2}n + 2, \ldots, \tfrac{1}{2}n. \qquad (3.243)$$

For $n = 4$ and $n = 6$, the zero and pole locations of the gain function $G(-s^2)$ are presented in Fig. 3.24. They have the same

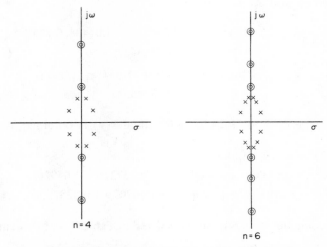

FIG. 3.24. The zero and pole locations of the fourth-order and sixth-order elliptic responses.

general configuration as in the odd case except that poles are not located on the real axis.

In Appendix C, the coefficients of the polynomials $r(y)$ are presented for various values of ϵ and k and for $n = 2, 4, \ldots, 10$.

EXAMPLE 3.13. Consider the same problem as in Example 3.12 except that we wish now to determine the fourth-order Hurwitz polynomial $r(y)$.

From Example 3.9, we have $k_1 = 0.0079183$ for $n = 4$ and $k = 0.71429$, giving $K = 1.86282$ and

$$K_1 = K(0.0079183) = 1.57082. \quad (3.244a)$$

Substituting these in (3.228b) yields

$$a = -j\frac{1.86282}{4 \times 1.57082} \operatorname{sn}^{-1}(j/0.34931, 0.0079183) = 1.01488. \quad (3.244b)$$

Thus, from (3.243) the LHS poles of the gain function can be determined and are located at

$$y_{p(-1)}, y_{p2} = j \operatorname{sn}(\mp 3 \times 1.86282/4 + j1.01488, 0.71429)$$
$$= -0.11860 \mp j1.02234, \quad (3.245a)$$

$$y_{p0}, y_{p1} = j \operatorname{sn}(\mp 1.86282/4 + j1.01488, 0.71429)$$
$$= -0.46668 \mp j0.53045. \quad (3.245b)$$

Finally, the desired Hurwitz polynomial $r(y)$ is obtained as

$$r(y) = (y - y_{p1})(y - \bar{y}_{p1})(y - y_{p2})(y - \bar{y}_{p2})$$
$$= (y^2 + 0.93336y + 0.49917)(y^2 + 0.23721y + 1.05926)$$
$$= y^4 + 1.17056y^3 + 1.77982y^2 + 1.10707y + 0.52875, \quad (3.246)$$

which coincides with the numbers listed in Table C.6 of Appendix C.

4.4. Elliptic networks

Having obtained the gain function, we now proceed to design a lossless two-port network having this gain characteristic.

Referring again to Fig. 3.4, let $S(s)$ of (3.22) be the scattering matrix of N normalizing to R_1 and R_2. The network possesses an nth-order elliptic transducer power-gain characteristic

$$|S_{21}(j\omega)|^2 = G(\omega^2) = \frac{H_n}{1 + \epsilon^2 F_n^2(\omega/\omega_c)}, \quad (3.247)$$

APPROXIMATION AND LADDER REALIZATION 191

H_n as before being bounded between 0 and 1. Proceeding as in (3.84), we obtain the input reflection coefficient $S_{11}(s)$ of N as

$$S_{11}(s)S_{11}(-s) = (1 - H_n)\frac{1 + \hat{\epsilon}^2 F_n^2(-jy)}{1 + \epsilon^2 F_n^2(-jy)}, \quad (3.248a)$$

where

$$\hat{\epsilon} = \epsilon(1 - H_n)^{-1/2}. \quad (3.248b)$$

As in the Butterworth and Chebyshev cases, the denominator of (3.248a) can be uniquely decomposed, while its numerator can have many permissible decompositions. Let

$$S_{11}(s) = \pm \lambda \frac{\hat{r}(y)}{r(y)} \quad (3.249a)$$

be a minimum-phase decomposition of (3.248a) with

$$\hat{r}(y) = \hat{c}_0 + \hat{c}_1 y + \cdots + \hat{c}_{n-1} y^{n-1} + \hat{c}_n y^n = \sum_{m=0}^{n} \hat{c}_m y^m, \quad (3.249b)$$

$\hat{c}_n = 1$, being the Hurwitz polynomial formed by the LHS zeros of the equation $1 + \hat{\epsilon}^2 F_n^2(-jy) = 0$, where

$$\lambda = 1, \quad n \text{ odd} \quad (3.250a)$$

$$= \left[1 - \frac{H_n}{1 + (\epsilon/k_1)^2}\right]^{1/2}, \quad n \text{ even}, \quad (3.250b)$$

which follows directly from the fact that $G(\infty) = 0$ for n odd and $F_n(\infty) = 1/k_1$ for n even. Then $S_{11}(s)$ is a bounded-real reflection coefficient. From (2.112), (1.105) and (1.106), the input impedance $Z_{11}(s)$ of N with the output port terminating in R_2 is a positive-real function and is given by

$$Z_{11}(s) = R_1 \frac{r(y) \pm \lambda \hat{r}(y)}{r(y) \mp \lambda \hat{r}(y)}. \quad (3.251)$$

Substituting $y = 0$ in (3.251) yields a relation between the two

192   BROADBAND MATCHING NETWORKS

terminating resistances for an LC two-port network with $H_n \neq 0$:

$$\frac{R_2}{R_1} = \left[\frac{c_0 + \lambda \hat{c}_0}{c_0 - \lambda \hat{c}_0}\right]^{\pm 1} \qquad (3.252a)$$

$$= \left[\frac{1 + (1 - H_n)^{1/2}}{1 - (1 - H_n)^{1/2}}\right]^{\pm 1}, \quad n \text{ odd} \qquad (3.252b)$$

$$= \left[\frac{(1 + \epsilon^2)^{1/2} + (1 + \epsilon^2 - H_n)^{1/2}}{(1 + \epsilon^2)^{1/2} - (1 + \epsilon^2 - H_n)^{1/2}}\right]^{\pm 1}, \quad n \text{ even} \qquad (3.252c)$$

the $\pm$ signs being determined respectively according to $R_2 \geq R_1$ and $R_2 \leq R_1$. Thus, as in (3.87) if R_1 and R_2 are both specified in advance, then if n is odd, the dc gain is determined from (3.252b) and the passband tolerance may be chosen arbitrarily; and if n is even, the minimum passband gain is fixed by (3.252c) in accordance with (3.197) and the passband tolerance can be selected arbitrarily.

We illustrate the above results by the following examples for both n odd and even.

EXAMPLE 3.14. Design a low-pass filter that satisfies all the specifications described in Example 3.10 and that achieves a maximum dc gain.

For this we first compute the input reflection coefficient $S_{11}(s)$. Since n is odd, $\lambda = 1$ and the maximum permissible dc gain $H_3 = 1$. For this dc gain, it is clear from (3.248a) that $\hat{r}(y)$ is the polynomial formed by the simple zeros of $F_3^2(-jy)$, which according to (3.179) is given by

$$\hat{r}(y) = y(y^2 + 0.81206). \qquad (3.253a)$$

The denominator polynomial $r(y)$ was computed earlier in Example 3.12, and is given in (3.236). The input reflection coefficient is obtained as

$$S_{11}(s) = \pm \frac{y^3 + 0.81206y}{y^3 + 1.21995y^2 + 1.48376y + 0.91865}. \qquad (3.253b)$$

The terminating resistance R_2 can be determined in advance from (3.252b). Since $H_3 = 1$, we have $R_2 = R_1$, indicating an equal termination for an LC two-port network N.

APPROXIMATION AND LADDER REALIZATION

Depending upon the choice of the signs in (3.253b), two input impedance functions are possible, apart from R_1 one being the reciprocal of the other. Selecting the plus sign gives

$$\frac{Z_{11}(s)}{R_1} = \frac{2y^3 + 1.21995y^2 + 2.29582y + 0.91865}{1.21995y^2 + 0.67170y + 0.91865}$$

$$= 1.63941y + \frac{0.11876y^2 + 0.78978y + 0.91865}{1.21995y^2 + 0.67170y + 0.91865}. \quad (3.254)$$

Using Darlington's technique (Van Valkenburg, 1960), the second term can be realized as the input impedance of a lossless two-port network terminated in a 1-Ω resistor.

EXAMPLE 3.15. Let

$$k = 0.71429, \quad (3.255a)$$

$$n = 4, \quad (3.255b)$$

$$\epsilon = 0.32261, \quad (0.43\text{-dB ripple}), \quad (3.255c)$$

$$H_4 = 0.9814. \quad (3.255d)$$

Determine the input impedance function $Z_{11}(y)$ of N in terms of the normalized complex frequency y.

From the specifications, we obtain

$$K = K(0.71429) = 1.86282, \quad (3.255e)$$

$$k_1 = 0.0079183, \quad (3.255f)$$

$$K_1 = K(0.007918) = 1.57082, \quad (3.255g)$$

$$k_1' = (1 - k_1^2)^{1/2} = 0.99997, \quad (3.255h)$$

$$\hat{\epsilon} = \epsilon(1 - H_4)^{-1/2} = 2.36571. \quad (3.255i)$$

where K and k_1 were used in Example 3.9. First of all, we must determine the polynomials $r(y)$ and $\hat{r}(y)$ of (3.249). For this we can use formula (3.243) to compute their roots. The zeros of $r(y)$ will be computed first, as follows:

$$a = -j\frac{1.86282}{4 \times 1.57082} \text{sn}^{-1}(j/0.32261, 0.007918)$$
$$= 0.29647 F(\sin^{-1} 0.95170, 0.99997) = 0.54829, \quad (3.256a)$$

$$y_{p(-1)} = j \text{ sn}(-1.39711 + j0.54829, 0.71429)$$
$$= -0.12288 - j1.02888, \quad (3.256b)$$

$$y_{p0} = j \text{ sn}(-0.46571 + j0.54829, 0.71429)$$
$$= -0.48821 - j0.53904, \quad (3.256c)$$

$$y_{p1} = j \text{ sn}(0.46571 + j0.54829, 0.71429)$$
$$= -0.48821 + j0.53904 = \bar{y}_{p0}, \quad (3.256d)$$

$$y_{p2} = j \text{ sn}(1.39711 + j0.54829, 0.71429)$$
$$= -0.12288 + j1.02888 = \bar{y}_{p(-1)}. \quad (3.256e)$$

This gives the denominator polynomial

$$r(y) = (y^2 + 0.24576y + 1.07368)(y^2 + 0.97642y + 0.52891)$$
$$= y^4 + 1.22218y^3 + 1.84256y^2 + 1.17835y + 0.56789. \quad (3.257)$$

Similarly, we compute the numerator polynomial $\hat{r}(y)$ by repeating the above process for $\hat{\epsilon}$:

$$a = -j0.29647 \text{ sn}^{-1}(j/2.36571, 0.007918)$$
$$= 0.12186, \quad (3.258a)$$

$$y_{p(-1)} = j \text{ sn}(-1.39711 + j0.12186, 0.71429)$$
$$= -0.02919 - j0.94964, \quad (3.258b)$$

$$y_{p0} = j \text{ sn}(-0.46571 + j0.12186, 0.71429)$$
$$= -0.10398 - j0.44605, \quad (3.258c)$$

$$y_{p1} = j \text{ sn}(0.46571 + j0.12186, 0.71429)$$
$$= -0.10398 + j0.44605, \quad (3.258d)$$

$$y_{p2} = j \text{ sn}(1.39711 + j0.12186, 0.71429)$$
$$= -0.02919 + j0.94964, \quad (3.258e)$$

APPROXIMATION AND LADDER REALIZATION

yielding

$$\hat{r}(y) = (y^2 + 0.05838y + 0.90267)(y^2 + 0.20795y + 0.20977)$$
$$= y^4 + 0.26633y^3 + 1.12458y^2 + 0.19996y + 0.18935. \quad (3.259)$$

Since n is even, the constant λ is not unity and is determined by (3.250b), giving

$$\lambda = \left[1 - \frac{0.9814}{1 + (0.32261/0.0079183)^2}\right]^{1/2} = 0.99971. \quad (3.260)$$

Thus, the input reflection coefficient is given by

$$S_{11}(y) = \pm 0.99971 \frac{y^4 + 0.26633y^3 + 1.12458y^2 + 0.19996y + 0.18935}{y^4 + 1.22218y^3 + 1.84256y^2 + 1.17835y + 0.56789}. \quad (3.261)$$

Substituting it in (3.251) and selecting the plus sign in (3.261) yield the input impedance function

$$\frac{Z_{11}(y)}{R_1} = \frac{1.99971y^4 + 1.48843y^3 + 2.96681y^2 + 1.37825y + 0.75719}{0.00029y^4 + 0.95593y^3 + 0.71831y^2 + 0.97845y + 0.378595}. \quad (3.262)$$

For the choice of the minus sign in (3.261), the normalized input admittance is given in (3.262). This positive-real function can then be realized as a lossless two-port terminating in a resistor. By letting $y = 0$ in (3.262) yields the terminating resistance $R_2 = 2R_1$, which can be determined in advance by (3.252c) for an LC two-port network.

Suppose that $R_2 = 2R_1$ is specified in advance for an LC two-port network. Then from (3.252c) the quantity

$$\frac{H_4}{1 + \epsilon^2} = \frac{8}{9} \quad (3.263)$$

is fixed. By specifying the ripple factor $\epsilon = 0.32261$ as in (3.255c), the constant H_4 is determined from (3.263) to be 0.9814, as given in (3.255d), and cannot be chosen arbitrarily. On the other hand, if we set $H_4 = 1$ then the passband tolerance is fixed by (3.263) with $\epsilon = 0.35355$, which corresponds to a peak-to-peak ripple of 0.51 dB.

As a comparison to the case discussed in Example 3.14 for $n = 3$, let $H_4 = 1$ and $\epsilon = 0.34931$ (0.5-dB ripple). Then from (3.193), (3.246) and (3.250b), the input reflection coefficient is obtained as

$$S(s) = \pm 0.99975 \frac{y^4 + 1.08899 y^2 + 0.17441}{y^4 + 1.17056 y^3 + 1.77982 y^2 + 1.10707 y + 0.52875},$$
(3.264a)

where $\lambda = 0.99975$. Substituting it in (3.251) and selecting the plus sign in (3.264a) result in the input impedance

$$\frac{Z_{11}(s)}{R_1} = \frac{1.99975 y^4 + 1.17056 y^3 + 2.86854 y^2 + 1.10707 y + 0.70312}{0.00025 y^4 + 1.17056 y^3 + 0.69110 y^2 + 1.10707 y + 0.35438},$$
(3.264b)

which, by Darlington theory, can then be realized as the input impedance of a lossless two-port network terminated in a resistor. If N is an LC two-port network, the terminating resistance R_2 can be determined from (3.264b) by setting $y = 0$, giving $R_2 = 1.98409 R_1$, which can also be computed in advance by (3.252c).

5. Frequency transformations

So far we have concentrated exclusively on approximating the ideal low-pass transducer power-gain characteristic of Fig. 3.1. As mentioned at the beginning of this chapter, this should not be deemed to be a restriction. In the present section, we show that the results obtained in the foregoing for the low-pass case can readily be adapted to other cases such as high-pass, band-pass, band-elimination, etc., by means of transformations of the frequency variable.

It is evident that if we could find a frequency transformation which would map the desired passband and stopband onto the corresponding passband and stopband of the low-pass characteristic, we could then solve the equivalent low-pass problem. Applying the inverse transformation would lead to the solution of the original problem.

To discuss the transformation, let $s' = \sigma' + j\omega'$ be the complex frequency for the low-pass function and $s = \sigma + j\omega$ be the new complex frequency variable. Then a frequency transformation is a function

$$s' = f(s), \qquad (3.265)$$

which maps one or several frequency ranges of interest to the frequency range of the passband of the low-pass characteristic. Thus, in moving the passband, we obtain the different types of filter characteristics.

After realizing the network of the equivalent low-pass problem, the desired network is obtained from this low-pass network by replacing each inductance L by a one-port whose impedance is $Lf(s)$, and each capacitance C by a one-port whose admittance is $Cf(s)$. Since the impedance of a resistor is not a function of frequency, the resistances are not altered by the transformation. Thus, the low-pass filter and its derived network have the same value of gain at the corresponding frequencies defined by (3.265).

We now proceed to discuss three important cases that we are most interested in: *high-pass*, *band-pass* and *band-elimination*. To facilitate our discussion, it is convenient to normalize the low-pass characteristic to its cutoff frequency. With this, s' denotes the normalized low-pass frequency variable.

5.1. Transformation to high-pass

Consider the transformation

$$s' = \frac{\omega_0}{s}, \qquad (3.266)$$

which maps the interval from $j\omega_0$ to $+\infty$ in the s-plane to the interval $-j1$ to 0 in the s'-plane and from $-j\omega_0$ to $-\infty$ in the s-plane to $j1$ to 0 in the s'-plane, and vice versa, as indicated in Fig. 3.25.

Suppose now that a gain characteristic is prescribed over a high-pass interval. This characteristic can be transformed into a

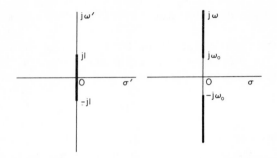

FIG. 3.25. The transformation from low-pass to high-pass.

corresponding low-pass characteristic by means of (3.266). We now solve the low-pass problem and obtain its network realization. To obtain the desired high-pass network realization, we replace each branch of the low-pass network by a branch whose impedance at a point in the high-pass interval is the same as the impedance of the replaced branch at the corresponding point in the low-pass interval. Thus, if L and C represent the inductance and capacitance in the low-pass network, we require that

$$Ls' = L\frac{\omega_0}{s} \equiv \frac{1}{C_h s}, \qquad (3.267a)$$

$$\frac{1}{Cs'} = \frac{s}{C\omega_0} \equiv L_h s, \qquad (3.267b)$$

where

$$C_h = \frac{1}{L\omega_0}, \qquad (3.268a)$$

$$L_h = \frac{1}{C\omega_0}. \qquad (3.268b)$$

This shows that, to obtain a high-pass network from its corresponding normalized low-pass realization, we simply replace each inductance by a capacitance and each capacitance by an inductance with the element values as given in (3.268). This is illustrated in Fig. 3.26. Clearly, the radian cutoff frequency of the

APPROXIMATION AND LADDER REALIZATION

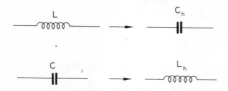

FIG. 3.26. The replacement of elements in the low-pass to high-pass transformation.

high-pass network is ω_0. For $\omega_0 = 1$, the two cutoff frequencies are the same, both being at $\omega = 1$.

EXAMPLE 3.16. Suppose that we wish to design a high-pass filter whose radian cutoff frequency is 10^5 rad/s. The filter is to be operated between a resistive generator of internal resistance 100 Ω and a 200-Ω load and is required to have an equiripple transducer power-gain characteristic. The peak-to-peak ripple in the passband must not exceed 1 dB and at $\omega = 2 \times 10^4$ rad/s or less the gain must be at least 60 dB down from its peak value in the passband.

The first step is to translate the high-pass specifications into the equivalent low-pass requirements. To this end, let $\omega_0 = 10^5$ rad/s. From (3.266) it is clear that, in terms of the low-pass specifications, the peak-to-peak ripple in the passband must not exceed 1 dB and at $\omega' = 5$ and beyond the gain must be at least 60 dB down from its peak value in the passband. Since the low-pass characteristic is normalized to its cutoff frequency, the low-pass problem was essentially solved in Example 3.6. Referring to the ladder network of Fig. 3.15, we frequency-scale the network by a factor of 10^{-4}. This gives the normalized values of the elements of the low-pass network as

$$L_1 = 16.82 \times 10^{-3}/10^{-4} = 168.2 \text{ H}, \tag{3.269a}$$

$$C_2 = 1.19 \times 10^{-6}/10^{-4} = 11.9 \times 10^{-3} \text{ F}, \tag{3.269b}$$

$$L_3 = 23.79 \times 10^{-3}/10^{-4} = 237.9 \text{ H}, \tag{3.269c}$$

$$C_4 = 0.84 \times 10^{-6}/10^{-4} = 8.4 \times 10^{-3} \text{ F}, \tag{3.269d}$$

arranged from left to right. Using (3.268) yields the element values of the desired high-pass filter:

$$C_{h1} = 10^{-5}/L_1 = 0.0595 \ \mu F, \qquad (3.270a)$$

$$L_{h2} = 10^{-5}/C_2 = 0.84 \ mH, \qquad (3.270b)$$

$$C_{h3} = 10^{-5}/L_3 = 0.042 \ \mu F, \qquad (3.270c)$$

$$L_{h4} = 10^{-5}/C_4 = 1.19 \ mH. \qquad (3.270d)$$

The high-pass ladder together with its terminations is presented in Fig. 3.27.

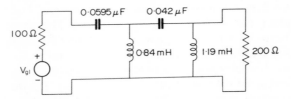

FIG. 3.27. A high-pass filter obtained by means of the low-pass to high-pass transformation.

5.2. Transformation to band-pass

The transformation from low-pass to band-pass can be handled in a similar manner. Consider the transformation

$$s' = \frac{\omega_0}{B}\left[\frac{s}{\omega_0} + \frac{\omega_0}{s}\right], \qquad (3.271a)$$

where

$$\omega_0^2 = \omega_1\omega_2, \qquad (3.271b)$$

$$B = \omega_2 - \omega_1. \qquad (3.271c)$$

This transformation maps the intervals $j\omega_1$ to $j\omega_2$ and $-j\omega_1$ to $-j\omega_2$ in the s-plane to the interval $-j1$ to $j1$ in the s'-plane, as indicated in Fig. 3.28. To see this, we substitute $s' = \pm j1$ and $s = j\omega$ in (3.271a), giving

$$\omega^2 \pm B\omega - \omega_0^2 = 0, \qquad (3.272)$$

APPROXIMATION AND LADDER REALIZATION 201

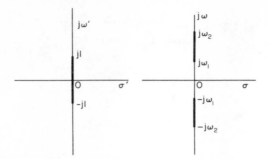

FIG. 3.28. The transformation from low-pass to band-pass.

whose solutions are given by $-\omega_1$, $-\omega_2$ and

$$\omega_1, \omega_2 = \mp B/2 + (B^2/4 + \omega_0^2)^{1/2}. \tag{3.273}$$

From this it is easy to confirm (3.271b) and (3.271c). Thus, B is the bandwidth of the band-pass filter and ω_0 is the geometric mean of the cutoff frequencies ω_1 and ω_2 and is referred to as the *mid-band frequency*. The reason for this name is that if frequency is plotted on a logarithmic scale, ω_0 falls midway between ω_1 and ω_2. In a similar way, we recognize that this transformation yields a gain characteristic that is geometrically symmetric with respect to the mid-band frequency.

We remark that s' is a double-valued function of s. For a point in the low-pass interval in the s'-plane, there correspond two points in the s-plane, one in the positive passband and one in the negative passband.

Using the transformation (3.271), we can translate the band-pass specifications to those of a low-pass. After realizing the low-pass filter, the required band-pass network is obtained by replacing each branch in the low-pass realization by a one-port whose impedance at a point in the band-pass interval is the same as the impedance of the replaced branch at the corresponding point in the low-pass interval. This requires that

$$Ls' = \frac{Ls}{B} + \frac{L\omega_0^2}{Bs} \equiv L_{b1}s + \frac{1}{C_{b1}s}, \tag{3.274a}$$

$$\frac{1}{Cs'} = \frac{1}{\dfrac{Cs}{B} + \dfrac{C\omega_0^2}{Bs}} \equiv \frac{1}{C_{b2}s + \dfrac{1}{L_{b2}s}}, \qquad (3.274b)$$

where

$$L_{b1} = L/B, \qquad (3.275a)$$

$$C_{b1} = B/L\omega_0^2, \qquad (3.275b)$$

$$L_{b2} = B/C\omega_0^2, \qquad (3.275c)$$

$$C_{b2} = C/B. \qquad (3.275d)$$

Thus, we conclude that, to obtain a band-pass network from its corresponding normalized low-pass realization, each inductance L in the low-pass realization is replaced by a series combination of an inductor with inductance L_{b1} and a capacitor with capacitance C_{b1} and each capacitance C in the low-pass realization is replaced by a parallel combination of an inductor with inductance L_{b2} and a capacitor with capacitance C_{b2}, as shown in Fig. 3.29.

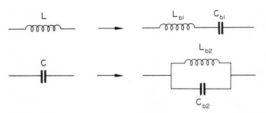

FIG. 3.29. The replacement of elements in the low-pass to band-pass transformation.

EXAMPLE 3.17. We wish to design a band-pass filter whose passband is from 10^5 rad/s to 4×10^5 rad/s. The filter is required to have an equiripple transducer power-gain characteristic and is to be operated between a resistive generator of internal resistance 100 Ω and a 200-Ω load. The peak-to-peak ripple in the passband must not exceed 1 dB and at $\omega = 15.263 \times 10^5$ rad/s the gain must be at least 60 dB down from its peak value in the passband.

As in the high-pass situation, we first translate the band-pass specifications to the equivalent low-pass requirements. For this we compute the bandwidth and the mid-band frequency as

$$B = \omega_2 - \omega_1 = 4 \times 10^5 - 10^5 = 3 \times 10^5 \text{ rad/s}, \quad (3.276a)$$

$$\omega_0 = (\omega_2 \omega_1)^{1/2} = (4 \times 10^5 \times 10^5)^{1/2} = 2 \times 10^5 \text{ rad/s}. \quad (3.276b)$$

We now consider a low-pass network which has been normalized to its cutoff frequency ω_c. At the frequency $\omega = 15.263$ rad/s, the corresponding frequency for the low-pass characteristic can be determined by (3.271a), and is given by

$$\omega' = \frac{2 \times 10^5}{3 \times 10^5}\left[\frac{15.263 \times 10^5}{2 \times 10^5} - \frac{2 \times 10^5}{15.263 \times 10^5}\right] = 5.$$

Hence, our problem is first to design an equiripple low-pass network with 1-dB ripple in the passband and 60 dB down at five times the normalized frequency, which is one, and beyond. This is precisely the same problem solved in Example 3.6 except that filter must be normalized to its cutoff frequency.

Referring again to Fig. 3.15 and frequency-scaling the network by a factor of 10^{-4} yield the normalized values as shown in (3.269). Using these values in (3.275) in conjunction with (3.276) gives the element values of the desired band-pass network as shown in Fig. 3.30.

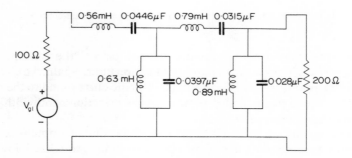

FIG. 3.30. A band-pass filter obtained by means of the low-pass to band-pass transformation.

5.3. Transformation to band-elimination

The transformation required in this case is given by

$$s' = \frac{1}{\dfrac{\omega_0}{B}\left[\dfrac{s}{\omega_0} + \dfrac{\omega_0}{s}\right]}, \qquad (3.277a)$$

where

$$\omega_0^2 = \omega_1 \omega_2, \qquad (3.277b)$$

$$B = \omega_2 - \omega_1, \qquad (3.277c)$$

which maps the desired intervals in the s-plane to the interval $-j1$ to $j1$ in the s'-plane as indicated in Fig. 3.31. The result is rather obvious

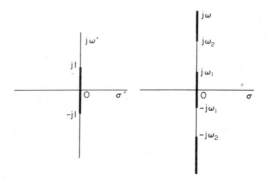

FIG. 3.31. The transformation from low-pass to band-elimination.

since the transformation relating the low-pass to the band-pass will transform a high-pass to a band-elimination filter. Again, we can show that the points $\pm j\omega_1$ and $\pm j\omega_2$ in the s-plane correspond to the points $\pm j1$ in the s'-plane, and B represents the rejection bandwidth of the band-elimination filter.

Proceeding as in (3.274), we can show that, to obtain a band-elimination network from its corresponding normalized low-pass realization, each inductance L in the low-pass realization is replaced by a parallel combination of an inductor with inductance L_{e1} and a capacitor with capacitance C_{e1} and each capacitance C in the low-pass realization is replaced by a series combination of an inductor with inductance L_{e2} and a capacitor with capacitance C_{e2}, whose values are given by (Problem 3.65)

$$L_{e1} = LB/\omega_0^2, \tag{3.278a}$$

$$C_{e1} = 1/LB, \tag{3.278b}$$

$$L_{e2} = 1/CB, \tag{3.278c}$$

$$C_{e2} = CB/\omega_0^2, \tag{3.278d}$$

This is also indicated in Fig. 3.32.

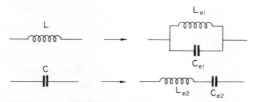

FIG. 3.32. The replacement of elements in the low-pass to band-elimination transformation.

We remark that the transformation (3.277a) can be put in a slightly different form as

$$s' = \frac{Bs}{s^2 + \omega_0^2}, \tag{3.279}$$

which is recognized as a reactance function. In fact, the transformations (3.271a), being the reciprocal of (3.279), and (3.266), being a special case of (3.279) with $\omega_0 = 0$, are all reactance functions. A little thought will show that if we are asked to provide a transformation that maps several segments on the imaginary axis of the s-plane onto the interval of the passband of the low-pass plane, a proper choice of the zeros and poles of a reactance function will perform the desired transformation. A detailed discussion of the reactance transformation is given in Cauer (1958), which also includes a discussion on transformations yielding functions without geometric symmetry versus the frequency.

6. Conclusions

In this chapter, we considered three popular rational function approximation schemes for the ideal low-pass brick-wall type of gain response. They are the Butterworth (maximally-flat) response, the Chebyshev (equiripple) response, and the elliptic (Cauer-parameter) response. The Butterworth response yields a maximally-flat gain near the origin and infinity and monotonic throughout, while Chebyshev response gives an equiripple characteristic in the passband and maximally-flat in the stopband. In each of these two cases, we presented explicit formulas for the element values of the corresponding ladder network realizations. The ladder networks are attractive from an engineering viewpoint in that they are unbalanced and contain no coupling coils. Considering the advantages of the Chebyshev response over the Butterworth, one is led to the development of the elliptic response, which provides equiripple characteristics in both the passband and stopband. The elliptic response is achieved by the use of elliptic functions. For this, a section on the theory of elliptic functions and some of their fundamental properties is included.

As to the distribution of the poles and zeros of these functions, it was shown that the poles obtained from a Butterworth response lie on a circle of radius equal to its cutoff frequency while those obtained from a Chebyshev response lie on an ellipse. All the zeros of the Butterworth and the Chebyshev responses are at the infinity while those of an elliptic response lie on the $j\omega$-axis, infinity included.

In order to get more flexibility, we started our design from the Butterworth characteristic to the Chebyshev characteristic and finally ended up with the elliptic characteristic. This flexibility is achieved by the increase in the number of parameters available to specify the desired filter, and thus increases the subsequent complexity of the design. In the case of Butterworth response, one parameter—the degree of the polynomial—is sufficient. For a Chebyshev response, two parameters are required, one being the degree of the Chebyshev polynomial and the other being determined by the amplitude of the passband ripple. To obtain an elliptic response, we must specify any three parameters of the four that can be specified: the order of the characteristic function, the passband ripple, the stopband ripple, and the width of the transitional frequency range.

Finally, we mentioned that confining attention to the low-pass characteristic is not so restrictive as it appears at first glance. We demonstrated this by considering frequency transformations which permit a low-pass gain characteristic to be converted to a high-pass, band-pass or band-elimination characteristic.

Problems

3.1. Consider the Butterworth power-gain characteristic $G(\omega^2)$ of (3.1). Obtain the Maclaurin series expansion of $G(1/\omega^2)$ and show that its first $2n - 1$ derivatives vanish at $\omega = 0$.

3.2. Show that the distinct roots of the polynomial (3.9) are given by the equation (3.10).

3.3. To obtain additional flexibility, consider the gain characteristic

$$G(\omega^2) = \frac{K_n}{1 + \epsilon^2(\omega/\omega_c)^{2n}}. \tag{3.280}$$

Determine its pole locations, and compare it with (3.44) at $\omega = \omega_c$.

3.4. Show that the phase of the Butterworth response (3.1) is given by the formula

$$-\sum_{m=0}^{\infty} \frac{(\omega/\omega_c)^{2m+1}}{(2m+1)\sin(2m+1)\gamma_1}, \tag{3.281}$$

where γ_1 is defined in (3.34c). The *time-delay function* $T_d(\omega)$ is defined as the negative derivative of the phase. From the phase function, show that the time-delay function can be put in the form

$$T_d(\omega) = \frac{\sum_{m=0}^{n-1} \frac{(\omega/\omega_c)^{2m}}{\sin(2m+1)\gamma_1}}{1 + (\omega/\omega_c)^{2n}}. \tag{3.282}$$

[*Hint.* Use the identity

$$\sum_{u=0}^{n-1} \beta^{-m(2u+1)} = (-j)^m \frac{\sin m\pi/2}{\sin \gamma_m}, \tag{3.283}$$

β and γ_m being given in (3.14b) and (3.34c).]

3.5. Derive the identities (3.36a) and (3.36b).

3.6. Design a low-pass filter having a maximally-flat transducer power-gain characteristic and operating between a resistive generator of internal resistance 70 Ω and a 200-Ω load. The filter must give at least 50 dB attenuation in gain at the frequency five times the radian cutoff frequency $\omega_c = 10^5$ rad/s and beyond, and has a maximum permissible dc gain.

3.7. In Problem 3.6, suppose that the load resistance is not specified. Design a Butterworth LC ladder filter having a maximum dc gain.

3.8. Show that the Chebyshev polynomial (3.45) can be put into the form

$$C_n(\omega) = \omega^n - \binom{n}{2}\omega^{n-2}(1-\omega^2) + \binom{n}{4}\omega^{n-4}(1-\omega^2)^2 - \cdots, \qquad (3.284)$$

which is recognized as a polynomial in ω.

3.9. Show that the Chebyshev polynomial of degree $2n$ and the square of the Chebyshev polynomial of degree n are related by

$$2C_n^2(\omega) = C_{2n}(\omega) + 1. \qquad (3.285)$$

Using this formula, reformulate the Chebyshev response (3.44) by avoiding the square function. [*Hint.* $2\cos^2 n\gamma = \cos 2n\gamma + 1$.]

3.10. For $\omega > 1$, the Chebyshev polynomial can be expressed equivalently as

$$2C_n(\omega) = [\omega + (\omega^2-1)^{1/2}]^n + [\omega + (\omega^2-1)^{1/2}]^{-n} \qquad (3.286a)$$

$$= [\omega - (\omega^2-1)^{1/2}]^n + [\omega - (\omega^2-1)^{1/2}]^{-n}. \qquad (3.286b)$$

Using the definitions of the hyperbolic functions, derive these two identities.

3.11. Using (3.74), show that $\sinh a$ and $\cosh a$ can be expressed explicitly in terms of the ripple factor ϵ as

$$\sinh a = \tfrac{1}{2}(\hat{k} - \hat{k}^{-1}), \qquad (3.287a)$$

$$\cosh a = \tfrac{1}{2}(\hat{k} + \hat{k}^{-1}), \qquad (3.287b)$$

where

$$\hat{k} = [(1 + 1/\epsilon^2)^{1/2} + 1/\epsilon]^{1/n}. \qquad (3.287c)$$

3.12. Applying (3.17), derive the coefficient formulas

$$a_{n-1} = \frac{1}{\sin \gamma_1}, \qquad (3.288a)$$

$$a_{n-2} = \frac{1}{2\sin^2 \gamma_1}, \qquad (3.288b)$$

$$a_{n-3} = \frac{\cos \gamma_1 \cos \gamma_2}{\sin \gamma_1 \sin \gamma_2 \sin \gamma_3}, \qquad (3.288c)$$

γ_m being given in (3.34c).

3.13. Show that b_0 given in (3.80d) can be expressed equivalently as

$$b_0 = j^n 2^{1-n} C_n(-j \sinh a). \qquad (3.289)$$

3.14. Show that the phase function of the Chebyshev response (3.44) is given by the formula

$$-\sum_{m=0}^{\infty} \frac{2e^{-(2m+1)a} C_{2m+1}(\omega/\omega_c)}{(2m+1)\sin(2m+1)\gamma_1}. \quad (3.290)$$

Using this phase function, derive the time-delay function (see Problem 3.4)

$$T_d(\omega) = \frac{\sum_{m=0}^{n-1} \dfrac{U_{2m}(\omega/\omega_c)\sinh(2n-2m-1)a}{\epsilon^2 \sin \gamma_{2m+1}}}{1 + \epsilon^2 C_n^2(\omega/\omega_c)}, \quad (3.291a)$$

where

$$U_n(\omega) = \frac{\sin[(n+1)\cos^{-1}\omega]}{\sin(\cos^{-1}\omega)} \quad (3.291b)$$

is the *Chebyshev polynomial of the second kind*. [*Hint.* Make use of

$$\frac{dC_n(\omega)}{d\omega} = nU_{n-1}(\omega) \quad (3.292)$$

and express $\sin[(n+1)\cos^{-1}\omega]$ in exponential form.]

3.15. Repeat Problem 3.6 for a Chebyshev transducer power-gain characteristic having a 1-dB peak-to-peak ripple in the passband.

3.16. In Problem 3.15, suppose that the load resistance is not specified. Design a Chebyshev *LC* ladder filter having a maximum attainable K_n.

3.17. Design a low-pass Chebyshev filter having the following specifications:
  (i) Peak-to-peak ripple in the passband must not exceed 1.5 dB.
  (ii) The minimum attenuation at three times the cutoff frequency, which is 50 MHz, and beyond is 40 dB.
  (iii) A resistive generator of internal resistance 150 Ω is the excitation, and the load resistance is 470 Ω.

3.18. Consider the transducer power-gain characteristic

$$G(\omega^2) = \frac{\epsilon^2 C_n^2(1/\omega)}{1 + \epsilon^2 C_n^2(1/\omega)}. \quad (3.293)$$

Show that it has a maximally-flat characteristic in the passband, and equal ripples in the stopband. This is known as an *inverse Chebyshev characteristic*.

3.19. Show that the frequencies for the zeros and the maxima of gain in the stopband of the inverse Chebyshev characteristic (3.293) alternate and are given by

$$\omega_k = \sec k\pi/2n, \qquad k = 0, 1, 2, \ldots, n, \quad (3.294)$$

the odd k corresponding to zeros and even k to peaks.

3.20. A Butterworth or Chebyshev low-pass filter is required that must be at least α_0 dB down from its passband maximum at k times the cutoff frequency and beyond. Show that the order of the filter can be determined by the formulas

$$n \geq \frac{\tfrac{1}{2}\log(A-1)}{\log k} \approx \frac{0.05\,\alpha_0}{\log k} \quad (3.295)$$

for a Butterworth filter, and

$$n \geq \frac{\cosh^{-1}[(A-1)^{1/2}/\epsilon]}{\cosh^{-1} k} \approx \frac{\frac{1}{2}\ln(4A/\epsilon)}{\cosh^{-1} k} \qquad (3.296)$$

for a Chebyshev filter; where $\alpha_0 = 10 \log A$. The approximation formulas are valid when $A \gg 1$.

3.21. Show that the zeros and poles of the function (3.84a) can be expressed as

$$y_u = \tfrac{1}{2}(w\beta^{2u+n+1} - w^{-1}\beta^{-(2u+n+1)}), \qquad u = 0, 1, 2, \ldots, 2n-1, \qquad (3.297a)$$

where $\beta = \exp(j\pi/2n)$ as in (3.14b), and

$$w = \{[(1 - K_n + \epsilon^2)^{1/2} + (1 - K_n)^{1/2}]/\epsilon\}^{1/n} \qquad (3.297b)$$

for the zeros and $w = \hat{k}$ of (3.287c) for the poles.

3.22. Repeat the problem stated in Example 3.3 for a Chebyshev transducer power-gain characteristic having a 3-dB peak-to-peak ripple in the passband. Compare your result with the Butterworth case.

3.23. Repeat Problem 3.17 for 1 dB ripple in the passband.

3.24. Let

$$K_{r,r+1} = L_r C_{r+1} \qquad \text{for } r \text{ odd}, \qquad (3.298a)$$

$$K_{r,r+1} = C_r L_{r+1} \qquad \text{for } r \text{ even}. \qquad (3.298b)$$

Show that the formulas (3.91) can be replaced by a single recurrence formula

$$K_{r,r+1} = \frac{16 \sin \gamma_{2r-1} \sin \gamma_{2r+1}}{\omega_c^2 f_r (\sinh a, \sinh \hat{a})}, \qquad r = 1, 2, \ldots, n-1. \qquad (3.299)$$

3.25. Let

$$K_{r,r+1} = C_r L_{r+1} \qquad \text{for } r \text{ odd}, \qquad (3.300a)$$

$$K_{r,r+1} = L_r C_{r+1} \qquad \text{for } r \text{ even}. \qquad (3.300b)$$

Show that the formulas (3.104b) and (3.104c) can be replaced by a single recurrence formula (3.299).

3.26. Derive expressions similar to those given in Problems 3.24 and 3.25 for the recurrence formulas (3.34) and (3.42) of the Butterworth response.

3.27. Prove the recurrence formulas (3.34) for $n = 2$ and $n = 3$.

3.28. Derive the coefficient formulas (3.80) for $n = 2$ and $n = 3$.

3.29. Prove the recurrence formulas (3.91) for $n = 2$ and $n = 3$.

3.30. Repeat the problem given in Example 3.6 for a passband tolerance of $\tfrac{1}{2}$ dB.

3.31. Show that

$$\operatorname{sn}^2(u, k) + \operatorname{cn}^2(u, k) = 1, \qquad (3.301a)$$

$$\operatorname{dn}^2(u, k) + k^2 \operatorname{sn}^2(u, k) = 1. \qquad (3.301b)$$

APPROXIMATION AND LADDER REALIZATION

3.32. Prove the identities given in (3.111).

3.33. Using (3.153a), confirm that

$$\text{sn}\,(\pm 1.24188 + j0.70012, 0.71429) = \pm 1.05066 + j0.20983. \quad (3.302)$$

3.34. Show that
$$\text{sn}\,(0, k) = 0, \qquad \text{cn}\,(0, k) = 1, \qquad \text{dn}\,(0, k) = 1. \quad (3.303)$$

3.35. Compute
$$\text{sn}^{-1}\,(j/0.32261, 0.03753), \quad (3.304)$$

and compare your result with (3.237a).

3.36. Using (3.153a), compute

$$\text{sn}\,(1.24188 + j0.72972, 0.71429), \quad (3.305)$$

and compare your result with (3.237c).

3.37. Determine the Hurwitz polynomial (3.231) for $n = 3$, $1/k = 1.25$ and $\epsilon = 0.35$.

3.38. Prove the identities given in (3.121).

3.39. Repeat Problem 3.37 for $n = 4$.

3.40. Determine the characteristic function for $n = 3$ and $1/k = 1.1$.

3.41. Determine the characteristic function for $n = 4$ and $1/k = 1.1$.

3.42. Derive the identities given in (3.132).

3.43. Suppose that we wish to design a low-pass filter having an elliptic transducer power-gain characteristic. The peak-to-peak ripple within the passband, which extends from 0 to 100 MHz, must not exceed 1 dB, and at 120 MHz the gain must be attenuated by at least 30 dB. Determine the transducer power-gain characteristic.

3.44. In Problem 3.43, suppose that the attenuation at 120 MHz is increased to at least 50 dB. Determine the transducer power-gain characteristic.

3.45. Show that

$$\text{sn}\,(K, k) = 1, \qquad \text{sn}\,(2K, k) = 0, \qquad \text{sn}\,(4K, k) = 0. \quad (3.306)$$

3.46. Consider the transducer power-gain characteristic obtained in Problem 3.43. Determine the maximum and minimum values of gain in both the passband and the stopband and the frequencies at which these maxima and minima occur.

3.47. Repeat Problem 3.46 for the transducer power-gain characteristic obtained in Problem 3.44.

3.48. Prove that
$$\text{sn}\,(u \pm j4K', k) = \text{sn}\,(u, k). \quad (3.307)$$

3.49. Show that

$$\text{cn}\,(u \pm j2K', k) = -\text{cn}\,(u, k), \quad (3.308a)$$
$$\text{cn}\,(u \pm j4K', k) = \text{cn}\,(u, k). \quad (3.308b)$$

3.50. Show that

$$\mathrm{dn}\,(u \pm j2K', k) = -\mathrm{dn}\,(u, k), \tag{3.309a}$$

$$\mathrm{dn}\,(u \pm j4K', k) = \mathrm{dn}\,(u, k). \tag{3.309b}$$

3.51. Using the specifications stated in Problem 3.43, compute the input impedance function $Z_{11}(s)$ of the filter.

3.52. Using the specifications stated in Problem 3.44, compute the input impedance function $Z_{11}(s)$ of the filter.

3.53. Show that (3.176) and (3.190) can be expressed equivalently as

$$k_1 = k^n \left[\prod_{m=1}^{v} \frac{\mathrm{cn}\,(uK/n, k)}{\mathrm{dn}\,(uK/n, k)} \right]^4, \tag{3.310}$$

where $v = \frac{1}{2}(n-1)$ and $u = 2m$ for n odd, and $v = \frac{1}{2}n$ and $u = 2m - 1$ for n even.

3.54. Derive the identities given in (3.144).

3.55. In (3.161), show that for k near 1 and k_1 small the integer n can be approximated by the formula

$$n \approx \frac{1}{\pi^2} \ln\,(16A/\epsilon^2) \ln\,[8k/(1-k)], \tag{3.311}$$

where $\alpha_0 = 10 \log A$ is the attenuation in the stopband from its passband maximum.

3.56. Derive the identities given in (3.148).

3.57. The formula (3.199a) is valid for all odd $n > 1$. For $n = 1$, is there any frequency at which the gain is minimum within the passband?

3.58. Suppose that we wish to design a filter having a transducer power-gain characteristic that gives at most 1-dB passband ripple and at least 30-dB attenuation in the stopband at 1.1 times the cutoff frequency. Determine the values of n for the Butterworth, Chebyshev and elliptic responses. Also compare your results.

3.59. Compute the value $\mathrm{sn}^{-1}\,(j/2.36571, 0.007918)$ and compare your result with (3.258a).

3.60. Show that the LHS zeros of (3.225) are given by (3.232) and (3.243).

3.61. Using (3.153a), confirm the computation given in (3.256).

3.62. Using (3.153a), confirm the computation given in (3.258).

3.63. Design a high-pass filter whose radian cutoff frequency is 10^6 rad/s. The filter is required to have an equiripple transducer power-gain characteristic and is to be operated between a resistive generator of internal resistance 50 Ω and a 150-Ω load. The peak-to-peak ripple in the passband must not exceed 1.5 dB and at 3×10^5 rad/s or less the gain must be at least 50 dB down from its peak value in the passband. Also plot the gain response versus ω.

3.64. Repeat Problem 3.63 for a maximally-flat high-pass filter.

APPROXIMATION AND LADDER REALIZATION 213

3.65. Derive the formulas (3.278) for the element values of a band-elimination filter.

3.66. Repeat Problem 3.63 for an elliptic high-pass filter.

3.67. Design a band-pass filter whose passband is from 10^6 rad/s to 5×10^6 rad/s. The filter is required to have an equiripple transducer power-gain characteristic and is to be operated between a resistive generator of internal resistance 50 Ω and a 150-Ω load. The peak-to-peak ripple in the passband must not exceed 1-dB and at $\omega = 15 \times 10^6$ rad/s the gain must be at least 60 dB down from its peak value in the passband. Also plot the gain response versus ω.

3.68. Repeat Problem 3.67 for a maximally-flat band-pass filter.

3.69. Repeat Problem 3.67 for an elliptic band-pass filter. Also plot the gain response versus ω.

3.70. In Problem 3.67, suppose that from 10^6 rad/s to 5×10^6 rad/s is the rejection bandwidth. Design an equiripple band-elimination filter that gives at least 60 dB attenuation at 1.8×10^6 rad/s, everything else being the same.

3.71. Repeat Problem 3.70 for an elliptic band-elimination filter and plot its gain response versus ω.

3.72. Repeat Problem 3.70 for a maximally-flat band-elimination filter and plot its gain response versus ω.

3.73. Repeat Example 3.16 for a high-pass elliptic filter and plot its gain response as a function of ω.

3.74. Repeat Example 3.17 for a maximally-flat band-pass filter and plot its gain response as a function of ω.

3.75. Show that (3.45a) and (3.45b) are completely equivalent, each being valid for all ω.

3.76. For n even, show that the characteristic function $F_n(\omega)$ of (3.159b) has the property that

$$F_n(\infty) = 1/k_1. \tag{3.312}$$

[*Hint.* Appeal to (3.147) and (3.161).]

3.77. Compare (3.199) to (3.196) and indicate how they can be combined to give a single formula.

3.78. Show that the half-power radian frequency $\omega_{3\text{-dB}}$ of a low-pass Chebyshev transducer power-gain characteristic (3.44) is given by

$$\omega_{3\text{-dB}} = \omega_c \cosh\left(\frac{1}{n}\cosh^{-1}\frac{1}{\epsilon}\right) \text{ rad/s}. \tag{3.313}$$

3.79. The transducer power-gain characteristic

$$G(\omega^2) = \frac{K_n}{1 + \epsilon^2 \omega^{2m} C_{n-m}^2(\omega)} \tag{3.314}$$

is called the *transitional Butterworth–Chebyshev response*. For $0 < m < n$, show that (3.314) possesses both Butterworth-like and Chebyshev-like characteristics. Determine the dB attenuation produced in the stopband by this response.

3.80. Show that the passband maxima and minima of the low-pass Chebyshev gain response (3.44) occur at the frequencies

$$\omega_{\max} = \cos m\pi/2n, \quad m = 1, 3, 5, \ldots, q_1, \quad (3.315a)$$

$$\omega_{\min} = \cos m\pi/n, \quad m = 0, 1, 2, \ldots, q_2. \quad (3.315b)$$

Also determine the integers q_1 and q_2.

3.81. By introducing a pair of $j\omega$-axis zeros $\pm j\omega_0$ in the Butterworth response, we obtain a transducer power-gain characteristic

$$G(\omega^2) = \frac{(\omega_0^2 - \omega^2)^2 K_n}{(\omega_0^2 - \omega^2)^2 + (\omega_0^2 - 1)^2 \omega^{2n}}. \quad (3.316)$$

Discuss this approximation and sketch $G(\omega^2)$ versus ω.

3.82. Show that the term a as given in (3.74) can be expressed directly in terms of the passband ripple in dB by the relation

$$a = \frac{1}{n} \coth^{-1} 10^{0.05\alpha}, \quad (3.317)$$

where $\alpha = 10 \log (1 + \epsilon^2)$.

3.83. For $n = 2$, show that the coefficient formulas (3.80) become

$$b_0 = \tfrac{1}{2} + \sinh^2 a, \quad (3.318a)$$

$$b_1 = 2^{1/2} \sinh a. \quad (3.318b)$$

3.84. For $n = 3$, show that the coefficient formulas (3.80) become

$$b_0 = (\tfrac{3}{4} + \sinh^2 a) \sinh a, \quad (3.319a)$$

$$b_1 = \tfrac{3}{4} + 2 \sinh a, \quad (3.319b)$$

$$b_2 = 2 \sinh a. \quad (3.319c)$$

3.85. For $n = 4$, show that the coefficient formulas (3.80) become

$$b_0 = \tfrac{1}{8} + \sinh^2 a + \sinh^4 a, \quad (3.320a)$$

$$b_1 = 2^{1/2}(2 + 2^{1/2})^{1/2}(1 - 2^{-3/2} + \sinh^2 a) \sinh a, \quad (3.320b)$$

$$b_2 = 1 + (2 + 2^{1/2}) \sinh^2 a, \quad (3.320c)$$

$$b_3 = 2^{1/2}(2 + 2^{1/2})^{1/2} \sinh a. \quad (3.320d)$$

3.86. Show that (3.80d) can be expressed equivalently as

$$b_0 = 2^{1-n}/\epsilon, \quad n \text{ odd} \quad (3.321a)$$

$$= 2^{1-n}(1 + \epsilon^2)^{1/2}/\epsilon, \quad n \text{ even}. \quad (3.321b)$$

3.87. Show that for n even the constant λ given in (3.250b) can be expressed as

$$\lambda = \frac{c_0}{\hat{c}_0}\left[1 - \frac{H_n}{1+\epsilon^2}\right]^{1/2}. \tag{3.322}$$

References

1. Abramowitz, M. and Stegun, I. A. (1965) *Handbook of Mathematical Functions*, New York: Dover.
2. Atiya, F. S. (1953) Theorie der Maximal-geebneten und quasi-Tschebyscheffschen Filter. *Arch. elek. Übertragung*, vol. 7, pp. 441–450.
3. Belevitch, V. (1952) Tchebyshev filters and amplifier networks. *Wireless Engr.*, vol. 29, pp. 106–110.
4. Bennett, W. R. (1932) Transmission network, U.S. Patent 1,849,656.
5. Bossé, G. (1951) Siebketten ohne Dämpfungsschwankungen im Durchlassbereich (Potenzketten). *Frequenz*, vol. 5, no. 10, pp. 279–284.
6. Butterworth, S. (1930) On the theory of filter amplifiers. *Wireless Engr.*, vol. 7, pp. 536–541.
7. Cauer, W. (1931) Ein Reaktanztheorem. *Sitzber. preuss. Akad. Wiss.*, vol. 30–32, pp. 673–681.
8. Cauer, W. (1958) *Synthesis of Linear Communication Networks*, New York: McGraw-Hill. (Translated from the German second edition by G. E. Knausenberger and J. N. Warfield.)
9. Chebyshev, P. L. (1899) Théorie des mécanismes connus sous le nom de parallélogrammes. *Oeuvres*, vol. 1, St. Petersburg.
10. Chen, W. H. (1964) *Linear Network Design and Synthesis*, New York: McGraw-Hill.
11. Chen, W. K. (1975) Design considerations of the impedance-matching networks having the elliptic gain response, *Proc. 18th Midwest Symp. Circuits and Systems*, Concordia University, Montreal, Canada, pp. 322–326, August 11–12.
12. Chen, W. K. (1975) On the minimum-phase reflection coefficient in broadband equalizers. *Int. J. Electronics*, vol. 39, no. 3, pp. 357–360.
13. Doyle, W. (1958) Lossless Butterworth ladder networks operating between arbitrary resistance. *J. Math. Phys.*, vol. 37, no. 1, pp. 29–37.
14. Fano, R. M. (1950) A note on the solution of certain approximation problems in network synthesis. *J. Franklin Inst.*, vol. 249, no. 3, pp. 189–205.
15. Fubini, E. G. and Guillemin, E. A. (1959) Minimum insertion loss filters. *Proc. IRE*, vol. 47, no. 1, pp. 37–41.
16. Green, E. (1954) Synthesis of ladder networks to give Butterworth or Bhebȳshev response in the pass band. *Proc. IEE* (London), vol. 101, pt. IV, pp. 192–203.
17. Hancock, H. (1958) *Theory of Elliptic Functions*, New York: Dover.
18. Herrero, J. L. and Willoner, G. (1966) *Synthesis of Filters*, Englewood Cliffs, N.J.: Prentice-Hall.
19. Jahnke, E. and Emde, F. (1945) *Tables of Functions*, New York: Dover.
20. Landon, V. D. (1941) Cascade amplifiers with maximal flatness. *RCA Rev.*, vol. 5, no. 3–4, pp. 347–362.
21. Meinguet, J. and Belevitch, V. (1958) On the realizability of ladder filters. *IRE Trans. Circuit Theory*, vol. CT-5, no. 4, pp. 253–255.

22. Norton, E. L. (1937) Constant resistance networks with applications to filter groups. *Bell Sys. Tech. J.*, vol. 16, no. 2, pp. 178–193.
23. Orchard, H. J. (1953) Formulae for ladder filters. *Wireless Engr.*, vol. 30, pp. 3–5.
24. Orchard, H. J. (1958) Computation of elliptic functions of rational fractions of a quarterperiod. *IRE Trans. Circuit Theory*, vol. CT-5, no. 4, pp. 352–355.
25. Rhodes, J. D. (1971) Explicit formulas for element values in elliptic function prototype networks. *IEEE Trans. Circuit Theory*, vol. CT-18, no. 2, pp. 264–276.
26. Rubini, R. (1961) Graphical methods for solving the approximation problem in the design of electric frequency filters. *Alta Frequenza*, vol. 30, no. 2, pp. 136–155.
27. Saal, R. and Ulbrich, E. (1958) On the design of filters by synthesis. *IRE Trans. Circuit Theory*, vol. CT-5, no. 4, pp. 284–327.
28. Saraga, W. (1952) Approximations in network design. *Wireless Engr.*, vol. 29, pp. 280–281.
29. Sharpe, C. B. (1954) A general Tchebycheff rational function. *Proc. IRE*, vol. 42, no. 2, pp. 454–457.
30. Skwirzynski, J. K. (1965) *Design Theory and Data for Electrical Filters*, Princeton, N.J.: Van Nostrand.
31. Skwirzynski, J. K. and Zdunek, J. (1958) Note on calculation of ladder coefficients for symmetrical and inverse impedance filters on a digital computer. *IRE Trans. Circuit Theory*, vol. CT-5, no. 4, pp. 328–333.
32. Storer, J. E. (1957) *Passive Network Synthesis*, New York: McGraw-Hill.
33. Takahasi, H. (1951) On the ladder-type filter network with Tchebysheff response. *J. Inst. Elec. Commun. Engrs. Japan*, vol. 34, no. 2, pp. 65–74.
34. Tuttle, W. N. (1959) The design of two-section symmetrical Zobel filters for Tchebycheff insertion loss. *Proc. IRE*, vol. 47, no. 1, pp. 29–36.
35. Van Valkenburg, M. E. (1960) *Modern Network Synthesis*, New York: Wiley.
36. Waldron, R. A. (1959) Coupling coefficients of ladder networks with maximally flat amplitude response. *J. Brit. IRE*, vol. 19, pp. 63–71.
37. Weinberg, L. (1956) Network design by use of modern synthesis techniques and tables. *Proc. Natl. Electronics Conf.*, vol. 12, pp. 794–817.
38. Weinberg, L. (1957) Additional tables for design of optimum ladder networks. *J. Franklin Inst.*, vol. 264, nos. 1 and 2, pp. 7–23 (pt. I) and pp. 127–138 (pt. II).
39. Weinberg, L. (1957) Explicit formulas for Tschebyscheff and Butterworth ladder networks. *J. Appl. Phys.*, vol. 28, no. 10, pp. 1155–1160; *IRE Natl. Conv. Record*, vol. 5, pt. 2, pp. 200–212.
40. Weinberg, L. (1957) Tables of networks whose reflection coefficients possess alternating zeros. *IRE Trans. Circuit Theory*, vol. CT-4, no. 4, pp. 313–320.
41. Weinberg, L. (1962) *Network Analysis and Synthesis*, New York: McGraw-Hill.
42. Weinberg, L. and Slepian, P. (1960) Takahasi's results on Tchebycheff and Butterworth ladder networks. *IRE Trans. Circuit Theory*, vol. CT-7, no. 2, pp. 88–101.
43. Zdunek, J. (1958) The network synthesis on the insertion-loss basis. *Proc. IEE* (London), vol. 105C, pp. 259–291.

CHAPTER 4

Theory of Broadband Matching: The Passive Load

IN CHAPTER 2, we have studied the properties of the scattering matrix associated with an n-port network, and indicated how it can be extended to complex normalization. In Chapter 3, we presented three popular rational function schemes for approximating the ideal low-pass brick-wall type of gain response. In the present chapter, we shall apply these results to the design of matching networks. As is well known, in the design of communication systems, a basic problem is to design a coupling network between a given source and a given load so that the transfer of power from the source to the load is maximized over a given frequency band of interest. A problem of this type invariably involves the design of a coupling network to transform a given load impedance into another specified one. We refer to this operation as *impedance matching* or *equalization*, and the resulting coupling network as *matching network* or *equalizer*. We recognize that the choice of a lossy equalizer would not only lessen the transducer power gain but also severely hamper our ability to manipulate since the scattering matrix of a lossy equalizer is not necessarily unitary. Hence, we shall deal exclusively with the design of lossless equalizers.

The matching problem was initiated by Bode (1945) for a class of very useful but restricted load impedance consisting of the parallel combination of a capacitor and a resistor. He established a fundamental gain-bandwidth limitation for this class of equalizers, but did not go further to investigate the additional limitations imposed on the lossless equalizers. Fano (1950) extended Bode's work and solved the problem of impedance matching between an arbitrary passive load and a resistive generator, in its full generality. Fano's results are expressed as a set of integral constraints with proper weighting functions depending on the load impedance. Recently, Youla (1964) developed a new theory based on the principle of complex normalization which circumvents some of the difficulties encountered in Fano's work. Further, Youla's theory can be generalized to the design of active equalizers, as will be demonstrated in the following chapter. Needless to say, Fano's and Youla's techniques have been extended and elaborated upon by many workers (see, for example, references 3–7, 9–18, 21, and 25).

In this chapter, we shall present Youla's theory in detail, and illustrate every phase of the theory with fully worked out examples. In particular, we establish the fundamental gain-bandwidth limitations for Bode's parallel RC load and Darlington's type-C load, in their full generality.

1. The Bode–Fano–Youla broadband matching problem

In most practical cases, the source can usually be represented as an ideal voltage source in series with a pure resistor, which may be the Thévenin equivalent of some other network. The load impedance is assumed to be strictly passive over a frequency band of interest, the reason being that the matching problem cannot be meaningfully defined if the load is purely reactive. The objective is to design an "optimum" lossless two-port network (equalizer) N to match out the load impedance $z_l(s)$ to a resistive generator with internal resistance r_g and to achieve a preassigned transducer power-gain characteristic $G(\omega^2)$ over the entire sinusoidal frequency spectrum, where the interpretation of "optimum" becomes evident as the study is developed. The arrangement is depicted schematically in Fig. 4.1.

Under this situation, the maximum power transfer is then achieved when the impedance presented to the generator is equal to the source resistance. With the exception that the load impedance is purely resistive, it will be shown later that it is not always possible to match an arbitrary passive load to a resistive generator over a frequency band of interest with a prescribed gain characteristic, and

THE PASSIVE LOAD

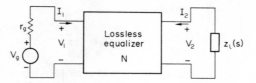

FIG. 4.1. The symbolic representation of the Bode–Fano–Youla broadband matching problem.

that the limitations originate from the physical realizability of the scattering parameters of the equalizer which, in turn, is dictated by the load impedance. Thus, any matching problem must include the maximum tolerance on the match as well as the minimum bandwidth within which the match is to be obtained.

2. Youla's theory of broadband matching: preliminary considerations

Consider the two-port network N of Fig. 4.1. Let $S(s)$ be its scattering matrix normalized to the reference impedances $z_1(s) = r_g$ and $z_2(s) = z_l(s)$. Then from §3.2 of Chapter 2, $S(s)$ can be expressed as

$$S(s) = h(s)S^I(s)h_*^{-1}(s), \quad (4.1)$$

where $S^I(s)$ is the current-based scattering matrix of N, and $h(s)h_*(s)$ is the para-hermitian part of the reference impedance matrix

$$z(s) = \begin{bmatrix} r_g & 0 \\ 0 & z_l(s) \end{bmatrix}. \quad (4.2)$$

Let

$$S(s) = [S_{ij}], \quad (4.3a)$$

$$S^I(s) = [S^I_{ij}], \quad (4.3b)$$

$$h(s) = \begin{bmatrix} h_1(s) & 0 \\ 0 & h_2(s) \end{bmatrix}. \quad (4.3c)$$

Then from (4.1), the current-based reflection coefficient at the output port

$$S_{22}^l(s) = \frac{Z_{22}(s) - z_l(-s)}{Z_{22}(s) + z_l(s)}, \qquad (4.4)$$

$Z_{22}(s)$ being the driving-point impedance looking into the output port when the input port is terminated in its reference impedance r_g, can be related to the normalized reflection coefficient $S_{22}(s)$ by the equation

$$S_{22}(s) = h_2(s)S_{22}^l(s)/h_2(-s). \qquad (4.5)$$

Recall the procedure for decomposing the para-hermitian part of $z(s)$ into factors $h(s)$ and $h_*(s)$ and the fact that each element of $h(s)$ is ambiguous within an all-pass function (Chapter 2, §2.3). We recognize that $h_2(s)/h_2(-s)$ is an all-pass function, whose poles include all the open LHS poles of $z_l(s)$. Thus it can be written as the product of the real all-pass function

$$A(s) = \prod_{i=1}^{\nu} \frac{s - s_i}{s + s_i} \qquad (4.6)$$

defined by the open RHS poles s_i ($i = 1, 2, \ldots, \nu$) of $z_l(-s)$ and another real all-pass function $\hat{\eta}(s)$, i.e.

$$h_2(s)/h_2(-s) = \hat{\eta}(s)A(s). \qquad (4.7)$$

Now observe that the open RHS poles of $S_{22}^l(s)$ are precisely those of $z_l(-s)$. The function defined by

$$\rho(s) = A(s)S_{22}^l(s) \qquad (4.8)$$

is analytic in the closed RHS. In other words, $\rho(s)$ is a bounded-real function, so is $S_{22}(s)$, which are also called the *bounded-real reflection coefficients*. Since $|A(j\omega)| = 1$ and $|\hat{\eta}(j\omega)| = 1$ and since the two-port network N is lossless whose scattering matrix $S(s)$ is para-unitary, the transducer power gain $G(\omega^2)$ can be expressed in terms of $\rho(j\omega)$ by the relation

$$G(\omega^2) = |S_{21}(j\omega)|^2 = 1 - |S_{22}(j\omega)|^2 = 1 - |\rho(j\omega)|^2. \qquad (4.9)$$

THE PASSIVE LOAD

Thus, to study the class of transducer power gains compatible with a prescribed load impedance $z_l(s)$, it suffices to consider the bounded-real reflection coefficient

$$\rho(s) = A(s) \frac{Z_{22}(s) - z_l(-s)}{Z_{22}(s) + z_l(s)}, \quad (4.10)$$

which together with (4.9) forms the cornerstone of Youla's theory on broadband matching. At times, we shall find it necessary throughout the remainder of this chapter to use

$$S_{22}(s) = \hat{\eta}(s)\rho(s) \quad (4.11)$$

instead of $\rho(s)$ in order to take advantage of the additional degrees of freedom introduced by the factor $\hat{\eta}(s)$. To proceed to the derivation of the limitations imposed on $\rho(s)$, we shall need two definitions based on $z_l(s)$.

DEFINITION 4.1. *Zero of transmission.* For a given load impedance $z_l(s)$, a closed RHS zero of multiplicity k of the function

$$w(s) \equiv \frac{r_l(s)}{z_l(s)}, \quad (4.12a)$$

where

$$r_l(s) = \tfrac{1}{2}[z_l(s) + z_l(-s)] \quad (4.12b)$$

is the even part of $z_l(s)$, is said to be a *zero of transmission of order k* of $z_l(s)$.

The term zero of transmission has its origin in Darlington synthesis; for if we realize $z_l(s)$ as the driving-point impedance of a lossless two-port network terminated in a 1-Ω resistor, the magnitude squared of the transfer impedance function on the $j\omega$-axis between the 1-Ω resistor and the input is equal to $r_l(j\omega)$. After analytic continuation, it is easy to see that the zeros of $r_l(s)$ would be the *zeros of transmission* of the lossless two-port network, which are defined as the complex frequencies at which zero output occurs for any finite input. We note that in Definition 4.1 not all of the zeros of $w(s)$ are called the zeros of transmission; only the closed RHS ones are.

For convenience, the zeros of transmission are divided into four mutually exclusive classes.

DEFINITION 4.2. *Classification of zeros of transmission.* Let $s_0 = \sigma_0 + j\omega_0$ be a zero of transmission of order k of a load impedance $z_l(s)$. Then s_0 belongs to one of the following four mutually exclusive classes depending on σ_0 and $z_l(s_0)$, as follows:

Class I: $\sigma_0 > 0$, which includes all the open RHS zeros of transmission.
Class II: $\sigma_0 = 0$ and $z_l(j\omega_0) = 0$.
Class III: $\sigma_0 = 0$ and $0 < |z_l(j\omega_0)| < \infty$.
Class IV: $\sigma_0 = 0$ and $|z_l(j\omega_0)| = \infty$.

With these definitions, we now proceed to derive constraints imposed on $\rho(s)$ by the load impedance $z_l(s)$.

3. Basic constraints on $\rho(s)$

The basic constraints imposed on the bounded-real reflection coefficient $\rho(s)$ by a given $z_l(s)$ will first be stated and illustrated in this section. Then in §5 we shall prove that they are necessary, and in §6 that they are sufficient, for the physical realizability of $\rho(s)$.

The restrictions on $\rho(s)$ are most conveniently formulated in terms of the coefficients of the Laurent series expansions of the following quantities about a zero of transmission $s_0 = \sigma_0 + j\omega_0$ of order k of $z_l(s)$:

$$\rho(s) = \sum_{m=0}^{\infty} \rho_m (s - s_0)^m, \tag{4.13a}$$

$$A(s) = \sum_{m=0}^{\infty} A_m (s - s_0)^m, \tag{4.13b}$$

$$F(s) \equiv 2r_l(s)A(s) = \sum_{m=0}^{\infty} F_m (s - s_0)^m. \tag{4.13c}$$

Basic constraints on $\rho(s)$. For each zero of transmission s_0 of order k of $z_l(s)$, one of the following four sets of coefficient conditions must be satisfied, depending on the classification of s_0:

(i) Class I: $A_x = \rho_x$ for $x = 0, 1, 2, \ldots, k - 1$.
(ii) Class II: $A_x = \rho_x$ for $x = 0, 1, 2, \ldots, k - 1$, and
$(A_k - \rho_k)/F_{k+1} \geqq 0$.
(iii) Class III: $A_x = \rho_x$ for $x = 0, 1, 2, \ldots, k - 2$, and
$(A_{k-1} - \rho_{k-1})/F_k \geqq 0$, where $k \geqq 2$.
(iv) Class IV: $A_x = \rho_x$ for $x = 0, 1, 2, \ldots, k - 1$, and
$F_{k-1}/(A_k - \rho_k) \geqq a_{-1}$, the residue of $z_l(s)$ at the pole $j\omega_0$.

We remark that for a preassigned transducer power-gain characteristic $G(\omega^2)$, the function $\rho(s)\rho(-s) = 1 - G(-s^2)$ is specified through (4.9) after appealing to the theory of analytic continuation of an analytic function. Our task, then, is to determine the function $\rho(s)$, knowing the function $\rho(s)\rho(-s)$, which must be the ratio of two even polynomials. Thus, the zeros and poles of $\rho(s)\rho(-s)$ must appear in quadrantal symmetry, i.e. be symmetric with respect to both the real and the imaginary axes. The question now is how to pick the zeros and poles of $\rho(s)$ from among those of $\rho(s)\rho(-s)$. For the poles, the answer is simple. Since $\rho(s)$ is a bounded-real function, being analytic in the closed RHS, and since the poles of $\rho(-s)$ are the negatives of the poles of $\rho(s)$, the poles of $\rho(s)\rho(-s)$ can be uniquely distributed: the open LHS poles of $\rho(s)\rho(-s)$ belong to $\rho(s)$, whereas those in the open RHS belong to $\rho(-s)$. Note that, for the lumped system considered in this chapter, $\rho(s)$ has no $j\omega$-axis poles.

As for the zeros, there are no unique ways to assign them. Since $\rho(s)$ may have closed RHS zeros, we need not assign all the LHS zeros of $\rho(s)\rho(-s)$ to $\rho(s)$. The only requirement is that the complex-conjugate pair of zeros must both be assigned to $\rho(s)$ or $\rho(-s)$. However, if it is specified that $\rho(s)$ is to be made a minimum-phase function, then all the open LHS zeros of $\rho(s)\rho(-s)$ are assigned to $\rho(s)$. The $j\omega$-axis zeros of $\rho(s)\rho(-s)$ are of even multiplicity, and thus they are divided equally between $\rho(s)$ and $\rho(-s)$. In other words, $\rho(s)$ is uniquely determined by the zeros and poles of $\rho(s)\rho(-s)$ only if $\rho(s)$ is required to be minimum-phase.

Let $\hat{\rho}(s)$ be the minimum-phase factorization of $\rho(s)\rho(-s)$ by the procedure outlined above. Then any reflection coefficient of the form $\pm \eta(s)\hat{\rho}(s)$, which is of the general type of (4.11), is admissible, where $\eta(s)$ is an arbitrary real all-pass function possessing the property that $\eta(s)\eta(-s) = 1$, as indicated in (2.31) of Chapter 2.

Having obtained the desired reflection coefficient $\rho(s)$, we must now require that it satisfy the basic constraints outlined above. The significance of these constraints is that, as will be shown in §5 and §6, they are both necessary and sufficient for the physical realizability of $\rho(s)$. However, before we justify this assertion, we first illustrate these basic constraints by considering the matching problem for Bode's parallel RC load.

4. Bode's parallel RC load

As mentioned at the beginning of this chapter, the matching problem for the parallel combination of an R-ohm resistor and a C-farad capacitor was first considered by Bode. In the present section, we shall apply the basic constraints on $\rho(s)$ discussed above to determine the gain-bandwidth restrictions imposed on the design of a lossless two-port network N that equalizes the parallel RC load (Fig. 4.2)

$$z_l(s) = \frac{R}{1 + RCs} \qquad (4.14)$$

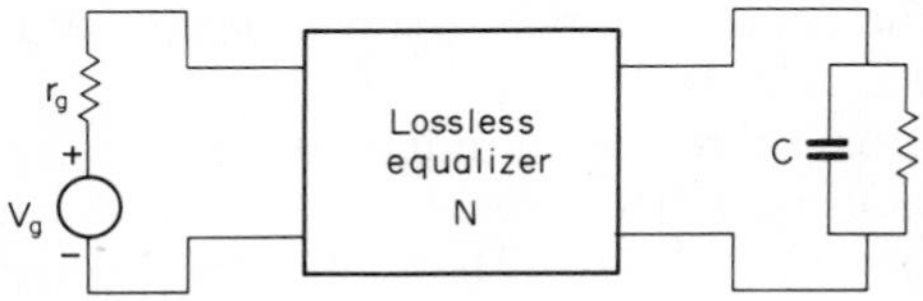

FIG. 4.2. The broadband matching problem for the parallel RC load.

to a resistive generator of internal resistance r_g, and to achieve the nth-order low-pass Butterworth, Chebyshev or elliptic transducer power-gain characteristic. To facilitate our discussion, the gain-bandwidth limitations for each of the three gain characteristics will be presented in a separate section.

THE PASSIVE LOAD

For the load $z_l(s)$ of (4.14), we first compute the following needed quantities:

$$r_l(s) = \tfrac{1}{2}[z_l(s) + z_l(-s)] = \frac{-\tau/C}{s^2 - \tau^2}, \qquad (4.15a)$$

$$A(s) = \frac{s - \tau}{s + \tau}, \qquad (4.15b)$$

$$F(s) = 2r_l(s)A(s) = \frac{-2\tau/C}{(s + \tau)^2}, \qquad (4.15c)$$

$$w(s) = \frac{r_l(s)}{z_l(s)} = \frac{-\tau}{s - \tau}, \qquad (4.15d)$$

where

$$\tau = \frac{1}{RC}. \qquad (4.15e)$$

Thus, from (4.15d) we see that the only zero of transmission of $z_l(s)$ is at $s = \infty$ and is of order 1. This means that $s_0 = \infty$ and $k = 1$. Since $z_l(s_0) = 0$, s_0 is a Class II zero of transmission of order 1.

We now proceed to discuss the gain-bandwidth limitations for the three gain characteristics in detail, as follows.

4.1. Butterworth transducer power-gain characteristic

We begin our discussion by considering the nth-order low-pass Butterworth transducer power-gain characteristic

$$G(\omega^2) = \frac{K_n}{1 + (\omega/\omega_c)^{2n}}, \qquad 0 \leq K_n \leq 1, \qquad (4.16)$$

as defined in (3.1), where ω_c is the 3-dB radian bandwidth and K_n is the dc gain.

Substituting (4.16) in (4.9) and appealing to analytic continuation give†

$$\rho(s)\rho(-s) = 1 - G(-s^2) = (1 - K_n)\frac{1 + (-1)^n x^{2n}}{1 + (-1)^n y^{2n}}, \qquad (4.17a)$$

where

$$x = (1 - K_n)^{-1/2n} s/\omega_c, \qquad (4.17b)$$

$$y = s/\omega_c. \qquad (4.17c)$$

†In the case $K_n = 1$, the numerator of (4.17a) becomes $(-1)^n y^{2n}$.

The numerator and denominator polynomials of (4.17a) can be factored in terms of the roots of the equation $(-1)^n s^{2n} + 1 = 0$. Like (3.12), let

$$q(s) = a_0 + a_1 s + \cdots + a_n s^n = \sum_{m=0}^{n} a_m s^m \qquad (4.18)$$

with $a_n = a_0 = 1$, be the nth-degree Hurwitz polynomial formed by the LHS zeros of $(-1)^n s^{2n} + 1 = 0$. To avoid the necessity of computing these coefficients they are tabulated in Appendix A. From (3.18) and Problem 3.12, we have

$$a_{n-1} = 1/\sin(\pi/2n). \qquad (4.19)$$

Proceeding as in (3.27) and for reason to be given shortly, let $\hat{\rho}(s)$ be the minimum-phase factorization of (4.17a). This yields a bounded-real reflection coefficient

$$\hat{\rho}(s) = (1 - K_n)^{1/2} \frac{q(x)}{q(y)}, \qquad (4.20)$$

which is devoid of zeros in the open RHS. Let

$$\rho(s) = \pm \eta(s)\hat{\rho}(s). \qquad (4.21)$$

where

$$\eta(s) = \prod_{i=1}^{u} \frac{s - \lambda_i}{s + \lambda_i}, \qquad \mathrm{Re}\,\lambda_i > 0 \ (i = 1, 2, \ldots, u) \qquad (4.22)$$

is an arbitrary real all-pass function, whose zeros or poles are to be chosen to maximize the dc gain K_n.

To apply the basic constraints, we must expand $\rho(s)$, $A(s)$ and $F(s)$ by Laurent series expansions about the zero of transmission s_0, which is at infinity. To this end, we shall use the binominal expansion formula

$$(s + c)^n = s^n + n s^{n-1} c + \frac{n(n-1)}{2!} s^{n-2} c^2 + \cdots, \qquad (4.23)$$

which is valid for all values of n if $|s| > |c|$; and is valid only for nonnegative integers n if $|s| \leq |c|$. That we can apply such procedures to obtain Laurent series expansions follows from the fact that the Laurent series expansion of an analytic function over a given annulus is unique. In other words, if an expansion of the Laurent type is found by any process, it must be *the* Laurent series expansion. The expansions about infinity are given by

$$A(s) = (s - \tau)(s + \tau)^{-1} = (s - \tau)(s^{-1} - s^{-2}\tau + s^{-3}\tau^2 + \cdots)$$
$$= 1 - 2\tau/s + 2\tau^2/s^2 + \cdots, \qquad (4.24a)$$

$$F(s) = (-2\tau/C)(s + \tau)^{-2} = (-2\tau/C)(s^{-2} - 2s^{-3}\tau + 3s^{-4}\tau^2 + \cdots)$$
$$= -2\tau/Cs^2 + 4\tau^2/Cs^3 + \cdots, \qquad (4.24b)$$

$$\pm \rho(s) = \eta(s)\hat{\rho}(s)$$
$$= (1 - K_n)^{1/2}(s^u + \tfrac{1}{2}\eta_1 s^{u-1} + \cdots)(s^u - \tfrac{1}{2}\eta_1 s^{u-1} + \cdots)^{-1}$$
$$\times (x^n + a_{n-1}x^{n-1} + \cdots)(y^n + a_{n-1}y^{n-1} + \cdots)^{-1}$$
$$= (1 - K_n)^{1/2}(s^u + \tfrac{1}{2}\eta_1 s^{u-1} + \cdots)(s^{-u} + \tfrac{1}{2}\eta_1 s^{-u-1} + \cdots)$$
$$\times (x^n + a_{n-1}x^{n-1} + \cdots)(y^{-n} - a_{n-1}y^{-n-1} + \cdots)$$
$$= (1 + \eta_1 s^{-1} + \cdots)(1 + a_{n-1}/x - a_{n-1}/y + \cdots)$$
$$= 1 + (\eta_1 + \hat{\rho}_1)/s + \eta_1\hat{\rho}_1/s^2 + \cdots, \qquad (4.24c)$$

where

$$\eta_1 = -2 \sum_{i=1}^{u} \lambda_i, \qquad (4.24d)$$

$$\hat{\rho}_1 = -[1 - (1 - K_n)^{1/2n}]\omega_c/\sin(\pi/2n). \qquad (4.24e)$$

Alternatively, the above expansions can be obtained by equating the known functions to their Laurent series expansions of unknown coefficients, as shown in (4.13). Multiplying both sides by the denominators of the known functions, we can solve the unknown coefficients by equating coefficients of like powers of s on the two sides of the equations. This procedure results, of course, in the same expansions (Problem 4.5).

Since the load impedance has only a Class II zero of transmission of order 1, the basic constraints on the coefficients of (4.24) are given by

$$A_0 = \rho_0, \qquad (4.25a)$$

$$(A_1 - \rho_1)/F_2 \geqq 0, \qquad (4.25b)$$

where $A_0 = 1$, $A_1 = -2\tau$, $\rho_0 = \pm 1$, $\rho_1 = \pm(\eta_1 + \hat{\rho}_1)$, and $F_2 = -2\tau/C$. Thus, to satisfy (4.25a) we must choose the plus sign in the expansion of $\rho(s)$. To satisfy the second constraint, we substitute the appropriate quantities in (4.25b), which yields the inequality

$$1 - (1 - K_n)^{1/2n} \leq 2[\sin(\pi/2n)]\left(\frac{1}{RC} - \sum_{i=1}^{u} \lambda_i\right)\Big/\omega_c. \quad (4.26)$$

In order that (4.26) possesses a solution K_n in the range $0 \leq K_n \leq 1$, the zeros λ_i of $\eta(s)$ are not completely arbitrary, and must be chosen in accordance with the requirement

$$\frac{1}{RC} \geq \sum_{i=1}^{u} \lambda_i, \quad (4.27)$$

and the dc gain is bounded by

$$(1 - K_n)^{1/2n} \geq 1 - 2(\sin \pi/2n)\left(\frac{1}{RC} - \sum_{i=1}^{u} \lambda_i\right)\Big/\omega_c. \quad (4.28)$$

For an adequate discussion, two cases are considered.

Case 1. $2(\sin \pi/2n)/RC\omega_c \geq 1$. Since $\Sigma \lambda_i$ is real and positive, by choosing the zeros λ_i appropriately, we can guarantee that the inequality (4.28) can be satisfied with $K_n = 1$ (Problem 4.7). Thus, in this case, we can achieve a dc gain of unity. In particular, this is true if we set all $\lambda_i = 0$.

Case 2. $2(\sin \pi/2n)/RC\omega_c < 1$. Then

$$2(\sin \pi/2n)\left(\frac{1}{RC} - \sum_{i=1}^{u} \lambda_i\right)\Big/\omega_c \leq 1. \quad (4.29)$$

Thus, if we choose $\lambda_i \neq 0$ ($i = 1, 2, \ldots, u$), we reduce the maximal obtainable dc gain K_n for a preassigned bandwidth ω_c. Setting all $\lambda_i = 0$ in (4.28) yields the inequality

$$K_n \leq 1 - [1 - 2(\sin \pi/2n)/RC\omega_c]^{2n}. \quad (4.30)$$

THE PASSIVE LOAD

In the limit as $n \to \infty$,

$$K_\infty \leq \lim_{n \to \infty} \{1 - [1 - 2(\sin \pi/2n)/RC\omega_c]^{2n}\}$$

$$= \lim_{n \to \infty} \{1 - [1 - \pi/nRC\omega_c]^{(-nRC\omega_c/\pi)(-2\pi/RC\omega_c)}\}$$

$$= 1 - \exp(-2\pi/RC\omega_c) = 1 - \exp(-1/RCf_c) \quad (4.31)$$

where $\omega_c = 2\pi f_c$, f_c being the cutoff frequency in hertz. The last line follows from the identity

$$\lim_{x \to \infty} \left(1 + \frac{1}{x}\right)^x = e. \quad (4.32)$$

At this point, we might be tempted to conclude that the maximum obtainable dc gain K_n would increase or decrease monotonically with n. That this is *not* the case follows directly from the curves of Fig. 4.3, in which the maximum attainable dc gain K_n of (4.30) is plotted as a function of n and RCf_c, the normalized bandwidth. It is clear from these plots that with the exception for $n = 1$, K_∞ is approached from the above (also see Problem 4.14). As an example, let us consider the case where $RCf_c = 1$. Then

$$K_{\infty \max} = 1 - e^{-1} = 1 - 0.37 = 0.63,$$

and for $n = 3$, we have

$$K_{3 \max} = 1 - \left[1 - \frac{1}{\pi} \sin \pi/6\right]^6 = 0.645.$$

Thus, there exists an optimum gain for a specified RCf_c that is attained at some finite n, as indicated in Fig. 4.3(c).

We remark that (4.31) also represents the maximum obtainable constant gain for the ideal brick-wall type of low-pass response for $G(\omega^2)$ with a preassigned bandwidth ω_c, i.e. (Problem 4.9)

$$G(\omega^2) \leq 1 - e^{-1/RCf_c}, \quad 0 \leq \omega \leq \omega_c, \quad (4.33a)$$

$$= 0, \quad \omega > \omega_c. \quad (4.33b)$$

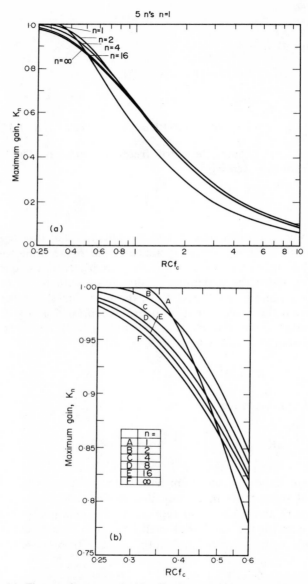

FIG. 4.3. (a) The maximum attainable K_n for equalizers having maximally-flat transducer power-gain characteristics. (b) An expanded view of Fig. 4.3(a).

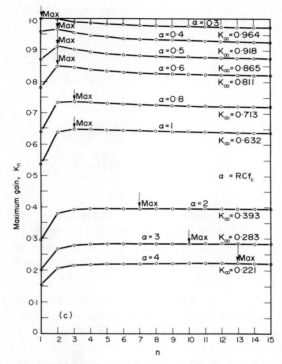

FIG. 4.3 (*cont.*). (c) The maximum attainable K_n plotted as a function of n.

We also note that for the parallel RC load, it is never necessary to insert any open RHS zeros in $\rho(s)$.

Before we work out a detailed numerical example on this load, we shall first justify an earlier statement that the bounded-real reflection coefficient $\hat{\rho}(s)$, as given in (4.20), should be devoid of zeros in the open RHS. To see this, let us replace $q(x)$ by $\hat{q}(x)$ in (4.20), where

$$\hat{q}(x) = \hat{a}_0 + \hat{a}_1 x + \cdots + \hat{a}_n x^n = \sum_{m=0}^{n} \hat{a}_m x^m, \quad (4.34)$$

$\hat{a}_0 = \hat{a}_n = 1$, is a factorization (not necessarily a Hurwitz polynomial) of the numerator polynomial of (4.17a). Proceeding in an entirely similar manner as in (4.21)–(4.24), we obtain the new coefficient ρ_1 of the Laurent series expansion of $\rho(s)$ as

$$\rho_1 = -[1 - d(1 - K_n)^{1/2n}]\omega_c/(\sin \pi/2n) + \eta_1, \qquad (4.35a)$$

where $d = \hat{a}_{n-1}/a_{n-1}$, and is bounded by (Problem 4.12)

$$-1 \leq d \leq 1. \qquad (4.35b)$$

Substituting (4.35) in (4.25b) yields the inequality

$$1 - d(1 - K_n)^{1/2n} \leq 2(\sin \pi/2n)\left(\frac{1}{RC} - \sum_{i=1}^{u} \lambda_i\right)\bigg/\omega_c. \qquad (4.36)$$

Thus, to maximize K_n, we choose $d = 1$ (Problem 4.13), which is equivalent to the minimum-phase factorization of (4.17a). An alternative approach is described in Problem 4.52.

With this digression, we now proceed to make some detailed computations on the design of an equalizer with a specific parallel RC load.

EXAMPLE 4.1. Consider the parallel combination of a 50-Ω resistor and a 100-pF capacitor. It is desired to equalize this load to a resistive generator of internal resistance 100 Ω, and to achieve the third-order low-pass Butterworth transducer power gain with a maximal attainable dc gain. The 3-dB bandwidth is required to be 10^8 rad/s. Realize the desired lossless equalizer.

To simplify the computation, the load impedance is first magnitude-scaled down by a factor 10^{-2}, and frequency-scaled down by 10^{-8}. This results in the normalized quantities $R = \frac{1}{2}\Omega$, $C = 1$ F, $r_g = 1 \Omega$, and $\omega_c = 1$ rad/s with

$$2(\sin \pi/2n)/RC\omega_c = 4 \sin \pi/6 = 2 > 1.$$

Thus, Case 1 applies and we can attain a maximal dc gain of unity, i.e. $K_3 = 1$. For $n = 3$, the Hurwitz polynomial $q(s)$ becomes

$$q(s) = 1 + 2s + 2s^2 + s^3 = (s + 1)(s^2 + s + 1). \qquad (4.37)$$

The minimum-phase factorization $\hat{\rho}(s)$, as indicated in (4.20), is obtained as

THE PASSIVE LOAD

$$\hat{\rho}(s) = \frac{s^3}{s^3 + 2s^2 + 2s + 1}. \tag{4.38}$$

The other required functions are computed from (4.15), as follows:

$$z_l(s) = \frac{1}{s+2}, \tag{4.39a}$$

$$A(s) = \frac{s-2}{s+2}, \tag{4.39b}$$

$$F(s) = \frac{-4}{(s+2)^2}. \tag{4.39c}$$

From (4.10), the driving-point impedance $Z_{22}(s)$ at the output port when the input port is terminated in $r_g = 1\,\Omega$ can be computed, and is given by

$$\begin{aligned} Z_{22}(s) &= \frac{F(s)}{A(s) - \hat{\rho}(s)} - z_l(s) \\ &= \frac{2s^3 + 6s^2 + 5s + 2}{(s+2)(2s^3 + 2s^2 + 3s + 2)} \\ &= \frac{2s^2 + 2s + 1}{2s^3 + 2s^2 + 3s + 2}, \end{aligned} \tag{4.40}$$

which, as will be shown in §6, must be positive-real. Thus, it is physically realizable, since according to Darlington's theory (Van Valkenburg, 1960) any positive-real function can be realized as the input impedance of a lossless two-port network terminated in a 1-Ω resistor. If the terminating resistance other than 1 Ω is required, an ideal transformer may be inserted in the final matching network to give the desired impedance levels. For the present case, $Z_{22}(s)$ can be realized as an LC ladder network terminating in a resistor by expanding it in a continued fraction, yielding

$$Z_{22}(s) = \cfrac{1}{s + \cfrac{1}{s + \cfrac{1}{2s + \cfrac{1}{0.5}}}}. \tag{4.41}$$

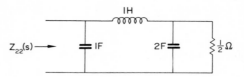

FIG. 4.4. A ladder realization of the impedance (4.40).

The corresponding network is given in Fig. 4.4. After denormalization, the equalizer together with its resistive generator and the parallel RC load is presented in Fig. 4.5. Since the generator resistance is 100 Ω, an ideal transfer of turns ratio

$$\left(\frac{100}{50}\right)^{1/2} = 1.414$$

is required at the input, as indicated in Fig. 4.5.

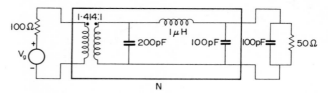

FIG. 4.5. An equalizer together with its terminations possessing the third-order Butterworth transducer power-gain characteristic.

For illustrative purposes, we compute the scattering matrix of the equalizer normalizing to the load impedances as indicated in Fig. 4.5. For simplicity, again the network is magnitude-scaled down by a factor 10^{-2} and frequency-scaled down by 10^{-8}. The open-circuit impedance matrix $Z(s)$ of the equalizer together with the reference impedance matrix is given by

$$Z(s) = \frac{1}{s(2s^2+3)} \begin{bmatrix} 2s^2+2 & \sqrt{2} \\ \sqrt{2} & 2s^2+1 \end{bmatrix}, \quad (4.42a)$$

$$z(s) = \begin{bmatrix} 1 & 0 \\ 0 & \dfrac{1}{s+2} \end{bmatrix}, \quad (4.42b)$$

THE PASSIVE LOAD 235

from which we can compute the current-based scattering matrix

$$S^I(s) = [Z(s)+z(s)]^{-1}[Z(s)-z_*(s)]$$
$$= \frac{(2s^2+3)(s+2)}{2(s+1)(2s^4+2s^3+5s^2+3s+3)} \begin{bmatrix} \frac{-2s^3}{s+2} & \frac{2\sqrt{2}}{4-s^2} \\ \sqrt{2} & \frac{2s^3}{s-2} \end{bmatrix}. \quad (4.43)$$

Finally, the normalized scattering matrix $S(s)$ is obtained as

$$S(s) = h(s)S^I(s)h_*^{-1}(s)$$
$$= \frac{2s^2+3}{2(s+1)(2s^4+2s^3+5s^2+3s+3)} \begin{bmatrix} -2s^3 & 2 \\ 2 & -2s^3 \end{bmatrix}$$
$$= \frac{1}{(s+1)(s^2+s+1)} \begin{bmatrix} -s^3 & 1 \\ 1 & -s^3 \end{bmatrix}, \quad (4.44a)$$

where

$$h(s) = \begin{bmatrix} 1 & 0 \\ 0 & \frac{\sqrt{2}}{s+2} \end{bmatrix}. \quad (4.44b)$$

Thus, the transducer power gain of the normalized lossless equalizer is given by

$$G(\omega^2) = |S_{21}(j\omega)|^2 = \frac{1}{1+\omega^6}, \quad (4.45)$$

confirming our design of the equalizer.

4.2. Chebyshev transducer power-gain characteristic

Consider the same problem discussed in the preceding section except now that we wish to achieve the nth-order low-pass Chebyshev transducer power-gain characteristic

$$G(\omega^2) = \frac{K_n}{1+\epsilon^2 C_n^2(\omega/\omega_c)}, \quad 0 \leq K_n \leq 1, \quad (4.46)$$

where the quantities K_n, ϵ and ω_c are defined the same as in (3.44).

We remark that the dc gain is

$$G(0) = K_n, \quad n \text{ odd} \tag{4.47a}$$

$$= \frac{K_n}{1+\epsilon^2}, \quad n \text{ even.} \tag{4.47b}$$

In the passband, (4.47a) is the value of the maxima while (4.47b) denotes the value of the minima.

Substituting (4.46) in (4.9) and appealing to the theory of analytic continuation give

$$\rho(s)\rho(-s) = (1-K_n)\frac{1+\hat{\epsilon}^2 C_n^2(-jy)}{1+\epsilon^2 C_n^2(-jy)}, \tag{4.48a}$$

where

$$\hat{\epsilon} = \epsilon(1-K_n)^{-1/2}, \tag{4.48b}$$

$$y = s/\omega_c, \tag{4.48c}$$

as in (4.17c). In the case $K_n = 1$, (4.48a) reduces to

$$\rho(s)\rho(-s) = \frac{\epsilon^2 C_n^2(-jy)}{1+\epsilon^2 C_n^2(-jy)}. \tag{4.49}$$

Like the Butterworth response, the numerator and denominator polynomials of (4.48a) can be factored in terms of the roots of

$$1 + \nu^2 C_n^2(-jy) = 0, \tag{4.50}$$

where $\nu = \epsilon$ or $\hat{\epsilon}$. Again, let $\hat{\rho}(s)$ be the minimum-phase decomposition of (4.48a) formed by its LHS zeros and poles (Problems 4.24 and 4.53). Write

$$\hat{\rho}(y) = \frac{y^n + \hat{b}_{n-1}y^{n-1} + \cdots + \hat{b}_1 y + \hat{b}_0}{y^n + b_{n-1}y^{n-1} + \cdots + b_1 y + b_0}, \tag{4.51}$$

as in (3.85a). These polynomials can be computed directly by means of the formula (3.79). For some values of n and ϵ, they are tabulated in Appendix B.

Explicit formula for the coefficient b_{n-1} or $\hat{b}_{n-1}$ of y^{n-1} in terms of n and ϵ is needed. From (3.80a), we have

$$b_{n-1} = \frac{\sinh a}{\sin \pi/2n}, \tag{4.52a}$$

$$\hat{b}_{n-1} = \frac{\sinh \hat{a}}{\sin \pi/2n}, \tag{4.52b}$$

where

$$a = \frac{1}{n} \sinh^{-1} \frac{1}{\epsilon}, \tag{4.53a}$$

$$\hat{a} = \frac{1}{n} \sinh^{-1} \frac{1}{\hat{\epsilon}}. \tag{4.53b}$$

Now consider $\rho(s)$ of (4.21) with $\hat{\rho}(s)$ being given by (4.51). Expanding $\rho(s)$ by Laurent series as in (4.24c) yields

$$\begin{aligned}\pm \rho(s) &= \eta(s)\hat{\rho}(s) \\ &= (1 + \eta_1 s^{-1} + \cdots)(1 + \hat{b}_{n-1} y^{-1} - b_{n-1} y^{-1} + \cdots) \\ &= 1 + [\eta_1 + (\hat{b}_{n-1} - b_{n-1})\omega_c]/s + \cdots,\end{aligned} \tag{4.54}$$

where η_1 is given in (4.24d). Thus, we have

$$\rho_0 = \pm 1, \tag{4.55a}$$

and from (4.52) and (4.53)

$$\rho_1 = \pm \left[\frac{(\sinh \hat{a} - \sinh a)\omega_c}{\sin \pi/2n} - 2 \sum_{i=1}^{u} \lambda_i \right]. \tag{4.55b}$$

Referring to the basic constraints (4.25) imposed on the coefficients ρ_0 and ρ_1, we see that we must first choose the plus sign in the expansion of $\rho(s)$ in (4.54), and that, in addition, the inequality

$$\sinh a - \sinh \hat{a} \leqq 2 (\sin \pi/2n)\left(\frac{1}{RC} - \sum_{i=1}^{u} \lambda_i\right)\bigg/ \omega_c \tag{4.56}$$

must be satisfied. Combining (4.48b), (4.53) and (4.56), we obtain the

238   BROADBAND MATCHING NETWORKS

bona fide restriction on the constant K_n (Problem 4.22):

$$(1 - K_n)^{1/2} \geq \epsilon \sinh\left\{n \sinh^{-1}\left[\sinh\left(\frac{1}{n}\sinh^{-1}\frac{1}{\epsilon}\right)\right.\right.$$
$$\left.\left. - \frac{2\sin \pi/2n}{\omega_c}\left(\frac{1}{RC} - \sum_{i=1}^{u}\lambda_i\right)\right]\right\}. \qquad (4.57)$$

In order that (4.57) possesses a solution K_n in the range $0 \leq K_n \leq 1$, the zeros λ_i of $\eta(s)$ must again be chosen in accordance with the requirement (4.27). But even so, two cases are distinguished.

Case 1. $2(\sin \pi/2n)/RC\omega_c \geq \sinh a$. Then by choosing the open RHS zeros λ_i appropriately, we can guarantee that the right-hand side of (4.57) is nonpositive, so that we can achieve $K_n = 1$. In particular, this is true if we set all $\lambda_i = 0$.

Case 2. $2(\sin \pi/2n)/RC\omega_c < \sinh a$. It is clear that if we choose $\lambda_i \neq 0$, it will only increase the right-hand side of (4.57), and thus lead to a reduction in K_n. Setting all $\lambda_i = 0$, the constant K_n simplifies to

$$K_n \leq 1 - \epsilon^2 \sinh^2\left\{n \sinh^{-1}\left[\sinh\left(\frac{1}{n}\sinh^{-1}\frac{1}{\epsilon}\right) - \frac{2\sin \pi/2n}{RC\omega_c}\right]\right\}. \qquad (4.58)$$

In the limit as $n \to \infty$ and $\epsilon \to 0$, we have

$$\lim_{\epsilon \to 0}\lim_{n \to \infty} K_n \leq \lim_{\epsilon \to 0}\lim_{n \to \infty}\left\{1 - \epsilon^2 \sinh^2\left[n \sinh^{-1}\left(\frac{1}{n}\sinh^{-1}\frac{1}{\epsilon} - \frac{\pi}{nRC\omega_c}\right)\right]\right\}$$
$$= \lim_{\epsilon \to 0}\left[1 - \epsilon^2 \sinh^2\left(\sinh^{-1}\frac{1}{\epsilon} - \frac{\pi}{RC\omega_c}\right)\right]$$
$$= \lim_{\epsilon \to 0}\left\{1 - \epsilon^2\left[\frac{1}{\epsilon}\cosh\frac{\pi}{RC\omega_c} - \cosh\left(\sinh^{-1}\frac{1}{\epsilon}\right)\sinh\frac{\pi}{RC\omega_c}\right]^2\right\}$$
$$= 1 - \left[\cosh\frac{\pi}{RC\omega_c} - \sinh\frac{\pi}{RC\omega_c}\right]^2 \quad \text{(See Problem 4.27.)}$$
$$= 1 - \exp\left(-1/RCf_c\right), \qquad (4.59)$$

as given in (4.31). Thus, the brick-wall type of flat-gain response is the limiting case of both the Butterworth and Chebyshev responses.

THE PASSIVE LOAD 239

The plots of the maximal attainable K_n of (4.58) as a function of RCf_c for various values of n and ϵ together with the limiting case of (4.59) are presented in Fig. 4.6.

We note that, like the Butterworth case discussed in the preceding section, it is never necessary to insert any open RHS zeros λ_i in $\rho(s)$ for the parallel RC load.

We shall illustrate the above procedure by the following detailed numerical examples.

EXAMPLE 4.2. It is desired to design a lossless two-port network N to equalize the load as shown in Fig. 4.7 to a generator with internal resistance of $100\,\Omega$ and to achieve the fourth-order Chebyshev transducer power gain having a maximal attainable K_4. The passband tolerance is $1\,\text{dB}$ and the cutoff frequency is $50/\pi$ MHz.

From specifications, we have

$$r_g = 100\,\Omega,$$
$$R = 100\,\Omega,$$
$$C = 200\,\text{pF},$$
$$\omega_c = 10^8\,\text{rad/s}.$$

For computational purposes, the network of Fig. 4.7 is first magnitude-scaled down by a factor 10^{-2}, and frequency-scaled down by 10^{-8}. This results in the normalized quantities $R = 1\,\Omega$, $r_g = 1\,\Omega$, $C = 2\,\text{F}$ and $\omega_c = 1\,\text{rad/s}$. From (3.62) the 1 dB peak-to-peak ripple in the passband corresponds to a ripple factor

$$\epsilon = (10^{0.1} - 1)^{1/2} = 0.50885.$$

Then we have

$$2(\sin \pi/2n)/RC\omega_c = \sin \pi/8 = 0.383 > \sinh a$$
$$= \sinh \left(\frac{1}{4} \sinh^{-1} \frac{1}{0.509}\right) = 0.365.$$

Thus, Case 1 applies and we can achieve the maximal attainable $K_4 = 1$. The Hurwitz polynomial formed by the open LHS zeros of

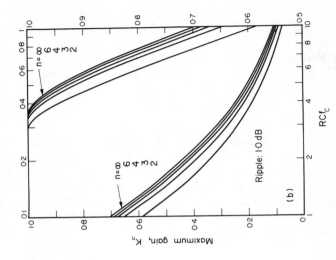
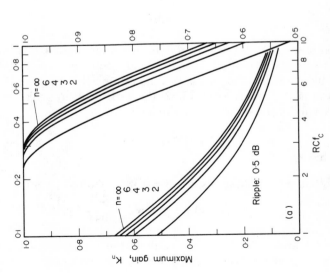

FIG. 4.6. The maximum attainable K_n for equalizers having equiripple transducer power-gain characteristics. (a) Passband ripple: $\frac{1}{2}$ dB. (b) Passband ripple: 1 dB. (Use the scale on the top and right for the upper-right set of curves.)

THE PASSIVE LOAD

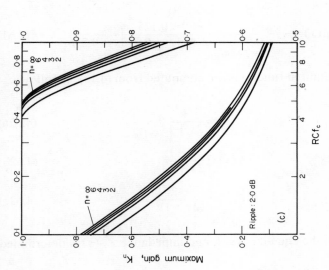

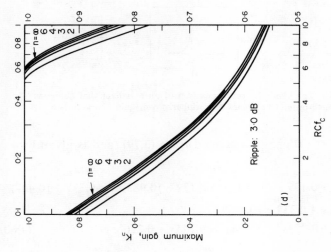

FIG. 4.6 (*cont.*). The maximum attainable K_n for equalizers having equiripple transducer power-gain characteristics. (c) Passband ripple: 2 dB. (d) Passband ripple: 3 dB. (Use the scale on the top and right for the upper-right set of curves.)

242 BROADBAND MATCHING NETWORKS

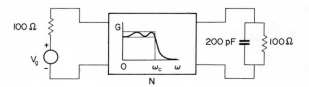

FIG. 4.7. The symbolic representation of an equalizer that achieves the fourth-order Chebyshev transducer power-gain characteristic.

$1 + \epsilon^2 C_4^2(-js) = 0$ can be obtained from (3.79), and is given by

$$\sum_{i=0}^{4} b_i s^i = (s + 0.139 + j0.983)(s + 0.139 - j0.983)(s + 0.337 + j0.407)$$
$$\times (s + 0.337 - j0.407)$$
$$= s^4 + 0.953s^3 + 1.454s^2 + 0.743s + 0.276. \quad (4.60)$$

Since $K_4 = 1$, the numerator polynomial of (4.48a) becomes $\epsilon^2 C_n^2(-js)$, which can easily be factored from (3.51). Thus, the desired bounded-real reflection coefficient $\hat{\rho}(s)$ is obtained as

$$\hat{\rho}(s) = \frac{s^4 + s^2 + 0.125}{s^4 + 0.953s^3 + 1.454s^2 + 0.743s + 0.276}. \quad (4.61)$$

The other required functions are computed from (4.15), as follows:

$$z_l(s) = \frac{1}{2s+1}, \quad (4.62a)$$

$$A(s) = \frac{2s-1}{2s+1}, \quad (4.62b)$$

$$F(s) = \frac{-2}{(2s+1)^2}. \quad (4.62c)$$

From (4.10), the equalizer back-end impedance $Z_{22}(s)$ is determined by

THE PASSIVE LOAD

$$Z_{22}(s) = \frac{F(s)}{A(s) - \hat{\rho}(s)} - z_l(s)$$

$$= \frac{1.906s^4 + 1.861s^3 + 1.940s^2 + 1.045s + 0.151}{(2s+1)(0.094s^4 + 0.045s^3 + 0.968s^2 + 0.441s + 0.401)}$$

$$= \frac{0.953s^3 + 0.454s^2 + 0.743s + 0.151}{0.094s^4 + 0.045s^3 + 0.968s^2 + 0.441s + 0.401}, \quad (4.63a)$$

which, as will be shown in §6, is guaranteed to be positive-real. In fact, it can be realized as a lossless ladder terminated in a resistor. This is accomplished by expanding $Z_{22}(s)$ in a continued fraction:

$$Z_{22}(s) = \cfrac{1}{0.099s + \cfrac{1}{1.064s + \cfrac{1}{2.83s + \cfrac{1}{0.79s + \cfrac{1}{2.66}}}}}. \quad (4.63b)$$

The corresponding network is realized in Fig. 4.8. Denormalizing the element values with regard to magnitude-scaling by a factor 100 and

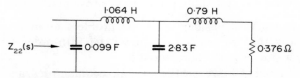

FIG. 4.8. A ladder realization of the impedance (4.63).

frequency-scaling by 10^8 gives the final design of the equalizer as shown in Fig. 4.9 together with the resistive generator and the parallel RC load.

For illustrative purposes, let us compute the transducer power gain. For simplicity, we again consider the normalized network of Fig. 4.8. The input impedance $Z_{11}(s)$ looking into the input port when the output port is terminated in its RC load is given by

$$Z_{11}(s) = 2.09 \frac{s^4 + 0.476s^3 + 1.225s^2 + 0.372s + 0.2}{s^3 + 0.477s^2 + 0.777s + 0.158}, \quad (4.64a)$$

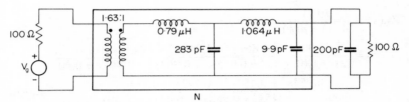

FIG. 4.9. An equalizer together with its terminations possessing the fourth-order Chebyshev transducer power-gain characteristic.

from which we can compute the transducer power gain $G(\omega^2)$, giving

$$G(\omega^2) = \frac{4 \operatorname{Re} Z_{11}(j\omega)}{|1 + Z_{11}(j\omega)|^2}$$

$$= \frac{0.06}{\omega^8 - 2.02\omega^6 + 1.244\omega^4 - 0.247\omega^2 + 0.076}. \quad (4.64b)$$

Alternatively, $G(\omega^2)$ can be obtained from (4.61) by the relation (4.9). However, using (4.64a) we can check the correctness of the element values of our design. A plot of (4.64b) as a function of ω together with the computer print out is presented in Fig. 4.10.

EXAMPLE 4.3. Consider the same problem as in Example 4.2 except now that we wish to achieve the fifth-order Chebyshev transducer power-gain characteristic.

As in Example 4.2, we use the normalized quantities, and compute

$$2(\sin \pi/2n)/RC\omega_c = \sin 18° = 0.309 > \sinh a = 0.2895,$$

where

$$a = \frac{1}{5} \sinh^{-1} \frac{1}{0.50885} = 0.2856.$$

Thus, Case 1 applies and we can attain $K_5 = 1$. Proceeding as in Example 4.2, the bounded-real reflection coefficient is obtained as

$$\hat{\rho}(s) = \frac{s^5 + 1.25s^3 + 0.312s}{s^5 + 0.937s^4 + 1.689s^3 + 0.974s^2 + 0.580s + 0.123}. \quad (4.65a)$$

THE PASSIVE LOAD

Substituting it in (4.10) in conjunction with (4.62) yields

$$Z_{22}(s) = \frac{0.937s^4 + 0.439s^3 + 0.974s^2 + 0.268s + 0.123}{0.126s^5 + 0.059s^4 + 0.991s^3 + 0.439s^2 + 0.647s + 0.123}$$

$$= \cfrac{1}{0.135s + \cfrac{1}{1.090s + \cfrac{1}{2.991s + \cfrac{1}{1.100s + \cfrac{1}{2.134s + \cfrac{1}{1}}}}}}, \quad (4.65b)$$

which can be identified as an LC ladder terminated in a 1-Ω resistor, as shown in Fig. 4.11. Denormalizing the element values with regard to magnitude-scaling by a factor 100 and frequency-scaling by 10^8 gives the final design of the equalizer as indicated in Fig. 4.12.

For illustrative purposes, we compute the transducer power gain in the normalized network of Fig. 4.11. For this, we compute the current-based reflection coefficient $S_{22}^I(s)$ at the output port by means of (2.11), giving

$$S_{22}^I(s) = \frac{-(1+2s)(2s^5 + 2.50s^3 + 0.625s)}{(1-2s)(2s^5 + 1.87s^4 + 3.378s^3 + 1.948s^2 + 1.16s + 0.246)}. \quad (4.66a)$$

Thus, the transducer power gain is given by

$$G(\omega^2) = |S_{21}(j\omega)|^2 = 1 - |S_{22}(j\omega)|^2 = 1 - |S_{22}^I(j\omega)|^2$$

$$= \frac{0.0151}{\omega^{10} - 2.5\omega^8 + 2.188\omega^6 - 0.781\omega^4 + 0.0975\omega^2 + 0.0151}. \quad (4.66b)$$

A plot of $G(\omega^2)$ as a function of the normalized frequency is presented in Fig. 4.13.

246 BROADBAND MATCHING NETWORKS

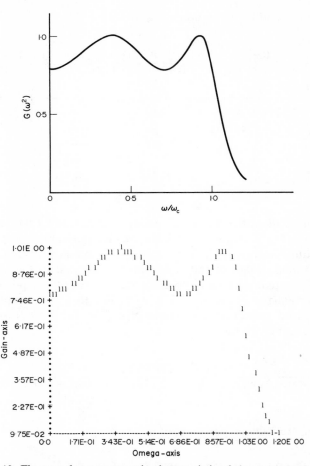

FIG. 4.10. The transducer power-gain characteristic of the network of Fig. 4.9.

4.3. Elliptic transducer power-gain characteristic

We now proceed to determine the gain-bandwidth limitations imposed on the design of a lossless equalizer that matches the parallel RC load to a resistive generator and achieves the nth-order

FIG. 4.10 (cont.)

Frequency in Omega	Gain	Frequency in Omega	Gain	Frequency in Omega	Gain
0.00	0.792	0.54	0.904	1.08	0.385
0.02	0.793	0.56	0.883	1.10	0.306
0.04	0.797	0.58	0.864	1.12	0.242
0.06	0.802	0.60	0.845	1.14	0.192
0.08	0.809	0.62	0.829	1.16	0.152
0.10	0.819	0.64	0.814	1.18	0.122
0.12	0.830	0.66	0.803	1.20	0.098
0.14	0.843	0.68	0.795		
0.16	0.857	0.70	0.791		
0.18	0.873	0.72	0.791		
0.20	0.890	0.74	0.795		
0.22	0.908	0.76	0.804		
0.24	0.925	0.78	0.818		
0.26	0.943	0.80	0.837		
0.28	0.959	0.82	0.861		
0.30	0.974	0.84	0.890		
0.32	0.987	0.86	0.922		
0.34	0.997	0.88	0.955		
0.36	1.003	0.90	0.983		
0.38	1.006	0.92	1.000		
0.40	1.005	0.94	0.997		
0.42	0.999	0.96	0.967		
0.44	0.990	0.98	0.905		
0.46	0.978	1.00	0.814		
0.48	0.962	1.02	0.704		
0.50	0.944	1.04	0.589		
0.52	0.924	1.06	0.481		

low-pass elliptic transducer power-gain characteristic

$$G(\omega^2) = \frac{H_n}{1 + \epsilon^2 F_n^2(\omega/\omega_c)}, \qquad 0 \leq H_n \leq 1, \qquad (4.67)$$

as given in (3.158). As before, since K denotes the complete elliptic integral of the first kind, to avoid possible confusion we use the symbol H_n instead of K_n in the numerator of (4.67). Note that k denotes either the order of a zero of transmission or the selectivity factor of the elliptic response. This should not create any difficulty, since in this section k denotes exclusively the selectivity factor.

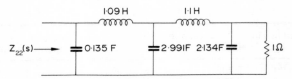

FIG. 4.11. A ladder realization of the impedance (4.65b).

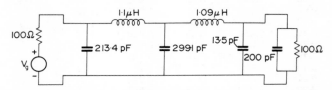

FIG. 4.12. An equalizer together with its terminations possessing the fifth-order Chebyshev transducer power-gain characteristic.

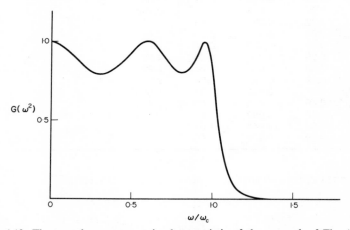

FIG. 4.13. The transducer power-gain characteristic of the network of Fig. 4.12.

Proceeding as in (4.48), we substitute (4.67) in (4.9) and invoke the theory of analytic continuation, giving

$$\rho(s)\rho(-s) = (1 - H_n)\frac{1 + \hat{\epsilon}^2 F_n^2(-jy)}{1 + \epsilon^2 F_n^2(-jy)}, \quad (4.68a)$$

THE PASSIVE LOAD   249

where $y = s/\omega_c$ as in (4.48c) and

$$\hat{\epsilon} = \epsilon(1 - H_n)^{-1/2}, \qquad (4.68b)$$

as in (3.248b). In the case $H_n = 1$, (4.68a) reduces to

$$\rho(s)\rho(-s) = \frac{\epsilon^2 F_n^2(-jy)}{1 + \epsilon^2 F_n^2(-jy)}. \qquad (4.69)$$

Like the Butterworth and Chebyshev cases, the denominator of (4.68a) can be uniquely decomposed while its numerator may have many permissible decompositions. Again, to maximize the gain (Problem 4.51), we choose $\hat{\rho}(s)$ to be the minimum-phase factorization of (4.68a). Write

$$\hat{\rho}(y) = \lambda \frac{y^n + \hat{c}_{n-1}y^{n-1} + \cdots + \hat{c}_1 y + \hat{c}_0}{y^n + c_{n-1}y^{n-1} + \cdots + c_1 y + c_0}, \qquad (4.70)$$

as in (3.249). These polynomials can be computed directly by means of the formula (3.228) or (3.243), depending upon whether n is odd or even. For some values of n, ϵ and k, they are tabulated in Appendix C.

Now consider $\rho(s)$ of (4.21) with $\hat{\rho}(s)$ being given by (4.70). Expanding $\rho(s)$ by Laurent series as in (4.54) yields

$$\begin{aligned}\pm \rho(s) &= \eta(s)\hat{\rho}(s) \\ &= (1 + \eta_1 s^{-1} + \cdots)(\lambda + \lambda \hat{c}_{n-1}y^{-1} - \lambda c_{n-1}y^{-1} + \cdots) \\ &= \lambda + \lambda[\eta_1 + (\hat{c}_{n-1} - c_{n-1})\omega_c]/s + \cdots, \end{aligned} \qquad (4.71)$$

where η_1 is given in (4.24d). Thus, we have

$$\rho_0 = \pm \lambda, \qquad (4.72a)$$

$$\rho_1 = \pm \lambda \left[(\hat{c}_{n-1} - c_{n-1})\omega_c - 2\sum_{i=1}^{u} \lambda_i\right]. \qquad (4.72b)$$

Referring to the basic constraints (4.25) imposed on the coefficients ρ_0 and ρ_1, we see again that we must choose the plus sign in the expansion of $\rho(s)$ in (4.71) with $\lambda = 1$, and that, in addition, the inequality

$$\hat{c}_{n-1} \geqq c_{n-1} - 2\left(\frac{1}{RC} - \sum_{i=1}^{u} \lambda_i\right)\bigg/\omega_c \tag{4.73}$$

must be satisfied. But from (3.250), $\lambda = 1$ if and only if n is odd. Thus, matching is possible only for elliptic response of odd orders. Appealing to (3.232) in conjunction with (3.228b), the coefficients c_{n-1} and $\hat{c}_{n-1}$, being the negative of the sums of the LHS roots, can be expressed explicitly in terms of n, k and ϵ, and are given by

$$c_{n-1} = u_0 + 2 \sum_{m=1}^{(n-1)/2} u_m, \tag{4.74a}$$

$$\hat{c}_{n-1} = \hat{u}_0 + 2 \sum_{m=1}^{(n-1)/2} \hat{u}_m, \tag{4.74b}$$

where n is odd and

$$-y_{pm} = u_m + jv_m = -j\,\text{sn}\,(2mK/n + ja, k), \tag{4.75a}$$

$$-\hat{y}_{pm} = \hat{u}_m + j\hat{v}_m = -j\,\text{sn}\,(2mK/n + j\hat{a}, k), \tag{4.75b}$$

$$m = 0, \pm 1, \pm 2, \ldots, \pm \tfrac{1}{2}(n-1),$$

$$a = -j\frac{K}{nK_1}\,\text{sn}^{-1}\,(j/\epsilon, k_1), \tag{4.76a}$$

$$\hat{a} = -j\frac{K}{nK_1}\,\text{sn}^{-1}\,[j(1-H_n)^{1/2}/\epsilon, k_1]. \tag{4.76b}$$

We remark that k_1 is determined by (3.176), once n and k are known. Since n is odd, according to (3.200a), H_n denotes the dc gain. To maximize H_n for a chosen set of parameters n, ϵ and k, we must, by (4.76b), minimize $\hat{a}$. From (4.75b) and (4.74b), this is equivalent to minimizing $\hat{c}_{n-1}$. For this, two cases are distinguished.

Case 1. $2/RC\omega_c \geqq c_{n-1}$. Then by selecting the open RHS zeros λ_i properly, we can guarantee that the right-hand side of (4.73) is nonpositive, so that $H_n = 1$ can always be achieved. In particular, this is true if we set all $\lambda_i = 0$.

THE PASSIVE LOAD   251

Case 2. $2/RC\omega_c < c_{n-1}$. It is evident that if we choose $\lambda_i \neq 0$, it will only increase the right-hand side of (4.73), and thus lead to a reduction in H_n. Setting all $\lambda_i = 0$ in (4.73) gives

$$\hat{c}_{n-1} \geqq c_{n-1} - \frac{2}{RC\omega_c}, \tag{4.77}$$

The maximum dc gain is achieved when the equality is attained. This equation together with (4.74)–(4.76), has been solved to yield a maximum H_n for various values of n, ϵ and k by a computer. In Fig. 4.14, the optimum dc gain H_n is plotted as a function of RCf_c for various odd orders n, passband ripples, and the steepness $1/k$ in the transitional frequency band.

We note that, like the previous two responses, it is never necessary to insert any open RHS zeros λ_i in $\rho(s)$ for this type of load.

We illustrate the above results by the following detailed numerical example.

EXAMPLE 4.4. It is desired to design a lossless two-port network N to equalize the load as shown in Fig. 4.15 to a resistive generator of internal resistance of 100 Ω and to achieve the third-order elliptic transducer power-gain characteristic with a maximum attainable dc gain. The peak-to-peak ripple within the passband, which extends from 0 to $50/\pi$ MHz, must not exceed 0.5 dB, and the edge of the stopband starts at $70/\pi$ MHz.

From specifications, we have

$$r_g = 100 \,\Omega, \quad R = 100 \,\Omega, \quad n = 3, \tag{4.78a}$$

$$C = 200 \text{ pF}, \quad \omega_c = 10^8 \text{ rad/s}, \tag{4.78b}$$

$$\omega_s = 1.4 \times 10^8 \text{ rad/s}, \quad \epsilon = 0.34931 \quad (0.5\text{-dB ripple}), \tag{4.78c}$$

giving

$$1/k = \frac{140 \times 10^6}{10^8} = 1.4 = 1/0.71429, \tag{4.79a}$$

$$RCf_c = 100 \times 200 \times 10^{-12} \times 50 \times 10^6/\pi = 0.3183. \tag{4.79b}$$

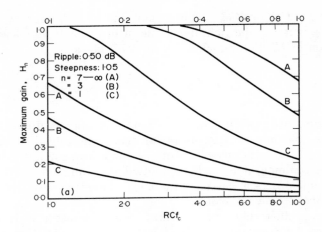

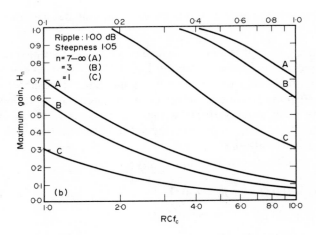

FIG. 4.14. (a) The maximum attainable H_n for equalizers having elliptic transducer power-gain characteristics. Steepness: 1.05. Passband ripple: $\frac{1}{2}$ dB. ($k = 0.95238$ and $\epsilon = 0.34931$.) (Use the scale on the top for the top three curves.) (b) The maximum attainable H_n for equalizers having elliptic transducer power-gain characteristics. Steepness: 1.05. Passband ripple: 1 dB. ($k = 0.95238$ and $\epsilon = 0.50885$.) (Use the scale on the top for the top three curves.)

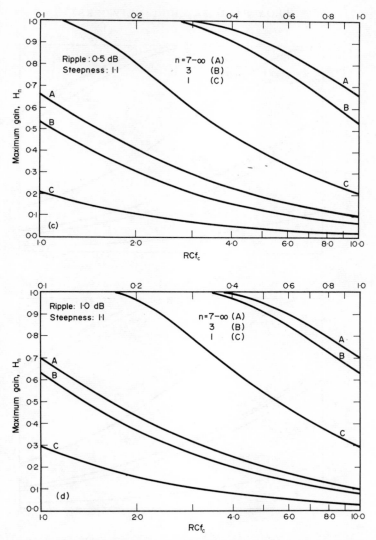

FIG. 4.14 (*cont.*). (c) The maximum attainable H_n for equalizers having elliptic transducer power-gain characteristics. Steepness: 1.10. Passband ripple: $\frac{1}{2}$ dB. ($k = 0.90909$ and $\epsilon = 0.34931$.) (Use the scale on the top for the top three curves.) (d) The maximum attainable H_n for equalizers having elliptic transducer power-gain characteristics. Steepness: 1.10. Passband ripple: 1 dB. ($k = 0.90909$ and $\epsilon = 0.50885$.) (Use the scale on the top for the top three curves.)

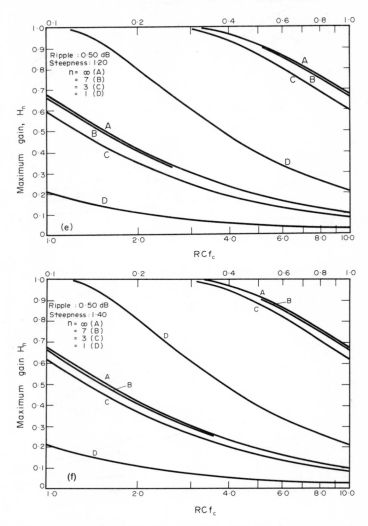

FIG. 4.14 (*cont.*). (e) The maximum attainable H_n for equalizers having elliptic transducer power-gain characteristics. Steepness: 1.20. Passband ripple: $\frac{1}{2}$ dB. ($k = 0.83333$ and $\epsilon = 0.34931$.) (Use the scale on the top for the top four curves.) (f) The maximum attainable H_n for equalizers having elliptic transducer power-gain characteristics. Steepness: 1.40. Passband ripple: $\frac{1}{2}$ dB. ($k = 0.71429$ and $\epsilon = 0.34931$.) (Use the scale on the top for the top four curves.)

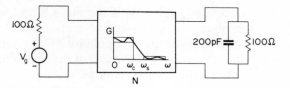

FIG. 4.15. The symbolic representation of an equalizer that achieves the third-order elliptic transducer power-gain characteristic.

For $k = 0.71429$, $n = 3$ and $\epsilon = 0.34931$, the denominator polynomial of (4.70) was computed in Example 3.12, and is given by

$$r(y) = y^3 + 1.21995 y^2 + 1.48376 y + 0.91865, \qquad (4.80)$$

which is also listed in Appendix C. Since

$$\frac{2}{RC\omega_c} = 1 < c_2 = 1.21995, \qquad (4.81)$$

Case 2 applies and $H_3 = 1$ cannot be attained. From Fig. 4.14(f), we choose

$$H_3 = 0.985, \qquad (4.82)$$

yielding

$$\hat{\epsilon} = \epsilon (1 - H_3)^{-1/2} = 2.85212. \qquad (4.83)$$

With this, we must now determine the numerator polynomial $\hat{r}(y)$ of (4.70) by the formulas (4.75b) and (4.76b), as follows:

$$k' = (1 - k^2)^{1/2} = 0.69985, \qquad (4.84a)$$

$$k_1 = 0.03753, \qquad (4.84b)$$

$$k'_1 = (1 - k_1^2)^{1/2} = 0.99930, \qquad (4.84c)$$

$$K = K(k) = K(0.71429) = 1.86282, \qquad (4.84d)$$

$$K' = K(k') = K(0.69985) = 1.84553, \qquad (4.84e)$$

$$K_1 = K(k_1) = K(0.03753) = 1.57135, \qquad (4.84f)$$

$$K'_1 = K(k'_1) = K(0.99930) = 4.67029, \qquad (4.84g)$$

where k_1 was computed earlier in Example 3.8 and is given in (3.178). The complete elliptic integrals are obtained from the standard mathematical tables by interpolation. For references, the reader is referred to those listed at the end of the preceding chapter. As a check, we use (3.161) which requires

$$\frac{K_1'}{K'} = 2.53060 = \frac{nK_1}{K}. \tag{4.85}$$

From (4.75b) and (4.76b), we have

$$\hat{a} = -j\frac{1.86282}{3 \times 1.57135} \operatorname{sn}^{-1}[j(1-0.985)^{1/2}/0.34931, 0.03753]$$
$$= 0.39516 \operatorname{tn}^{-1}(0.35062, 0.99930)$$
$$= 0.13586, \tag{4.86}$$

$$-\hat{y}_{p0} = \hat{u}_0 + j\hat{v}_0 = -j \operatorname{sn}(j0.13586, 0.71429)$$
$$= 0.13649, \tag{4.87a}$$

$$-\hat{y}_{p1}, -\hat{y}_{p(-1)} = \hat{u}_1 \pm j\hat{v}_1$$
$$= -j \operatorname{sn}(\pm 2 \times 1.86282/3 + j0.13586, 0.71429)$$
$$= 0.04494 \mp j0.90681. \tag{4.87b}$$

This leads to

$$\hat{r}(y) = (y + 0.13649)(y + 0.04494 + j0.90681)(y + 0.04494 - j0.90681)$$
$$= y^3 + 0.22636y^2 + 0.83659y + 0.11251. \tag{4.88}$$

Substituting (4.80) and (4.88) in (4.70) and selecting $\lambda = 1$ yield a bounded-real reflection coefficient

$$\hat{\rho}(y) = \frac{y^3 + 0.22636y^2 + 0.83659y + 0.11251}{y^3 + 1.21995y^2 + 1.48376y + 0.91865}. \tag{4.89}$$

The other required functions are computed from (4.14) and (4.15), giving

$$z_l(s) = \frac{100}{2y+1}, \tag{4.90a}$$

THE PASSIVE LOAD

$$A(s) = \frac{2y-1}{2y+1}, \tag{4.90b}$$

$$F(s) = \frac{-200}{(2y+1)^2}. \tag{4.90c}$$

Finally, from (4.10) the equalizer back-end impedance $Z_{22}(s)$ is determined by

$$Z_{22}(s) = \frac{F(s)}{A(s) - \hat{\rho}(s)} - z_l(s)$$
$$= 7750.312 \frac{y^2 + 0.65135y + 0.81134}{y^3 + 11.85413y^2 + 55.23167y + 80.43370}, \tag{4.91}$$

which, as will be shown in §6, is guaranteed to be positive-real. Using Darlington's technique, $Z_{22}(s)$ can be realized as the input impedance of a lossless two-port network terminated in a resistor.

4.4. Equalizer back-end impedance

In the foregoing, we have indicated that the equalizer back-end impedance $Z_{22}(s)$ can be determined directly from (4.10), which as will be shown in §6 is positive-real, and thus is physically realizable provided that the bounded-real reflection coefficient $\rho(s)$ satisfies the coefficient constraints outlined in §3. In the present section, we show that for the parallel RC load, the formula for $Z_{22}(s)$ may be simplified and can be expressed directly in terms of the coefficients of $\rho(s)$. For this, let

$$\rho(s) = \frac{\hat{d}_n s^n + \hat{d}_{n-1} s^{n-1} + \cdots + \hat{d}_1 s + \hat{d}_0}{d_n s^n + d_{n-1} s^{n-1} + \cdots + d_1 s + d_0}. \tag{4.92}$$

Substituting (4.92), (4.14) and (4.15b) in (4.10) and solving for $Z_{22}(s)$ give

$$Z_{22}(s) = \frac{F(s)}{A(s) - \rho(s)} - z_l(s)$$
$$= R \frac{\sum_{m=0}^{n-1} (d_m - \hat{d}_m) s^m}{\sum_{m=0}^{n} (d_m - RCd_{m-1} + \hat{d}_m + RC\hat{d}_{m-1}) s^m}, \tag{4.93}$$

where $d_{-1} = \hat{d}_{-1} = 0$. The significance of this formula is that we avoid the cancellation of the common factor $(RCs + 1)$ in $Z_{22}(s)$, as witnessed in the computation of $Z_{22}(s)$ in the preceding three examples.

Let us illustrate this by re-computing the impedances $Z_{22}(s)$ in Examples 4.1, 4.2 and 4.4.

EXAMPLE 4.5. Using (4.92), re-compute the equalizer back-end impedances $Z_{22}(s)$ of Examples 4.1, 4.2 and 4.4.

(i) From Example 4.1, we have $R = \frac{1}{2}\Omega$, $RC = \frac{1}{2}$ and

$$\rho(s) = \frac{s^3}{s^3 + 2s^2 + 2s + 1}, \quad (4.94)$$

giving

$$d_2 - \hat{d}_2 = 2 - 0 = 2, \quad (4.95a)$$

$$d_1 - \hat{d}_1 = 2 - 0 = 2, \quad (4.95b)$$

$$d_0 - \hat{d}_0 = 1 - 0 = 1; \quad (4.95c)$$

$$d_3 - RCd_2 + \hat{d}_3 + RC\hat{d}_2 = 1 - 1 + 1 + 0 = 1, \quad (4.95d)$$

$$d_2 - RCd_1 + \hat{d}_2 + RC\hat{d}_1 = 2 - 1 + 0 + 0 = 1, \quad (4.95e)$$

$$d_1 - RCd_0 + \hat{d}_1 + RC\hat{d}_0 = 2 - \tfrac{1}{2} + 0 + 0 = 1.5, \quad (4.95f)$$

$$d_0 + \hat{d}_0 = 1 + 0 = 1, \quad (4.95g)$$

confirming (4.40).

(ii) From Example 4.2, we have $R = 1\,\Omega$, $RC = 2$ and

$$\rho(s) = \frac{s^4 + s^2 + 0.125}{s^4 + 0.953s^3 + 1.454s^2 + 0.743s + 0.276}, \quad (4.96)$$

giving

$$d_3 - \hat{d}_3 = 0.953 - 0 = 0.953, \quad (4.97a)$$

$$d_2 - \hat{d}_2 = 1.454 - 1 = 0.454, \quad (4.97b)$$

$$d_1 - \hat{d}_1 = 0.743 - 0 = 0.743, \quad (4.97c)$$

$$d_0 - \hat{d}_0 = 0.276 - 0.125 = 0.151; \quad (4.97d)$$

THE PASSIVE LOAD

$$(d_4 + \hat{d}_4) - RC(d_3 - \hat{d}_3) = 2 - 2 \times 0.953 = 0.094, \quad (4.97\text{e})$$

$$(d_3 + \hat{d}_3) - RC(d_2 - \hat{d}_2) = 0.953 - 2 \times 0.454 = 0.045, \quad (4.97\text{f})$$

$$(d_2 + \hat{d}_2) - RC(d_1 - \hat{d}_1) = 2.454 - 2 \times 0.743 = 0.968, \quad (4.97\text{g})$$

$$(d_1 + \hat{d}_1) - RC(d_0 - \hat{d}_0) = 0.743 - 2 \times 0.151 = 0.441, \quad (4.97\text{h})$$

$$d_0 + \hat{d}_0 = 0.276 + 0.125 = 0.401, \quad (4.97\text{i})$$

confirming (4.63).

(iii) From Example 4.4, we have $R = 100\,\Omega$, $RC = 2 \times 10^{-8}$ and

$$\rho(y) = \frac{y^3 + 0.22636 y^2 + 0.83659 y + 0.11251}{y^3 + 1.21995 y^2 + 1.48376 y + 0.91865}, \quad (4.99\text{a})$$

giving $\quad (4.99\text{b})$

$$d_2 - \hat{d}_2 = (1.21995 - 0.22636) \times 10^{-16} = 0.99359 \times 10^{-16}, \quad (4.99\text{c})$$

$$d_1 - \hat{d}_1 = (1.48376 - 0.83659) \times 10^{-8} = 0.64717 \times 10^{-8},$$

$$d_0 - \hat{d}_0 = 0.91865 - 0.11251 = 0.80614,$$

$$(d_3 + \hat{d}_3) - RC(d_2 - \hat{d}_2) = (2 - 2 \times 0.99359) \times 10^{-24}$$
$$= 0.01282 \times 10^{-24}, \quad (4.99\text{d})$$

$$(d_2 + \hat{d}_2) - RC(d_1 - \hat{d}_1) = (1.44631 - 2 \times 0.64717) \times 10^{-16}$$
$$= 0.15197 \times 10^{-16}, \quad (4.99\text{e})$$

$$(d_1 + \hat{d}_1) - RC(d_0 - \hat{d}_0) = (2.32035 - 2 \times 0.80614) \times 10^{-8}$$
$$= 0.70807 \times 10^{-8}, \quad (4.99\text{f})$$

$$d_0 + \hat{d}_0 = 0.91865 + 0.11251 = 1.03116, \quad (4.99\text{g})$$

which are equivalent to those given in (4.91).

5. Proof of necessity of the basic constraints on $\rho(s)$

The basic constraints imposed on $\rho(s)$ by the load impedance $z_l(s)$ were stated and extensively illustrated in §3 and §4. In the present section, we show that these constraints are indeed necessary for the physical realizability of $\rho(s)$.

Referring to (4.12) and (4.13), let us consider the function

$$A(s) - \rho(s) = \frac{2r_i(s)A(s)}{Z_{22}(s) + z_l(s)} = \frac{F(s)}{Z_{22}(s) + z_l(s)}, \qquad (4.100)$$

which follows directly from (4.10) and (4.13c). Since, by assumption, $Z_{22}(s)$ and $z_l(s)$ are positive-real, their zeros and poles are restricted to the closed LHS, and, furthermore, the $j\omega$-axis zeros and poles, if they exist, must be simple. This implies that $r_i(s)$ has no poles on the $j\omega$-axis. For, if there were such poles, the partial fraction expansion of $r_i(s)$ would contain an odd term

$$\frac{\hat{k}_1}{s + j\omega_1} + \frac{\hat{k}_1}{s - j\omega_1} = \frac{2\hat{k}_1 s}{s^2 + \omega_1^2}, \qquad (4.101)$$

and $r_i(s)$ would be odd, a contradiction. However, $r_i(s)$ may possess the $j\omega$-axis zeros, but such zeros must be of even multiplicity if $z_l(s)$ is positive-real. Thus, from (4.100) we conclude that every zero of transmission s_0 of $z_l(s)$ is also a zero of $A(s) - \rho(s)$. In other words, regardless of the choice of the equalizer N, there exist points s_0 in the closed RHS, dictated solely by the choice of the load impedance $z_l(s)$, such that

$$A(s_0) = \rho(s_0), \qquad (4.102)$$

recalling that $A(s)$ is completely specified via (4.6) by the poles of $z_l(-s)$ in the open RHS. Equation (4.102) places fundamental limitations on the reflection coefficient $\rho(s)$ and the accompanying transducer power-gain characteristic.

Let $s_0 = \sigma_0 + j\omega_0$ be a zero of transmission of $z_l(s)$ of order k. The proof will now be completed by considering each class of zeros of transmission separately, as follows:

Class I. $\sigma_0 > 0$. Then s_0 is a zero of $A(s) - \rho(s)$ of multiplicity at least k. This requires that

$$\left\{ \frac{d^m}{ds^m}[A(s) - \rho(s)] \right\}_{s=s_0} = 0 \qquad (4.103)$$

THE PASSIVE LOAD

for $m = 0, 1, 2, \ldots, k - 1$. From (4.13), we get

$$\rho_m = A_m. \tag{4.104}$$

Class II. $\sigma_0 = 0$ and $z_l(s_0) = 0$. This implies that $F(s)$ has a zero at $s_0 = j\omega_0$ of multiplicity $k + 1$ since $z_l(s)$ can only have simple zeros on the $j\omega$-axis. Two subcases are considered.

Subcase 1. $Z_{22}(s_0) = 0$. Then $Z_{22}(s_0) + z_l(s_0) = 0$, and s_0 is a simple zero of the denominator of the right-hand side of (4.100). Thus, $s_0 = j\omega_0$ is a zero of $A(s) - \rho(s)$ of multiplicity k, and (4.103) applies. This shows that

$$\rho_m = A_m \tag{4.105}$$

for $m = 0, 1, 2, \ldots, k - 1$. Write

$$Z_{22}(s) = (s - j\omega_0)Z_{22}''(s), \tag{4.106a}$$

$$z_l(s) = (s - j\omega_0)z_l''(s). \tag{4.106b}$$

Then

$$Z_{22}''(j\omega_0) = \left.\frac{dZ_{22}(s)}{ds}\right|_{s=j\omega_0}, \tag{4.107a}$$

$$z_l''(j\omega_0) = \left.\frac{dz_l(s)}{ds}\right|_{s=j\omega_0}. \tag{4.107b}$$

Now we show that (4.107) are real and positive. It suffices to consider (4.107a); the other case can be proved in an entirely similar manner. To this end, we expand $Z_{22}(s)$ by Taylor series expansion about the point $j\omega_0$, which yields

$$\begin{aligned}Z_{22}(s) &= Z_{22}(j\omega_0) + Z_{22}''(j\omega_0)(s - j\omega_0) + \cdots \\ &= Z_{22}''(j\omega_0)(s - j\omega_0) + \cdots.\end{aligned} \tag{4.108}$$

As s approaches $j\omega_0$ from the RHS, in the limit, we shall find from (4.108) that

$$\arg_{s\to j\omega_0} Z_{22}(s) = \arg Z_{22}''(j\omega_0) + \arg_{s\to j\omega_0}(s - j\omega_0). \tag{4.109}$$

But the positive-real condition requires that $|\arg Z_{22}(s)| \leq \pi/2$ as long as $|\arg(s - j\omega_0)| \leq \pi/2$. Therefore, we conclude from (4.109) that $\arg Z_{22}''(j\omega_0) = 0$, and thus $Z_{22}''(j\omega_0)$ is real and positive.

Substituting (4.13) and (4.106) in (4.100) and comparing the coefficients of $(s - j\omega_0)^k$ result in

$$A_k - \rho_k = \frac{F_{k+1}}{Z_{22}''(j\omega_0) + z_l''(j\omega_0)} \qquad (4.110)$$

as s approaches $j\omega_0$. Since $j\omega_0$ is a zero of multiplicity $k + 1$ of $F(s)$, it follows that $F_{k+1} \neq 0$, and (4.110) becomes

$$\frac{A_k - \rho_k}{F_{k+1}} = \frac{1}{Z_{22}''(j\omega_0) + z_l''(j\omega_0)} > 0. \qquad (4.111)$$

Subcase 2. $Z_{22}(s_0) \neq 0$. Then $s_0 = j\omega_0$ is a zero of multiplicity at least $k + 1$ of $A(s) - \rho(s)$, and (4.103) and (4.104) apply for $m = 0, 1, 2, \ldots, k$.

The two subcases can be summarized as

$$A_m = \rho_m, \qquad m = 0, 1, 2, \ldots, k - 1, \qquad (4.112a)$$

$$\frac{A_k - \rho_k}{F_{k+1}} \geq 0, \qquad (4.112b)$$

in which the equality sign is attained if and only if $Z_{22}(j\omega_0) \neq 0$.

Class III. $\sigma_0 = 0$ and $0 < |z_l(s_0)| < \infty$. Then $s_0 = j\omega_0$ is a zero of multiplicity k of $r_l(s)$ or $F(s)$, k being even and $k \geq 2$. Two cases are distinguished.

Subcase 1. $Z_{22}(s_0) + z_l(s_0) = 0$. Then $j\omega_0$ is a zero of multiplicity $k - 1$ of $A(s) - \rho(s)$. Thus, from (4.103) we have

$$A_m = \rho_m, \qquad m = 0, 1, 2, \ldots, k - 2. \qquad (4.113)$$

From Class II, Subcase 1, we conclude that

$$A_{k-1} - \rho_{k-1} = \frac{F_k}{Z_{22}''(j\omega_0) + z_l''(j\omega_0)}, \qquad (4.114)$$

$Z''_{22}(j\omega_0)$ and $z''_l(j\omega_0)$ being defined in (4.107). Since $F_k \neq 0$, (4.114) becomes

$$\frac{A_{k-1} - \rho_{k-1}}{F_k} = \frac{1}{Z''_{22}(j\omega_0) + z''_l(j\omega_0)} > 0. \qquad (4.115)$$

Subcase 2. $Z_{22}(s_0) + z_l(s_0) \neq 0$. Then s_0 is a zero of multiplicity k of $A(s) - \rho(s)$. Hence, (4.113) applies for $m = 0, 1, 2, \ldots, k-1$.

The two subcases can be summarized as

$$A_m = \rho_m, \qquad m = 0, 1, 2, \ldots, k-2, k \geqq 2, \qquad (4.116a)$$

$$\frac{A_{k-1} - \rho_{k-1}}{F_k} \geqq 0, \qquad (4.116b)$$

in which the equality is attained if and only if $Z_{22}(s_0) + z_l(s_0) \neq 0$.

Class IV. $\sigma_0 = 0$ and $|z_l(s_0)| = \infty$. Then $s_0 = j\omega_0$ is a zero of multiplicity k of $A(s) - \rho(s)$, and (4.103) and (4.104) apply. Substituting (4.13) in (4.100) and invoking the known fact that $(s - j\omega_0)z_l(j\omega_0)$ is finite and nonzero as s approaches to $j\omega_0$, we can compare the coefficients of $(s - s_0)^k$ which yield

$$A_k - \rho_k = \left.\frac{F_{k-1}}{(s - j\omega_0)Z_{22}(s) + (s - j\omega_0)z_l(s)}\right|_{s \to j\omega_0}. \qquad (4.117)$$

Again, two subcases are considered.

Subcase 1. $|Z_{22}(j\omega_0)| \neq \infty$. Then (4.117) becomes

$$A_k - \rho_k = \frac{F_{k-1}}{a_{-1}(\omega_0)}, \qquad (4.118)$$

where

$$a_{-1}(\omega_0) = \lim_{s \to j\omega_0} (s - j\omega_0)z_l(s) \qquad (4.119)$$

is the residue of $z_l(s)$ evaluated at the pole $j\omega_0$.

Subcase 2. $|Z_{22}(j\omega_0)| = \infty$. Then since $A_k - \rho_k \neq 0$, (4.117) becomes

$$\frac{F_{k-1}}{A_k - \rho_k} = a_{-1}(\omega_0) + \lim_{s \to j\omega_0} (s - j\omega_0)Z_{22}(s). \qquad (4.120)$$

The limit in (4.120) is the residue of $Z_{22}(s)$ evaluated at the pole $j\omega_0$, which must be real and positive as $a_{-1}(\omega_0)$. Thus, we can write

$$\frac{F_{k-1}}{A_k - \rho_k} > a_{-1}(\omega_0). \tag{4.121}$$

The two subcases can be summarized as

$$A_m = \rho_m, \qquad m = 0, 1, 2, \ldots, k - 1, \tag{4.122a}$$

$$F_{k-1}/(A_k - \rho_k) \geqq a_{-1}(\omega_0), \tag{4.122b}$$

in which the equality is attained if and only if $|Z_{22}(s_0)|$ is finite.

6. Proof of sufficiency of the basic constraints on $\rho(s)$

In this section, we show that if the bounded-real reflection coefficient $\rho(s)$ satisfies the basic constraints stated in §3, then the impedance

$$Z_{22}(s) = \frac{2r_l(s)A(s)}{A(s) - \rho(s)} - z_l(s), \tag{4.123}$$

which is derived from (4.10), is a positive-real function. To prove this assertion, we appeal to Corollary 1.1 which states that a function $f(s)$ is positive-real if and only if $f(\sigma)$ is real and

(i) $f(s)$ is analytic in the open RHS,
(ii) $f(s)$ has at most simple poles with positive real residues on the $j\omega$-axis,
(iii) $\operatorname{Re} f(j\omega) \geqq 0$ for all ω.

To complete the proof, in the following we show that $Z_{22}(s)$ possesses the above three properties that are not presented in the listed order. We now divide our discussion into three parts, each corresponding to a property given above.

Part I. We show that $\operatorname{Re} Z_{22}(j\omega) \geqq 0$ for all ω. To this end, we consider the real part

THE PASSIVE LOAD

$$\text{Re } Z_{22}(j\omega) = \tfrac{1}{2}[Z_{22}(j\omega) + Z_{22}(-j\omega)]$$

$$= r_l(j\omega)\left[\frac{A(j\omega)}{A(j\omega) - \rho(j\omega)} + \frac{A(-j\omega)}{A(-j\omega) - \rho(-j\omega)} - 1\right]$$

$$= \frac{r_l(j\omega)[1 - |\rho(j\omega)|^2]}{|A(j\omega) - \rho(j\omega)|^2} \geqq 0, \qquad (4.124)$$

since $r_l(j\omega) \geqq 0$ and $|\rho(j\omega)| \leqq 1$. The third line follows from the fact that $A(s)A(-s) = 1$.

Part II. In this part, we show that $Z_{22}(s)$ is analytic in the open RHS. From (4.123), it suffices to show that the reciprocal of the function

$$f(s) = \frac{A(s) - \rho(s)}{2r_l(s)A(s)} \qquad (4.125)$$

is positive-real, being analytic in the open RHS. To this end, we shall show that $f(s)$ is positive-real since the reciprocal of a positive-real function is positive-real. Again, we show that $f(s)$ possesses the three properties described above. From (4.123), we have

$$\text{Re } \frac{1}{f(j\omega)} = \text{Re } Z_{22}(j\omega) + \text{Re } z_l(j\omega) \geqq 0, \qquad (4.126)$$

since from Part I, Re $Z_{22}(j\omega) \geqq 0$ and since $z_l(j\omega)$ by assumption is positive-real. This implies that (Problem 4.29)

$$\text{Re } f(j\omega) \geqq 0 \qquad (4.127)$$

for all ω.

Now we show that $f(s)$ has no poles in the open RHS. Knowing that $A(s)$ is an all-pass function and that $\rho(s)$ is bounded-real, $A(s)$ and $\rho(s)$ are analytic in the closed RHS. Thus, the open RHS poles of $f(s)$ can only be the open RHS zeros of $F(s) = 2r_l(s)A(s)$. Since the zeros of $A(s)$ are also the open RHS poles of $r_l(s)$, multiplicity included, the open RHS zeros of $F(s)$ and those of $r_l(s)/z_l(s)$ are identical, multiplicity included. Hence, the Class I zeros of transmission are precisely the open RHS zeros of $F(s)$. At such a zero, from hypothesis, we know that $A(s) - \rho(s)$ vanishes to at least the same order. Thus, $f(s)$ is analytic in the open RHS.

To complete our proof that $f(s)$ is positive-real, we must show that $f(s)$ has at most simple poles with real positive residues on the $j\omega$-axis. This is a bit complicated, and we shall consider three cases. Let $s_0 = j\omega_0$ be any $j\omega$-axis zero of transmission of order k of $z_l(s)$. Then the $j\omega$-axis poles of $f(s)$ can only be the $j\omega$-axis zeros of $F(s)$, and more precisely the $j\omega$-axis zeros of $r_l(s)$ or part of the $j\omega$-axis zeros of transmission of $z_l(s)$.

Case 1. $z_l(j\omega_0) = 0$. Then $j\omega_0$ is a zero of multiplicity $k+1$ of $F(s)$ since there is a single cancellation in $r_l(s)/z_l(s)$. From hypothesis, $j\omega_0$ is a zero of $A(s) - \rho(s)$ of multiplicity at least k, since it is a Class II zero of transmission. Therefore, such a zero is at most a simple pole of $f(s)$. Clearly, the residue at such a pole is

$$\frac{A_k - \rho_k}{F_{k+1}} \geqq 0, \tag{4.128}$$

which follows from the constraints on the Class II zeros of transmission.

Case 2. $0 < |z_l(j\omega_0)| < \infty$. Then the $j\omega$-axis zeros of $F(s)$ and the Class III zeros of transmission are identical, multiplicity included. Since by hypothesis at such a zero, $A(s) - \rho(s)$ vanishes to at least the order of $k - 1$, $j\omega_0$ is at most a simple pole of $f(s)$, whose residue is given by

$$(A_{k-1} - \rho_{k-1})/F_k \geqq 0, \tag{4.129}$$

which follows from hypothesis.

Case 3. $|z_l(j\omega_0)| = \infty$. Then $j\omega_0$ is a zero of multiplicity $k - 1$ of $F(s)$, and by hypothesis, $j\omega_0$ is a zero of $A(s) - \rho(s)$ of multiplicity at least k. Thus, $f(j\omega_0) = 0$.

We conclude from the above discussion that $f(s)$ and hence its reciprocal are positive-real. Therefore, $Z_{22}(s)$ is analytic in the open RHS.

Part III. In this part, we show that $Z_{22}(s)$ has at most simple $j\omega$-axis poles with real and positive residues at these poles.

From (4.123) and (4.125), we recognize that the $j\omega$-axis poles of $Z_{22}(s)$ are those or part of those of $1/f(s)$ and $z_l(s)$, which they themselves are positive-real. It follows that such poles, if they exist, must be simple. Let $j\omega_0$ be a pole of $Z_{22}(s)$. If $|z_l(j\omega_0)| \neq \infty$, the residue of $Z_{22}(s)$ evaluated at $j\omega_0$ is the same as that of $1/f(s)$ at $j\omega_0$, which must be real and positive. If $|z_l(j\omega_0)| = \infty$, the residue of $1/f(s)$ evaluated at the pole $j\omega_0$ is given by

$$F_{k-1}/(A_k - \rho_k) > 0, \tag{4.130}$$

since from Part II, Case 3, $j\omega_0$ is a simple zero of $f(s)$; and the corresponding residue of $Z_{22}(s)$ evaluated at this pole is therefore

$$\frac{F_{k-1}}{A_k - \rho_k} - a_{-1}(\omega_0) \geq 0, \tag{4.131}$$

which is obtained from (4.123). The inequality follows directly from hypothesis since $j\omega_0$ is also a Class IV zero of transmission. This completes the proof of sufficiency of the basic constraints on $\rho(s)$.

7. Design procedure for the equalizers

From the discussions presented in the preceding sections, we are now in a position to outline a simple procedure to design an optimum lossless equalizer N, as depicted in Fig. 4.1, to match out the load impedance $z_l(s)$ to a resistive generator with resistance r_g and to achieve a preassigned transducer power-gain characteristic $G(\omega^2)$ over the entire sinusoidal frequency spectrum. Without loss of generality, we assume that $r_g = 1$. The procedure is stated in eight steps, as follows:

Step 1. From a preassigned transducer power gain $G(\omega^2)$, verify that $G(\omega^2)$ is an even rational real function and satisfies the inequality

$$0 \leq G(\omega^2) \leq 1 \tag{4.132}$$

for all ω. The gain level is usually not specified to allow desired flexibility.

Step 2. From a prescribed passive load impedance $z_l(s)$, which has a positive real part over a frequency band of interest, compute

$$r_l(s) = \tfrac{1}{2}[z_l(s) + z_l(-s)], \qquad (4.133a)$$

$$A(s) = \prod_{i=1}^{\nu} \frac{s - s_i}{s + s_i}, \quad \text{Re } s_i > 0, \qquad (4.133b)$$

where s_i $(i = 1, 2, \ldots, \nu)$ are the open RHS poles of $z_l(-s)$, and

$$F(s) = 2r_l(s)A(s). \qquad (4.133c)$$

Step 3. Determine the locations and the orders of the zeros of transmission of $z_l(s)$, which are the closed RHS zeros of the function

$$w(s) = \frac{r_l(s)}{z_l(s)}, \qquad (4.134)$$

and divide them into respective classes according to Definition 4.2.

Step 4. Perform the unique factorization of the function

$$\hat{\rho}(s)\hat{\rho}(-s) = 1 - G(-s^2), \qquad (4.135)$$

in which the numerator of $\hat{\rho}(s)$ is a Hurwitz polynomial and the denominator of $\hat{\rho}(s)$ is a strictly Hurwitz polynomial. In other words, $\hat{\rho}(s)$ is a minimum-phase reflection coefficient.

Step 5. Obtain the Laurent series expansions of the functions $A(s)$, $F(s)$ and $\hat{\rho}(s)$ at each of the zeros of transmission s_0 of $z_l(s)$:

$$A(s) = \sum_{m=0}^{\infty} A_m (s - s_0)^m, \qquad (4.136a)$$

$$F(s) = \sum_{m=0}^{\infty} F_m (s - s_0)^m, \qquad (4.136b)$$

$$\hat{\rho}(s) = \sum_{m=0}^{\infty} \rho_m (s - s_0)^m. \qquad (4.136c)$$

The above expansions may be obtained by any available techniques.

Step 6. According to the classes of zeros of transmission, list the basic constraints imposed on the coefficients of (4.136). The gain level is determined from these constraints. If the constraints cannot all be satisfied, we consider the general bounded-real reflection coefficient (see Problems 4.51–4.53)

$$\rho(s) = \pm \eta(s)\hat{\rho}(s), \qquad (4.137)$$

$\eta(s)$ being an arbitrary real all-pass function. Then repeat Step 5 for $\rho(s)$. Of course, we should start with lower-order $\eta(s)$. If the constraints still cannot all be satisfied, then we must modify the preassigned transducer power-gain characteristic $G(\omega^2)$.

Step 7. Having successfully carried out Step 6, the equalizer back-end driving-point impedance is determined by

$$Z_{22}(s) = \frac{F(s)}{A(s) - \rho(s)} - z_l(s), \qquad (4.138)$$

where $\rho(s)$ may be $\hat{\rho}(s)$. $Z_{22}(s)$ is known to be positive-real.

Step 8. Using Darlington's procedure if necessary, realize the function $Z_{22}(s)$ as the driving-point impedance of a lossless two-port network terminated in a 1-Ω resistor. An ideal transformer may be required at the input port to compensate for the actual level of the generator resistance r_g. This completes the design of an equalizer.

We remark that it is sometimes convenient to use the magnitude and frequency scalings to simplify the numerical computation, as we did in Examples 4.1 and 4.2.

To illustrate the above procedure, we first work out a detailed numerical example.

EXAMPLE 4.6. Design a lossless two-port network to equalize the load as shown in Fig. 4.16 to a resistive generator of internal resistance of 100 Ω and to achieve the fourth-order Butterworth transducer power gain with a maximal dc gain. The cutoff frequency is $50/\pi$ MHz. Also compute the transducer power gain of the realized network.

To simplify our computation, the network of Fig. 4.16 is first magnitude-scaled by a factor of 10^{-2} and frequency-scaled by a factor of 10^{-8}. Thus, s denotes the normalized complex frequency

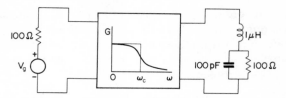

FIG. 4.16. The symbolic representation of an equalizer that achieves the fourth-order Butterworth transducer power-gain characteristic.

and ω the normalized real frequency. We now follow the eight steps outlined above to obtain an equalizer with desired specifications.

Step 1. From (4.16), we have

$$G(\omega^2) = \frac{K_4}{1+\omega^8}, \qquad 0 \leq K_4 \leq 1, \tag{4.139}$$

one of our objectives being to optimize the dc gain K_4.

Step 2. From (4.133), we compute

$$z_l(s) = \frac{s^2+s+1}{s+1}, \tag{4.140a}$$

$$r_l(s) = \frac{1}{1-s^2}, \tag{4.140b}$$

$$A(s) = \frac{s-1}{s+1}, \tag{4.140c}$$

$$F(s) = 2r_l(s)A(s) = \frac{-2}{(1+s)^2}. \tag{4.140d}$$

Step 3. The zero of transmission is defined by the closed RHS zero of the function

$$w(s) = \frac{r_l(s)}{z_l(s)} = \frac{1}{(s^2+s+1)(1-s)}, \tag{4.141}$$

indicating that $s = \infty$ is a Class IV zero of transmission of third order.

Step 4. Substituting (4.139) in (4.135) and appealing to the theory

THE PASSIVE LOAD

of analytic continuation give

$$\rho(s)\rho(-s) = (1 - K_4)\frac{1+x^8}{1+s^8}, \qquad (4.142a)$$

where

$$x = s/(1-K_4)^{1/8}. \qquad (4.142b)$$

Applying (3.13) or from the tables given in the Appendix A, the minimum-phase factorization of (4.142a) can be written as

$$\hat{\rho}(s) = (1-K_4)^{1/2}\frac{x^4 + 2.6131x^3 + 3.4142x^2 + 2.6131x + 1}{s^4 + 2.6131s^3 + 3.4142s^2 + 2.6131s + 1}. \qquad (4.143)$$

The constant K_4 will be determined from the coefficient constraints to be described below.

Step 5. Using any one of the procedures described below (4.23), the Laurent series expansions of $A(s)$, $F(s)$ and $\hat{\rho}(s)$ at the zero of transmission, which is at infinity, are given by

$$A(s) = 1 - \frac{2}{s} + \frac{2}{s^2} - \frac{2}{s^3} + \cdots, \qquad (4.144a)$$

$$F(s) = 0 + 0 - \frac{2}{s^2} + \frac{4}{s^3} + \cdots, \qquad (4.144b)$$

$$\hat{\rho}(s) = 1 + \frac{\rho_1}{s} + \frac{\rho_2}{s^2} + \frac{\rho_3}{s^3} + \cdots, \qquad (4.144c)$$

where

$$\rho_1 = 2.6131(\delta - 1), \qquad (4.145a)$$

$$\rho_2 = 3.4142(\delta^2 - 2\delta + 1), \qquad (4.145b)$$

$$\rho_3 = 2.6131(\delta^3 - 3.4142\delta^2 + 3.4142\delta - 1), \qquad (4.145c)$$

$$\delta = (1 - K_4)^{1/8}. \qquad (4.145d)$$

Step 6. For a Class IV zero of transmission of order 3, the coefficient constraints become

$$A_m = \rho_m, \qquad m = 0, 1, 2, \qquad (4.146a)$$

$$\frac{F_2}{A_3 - \rho_3} \geqq a_{-1}(\infty) = 1. \qquad (4.146b)$$

Substituting the coefficients of the Laurent series (4.144) in (4.146) yields the constraints imposed on K_4:

$$A_0 = \rho_0 = 1, \tag{4.147a}$$

$$A_1 = -2 = \rho_1 = 2.6131(\delta - 1), \tag{4.147b}$$

$$A_2 = 2 = \rho_2 = 3.4142(\delta - 1)^2, \tag{4.147c}$$

$$\frac{-2}{-2 - 2.6131(\delta - 1)(\delta^2 - 2.4142\delta + 1)} \geq 1, \tag{4.147d}$$

giving

$$\delta = 0.23463, \tag{4.148}$$

and from (4.145d), the optimum dc gain is

$$K_4 = 1 - \delta^8 = 0.99999. \tag{4.149}$$

Thus, the constraints are satisfied without inserting any open RHS zeros in $\rho(s)$, as indicated in (4.137).

Step 7. The equalizer back-end impedance is determined by

$$\begin{aligned}Z_{22}(s) &= \frac{F(s)}{A(s) - \hat{\rho}(s)} - z_l(s) \\ &= \frac{0.98s^4 + 2.556s^3 + 3.158s^2 + 2.576s + 1}{(s + 1)(1.02s^2 + 1.65s + 1)} \\ &= \frac{0.98s^3 + 1.576s^2 + 1.582s + 1}{1.02s^2 + 1.65s + 1},\end{aligned} \tag{4.150}$$

which is positive-real, as required.

Step 8. Expanding $Z_{22}(s)$ in a continued fraction expansion gives

$$Z_{22}(s) = 0.96s + \cfrac{1}{1.64s + \cfrac{1}{0.622s + \cfrac{1}{1}}}, \tag{4.151}$$

which is identified as an *LC* ladder terminated in a 1-Ω resistor as

THE PASSIVE LOAD

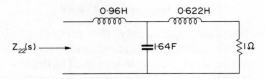

Fig. 4.17. A ladder realization of the impedance (4.150).

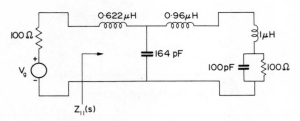

Fig. 4.18. An equalizer together with its terminations possessing the fourth-order Butterworth transducer power-gain characteristic.

shown in Fig. 4.17. Denormalizing the element values yields a final realization as indicated in Fig. 4.18.

To compute the transducer power gain from the realized network, we again consider the normalized network, and compute the input impedance when the output port is terminated in the load $z_l(s)$ as depicted in Fig. 4.18, giving

$$Z_{11}(s) = \frac{2s^4 + 2s^3 + 3.6s^2 + 2.58s + 1}{3.21s^3 + 3.21s^2 + 2.64s + 1}. \quad (4.152)$$

Using this, the transducer power gain can be computed, and is given by

$$G(\omega^2) = \frac{4 \operatorname{Re} Z_{11}(j\omega)}{|1 + Z_{11}(j\omega)|^2}$$
$$= \frac{1}{1 + \omega^8}, \quad (4.153)$$

confirming our design.

8. Darlington type-C load

In this section, we shall apply the preceding procedure to determine the gain-bandwidth limitations and discuss the design of a lossless matching network that equalizes the Darlington type-C load, as shown in Fig. 4.19, to a resistive generator and that achieves the

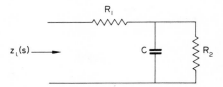

FIG. 4.19. The Darlington type-C load.

nth-order Butterworth, Chebyshev, or elliptic low-pass transducer power-gain characteristic. For illustrative purposes, we follow the steps outlined in the foregoing and present each response in a separate section. We begin our discussion by considering the Butterworth response.

8.1. Butterworth transducer power-gain characteristic

We follow the eight steps outlined in the preceding section.

Step 1. From (4.16) and (4.132), it is necessary that

$$0 \leq K_n \leq 1. \tag{4.154}$$

Step 2. According to (4.133), we compute

$$r_l(s) = \frac{(R_1 + R_2) - R_1 R_2^2 C^2 s^2}{1 - R_2^2 C^2 s^2}, \tag{4.155a}$$

$$A(s) = \frac{R_2 C s - 1}{R_2 C s + 1}, \tag{4.155b}$$

$$F(s) = \frac{2 R_1 R_2^2 C^2 s^2 - 2(R_1 + R_2)}{(1 + R_2 C s)^2}. \tag{4.155c}$$

THE PASSIVE LOAD

Step 3. Compute $w(s)$ from (4.134), giving

$$w(s) = \frac{R_2Cs^2 - (R_1 + R_2)/R_1R_2C}{(R_2Cs - 1)[s + (R_1 + R_2)/R_1R_2C]}, \quad (4.156)$$

which indicates that $z_l(s)$ has a Class I zero of transmission of order 1 at

$$s_0 = \sigma_0 = \frac{1}{R_2C}(1 + R_2/R_1)^{1/2}. \quad (4.157)$$

Step 4. From (4.17)–(4.20), we have the unique minimum-phase factorization

$$\hat{\rho}(s) = (1 - K_n)^{1/2} \frac{q(x)}{q(y)}, \quad (4.158)$$

where x and y are defined in (4.17b) and (4.17c).

Step 5. Since $z_l(s)$ has only a real Class I zero of transmission of order 1, no series expansions are needed in Step 6, and they are omitted here.

Step 6. For $k = 1$, the constraint for the coefficients for a Class I zero of transmission $s_0 = \sigma_0$ is simply $A_0 = \rho_0$, which is equivalent to $A(\sigma_0) = \hat{\rho}(\sigma_0)$ or more generally

$$A(\sigma_0) = \pm \eta(\sigma_0)\hat{\rho}(\sigma_0). \quad (4.159)$$

Substituting (4.155b) and (4.158) in (4.159) yields the basic limitation on the gain-bandwidth as (Problem 4.36)

$$\pm(1 - K_n)^{1/2}\eta(\sigma_0)q(x_0) = A(\sigma_0)q(y_0), \quad (4.160a)$$

where

$$x_0 = (1 - K_n)^{-1/2n}y_0, \quad (4.160b)$$

$$y_0 = \sigma_0/\omega_c. \quad (4.160c)$$

From (4.18), we show that

$$q(x_0) = \sum_{i=0}^{n} a_i x_0^i = x_0^n \sum_{i=0}^{n} a_i x_0^{i-n} = x_0^n \sum_{u=0}^{n} a_{n-u} x_0^{-u}$$

$$= x_0^n \sum_{u=0}^{n} a_u x_0^{-u} = x_0^n q(x_0^{-1}), \quad (4.161)$$

in which we have used the fact that the coefficients of $q(x)$ equi-distance from the ends are equal, as previously proved in (3.20) of Chapter 3. Substituting (4.161) in (4.160) results in the basic limitation on the gain-bandwidth imposed by the load $z_l(s)$:

$$\pm \eta(\sigma_0) q(\xi/y_0) = A(\sigma_0) q(1/y_0), \quad (4.162a)$$

where

$$\xi = (1 - K_n)^{1/2n}. \quad (4.162b)$$

Since $q(\sigma)$ increases monotonically for real $\sigma \geqq 0$, and $q(0) = 1$ for all n, we have the inequality

$$1 \leqq q(\xi/y_0) \leqq q(1/y_0), \quad y_0 \geqq 0, \quad (4.163)$$

since $0 \leqq \xi \leqq 1$. To facilitate our discussion, two cases are considered.

Case 1. $A(\sigma_0) q(1/y_0) < 1$. The condition (4.162a) cannot be satisfied if $\eta(\sigma_0) = 1$, since $q(\xi/y_0) \geqq 1$. Thus, we must consider nontrivial $\eta(s)$, and for simplicity let

$$\pm \eta(s) = \frac{s - \sigma_1}{s + \sigma_1} \quad (4.164)$$

be a first-order all-pass function with σ_1 to be determined. In order to achieve the maximum attainable dc gain, let $\xi = 0$, which corresponds to $K_n = 1$. Substituting these in (4.162a) yields

$$\sigma_1 = \sigma_0 \frac{1 - A(\sigma_0) q(\omega_c/\sigma_0)}{1 + A(\sigma_0) q(\omega_c/\sigma_0)} > 0, \quad (4.165)$$

which indicates that such σ_1 always exists. Therefore, it is never necessary to use higher-order all-pass functions $\eta(s)$; the first-order is sufficient.

Case 2. $A(\sigma_0) q(1/y_0) \geqq 1$. Since $|\eta(\sigma_0)|$ is less than unity, from (4.162) we conclude that the maximum dc gain is obtained if we choose $\pm \eta(s) = 1$. Under this condition, (4.162a) becomes

$$q(\xi_m/y_0) = A(\sigma_0) q(1/y_0), \quad (4.166)$$

THE PASSIVE LOAD

and the corresponding dc gain is given by

$$K_n = 1 - \xi_m^{2n}. \qquad (4.167)$$

Since $A(\sigma_0) < 1$, (4.166) implies that $q(1/y_0) > q(\xi_m/y_0)$, which after invoking the monotonic character of $q(\sigma)$ for real nonnegative σ, shows that

$$1/y_0 > \xi_m/y_0 \qquad (4.168)$$

or $\xi_m < 1$. Thus, the solution for K_n in (4.167) is always physical.

Step 7. Compute $Z_{22}(s)$ from (4.138).

Step 8. Realize $Z_{22}(s)$ as the driving-point impedance of a lossless two-port network terminated in a 1-Ω resistor.

We shall illustrate the above procedure by the following numerical examples.

EXAMPLE 4.7. Let

$$R_1 = 100 \ \Omega,$$
$$R_2 = 300 \ \Omega,$$
$$C = 200/3 \ \text{pF},$$
$$\omega_c = 10^8 \ \text{rad/s},$$
$$n = 2.$$

Then, we have

$$\sigma_0 = \frac{1}{R_2 C}(1 + R_2/R_1)^{1/2} = 10^8, \qquad (4.169a)$$

$$y_0 = \sigma_0/\omega_c = 1, \qquad (4.169b)$$

$$A(\sigma_0) = \frac{R_2 C \sigma_0 - 1}{R_2 C \sigma_0 + 1} = \frac{1}{3}. \qquad (4.169c)$$

For $n = 2$,

$$q(s) = 1 + 1.414s + s^2. \qquad (4.170)$$

We now compute

$$A(\sigma_0)q(1/y_0) = 1.138 > 1. \qquad (4.171)$$

Thus, Case 2 applies, and the constraint (4.166) becomes

$$\xi_m^2 + 1.414\xi_m - 0.138 = 0, \qquad (4.172)$$

which yields $\xi_m = -1.506$ and 0.092. Choosing $\xi_m = 0.092$, we obtain the dc gain

$$K_2 = 1 - \xi_m^4 = 0.999928. \qquad (4.173)$$

From (4.158), we can now compute the bounded-real reflection coefficient

$$\hat{\rho}(s) = \frac{(1-K_2)^{1/2} + \sqrt{2}(1-K_2)^{1/4}y + y^2}{1 + 1.414y + y^2}$$

$$= \frac{y^2 + 0.13y + 0.0085}{y^2 + 1.414y + 1}. \qquad (4.174)$$

Finally, from (4.155) and (4.174), the equalizer back-end impedance is obtained by (4.138) in terms of the normalized complex frequency $y = s/10^8$:

$$Z_{22}(y) = \frac{F(y)}{A(y) - \hat{\rho}(y)} - z_l(y)$$

$$= \frac{800 \dfrac{y^2 - 1}{(1+2y)^2}}{\dfrac{2y-1}{2y+1} - \dfrac{y^2 + 0.13y + 0.0085}{y^2 + 1.414y + 1}} - 200 \frac{y+2}{2y+1}$$

$$= 704 \frac{y^4 + 1.27y^3 - 0.39y^2 - 1.38y - 0.5}{y^3 + 1.29y^2 - 1.4y - 0.89}$$

$$= 704 \frac{y^2 + 1.77y + 0.992}{y + 1.77}, \qquad (4.175)$$

which can then be expanded in a continued fraction:

$$Z_{22}(s) = 7.04 \times 10^{-6}s + \cfrac{1}{14.3 \times 10^{-12}s + \cfrac{1}{395}}. \qquad (4.176)$$

THE PASSIVE LOAD

Equation (4.176) represents an LC ladder terminated in a resistor, as indicated in Fig. 4.20. The equalizer together with the resistive generator and the load is presented in Fig. 4.21, in which the ideal transformer of turns ratio

$$\left(\frac{395}{1}\right)^{1/2} = 19.87$$

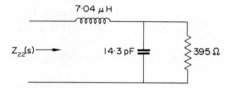

FIG. 4.20. A ladder realization of the impedance (4.175).

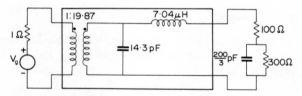

FIG. 4.21. An equalizer together with its terminations possessing the second-order Butterworth transducer power-gain characteristic.

performs the usual function of changing the resistance level at the generator end, which is specified as $1\,\Omega$.

As a check, we compute the transducer power gain from the realized network of Fig. 4.21. The driving-point impedance facing the resistive generator when the output port is terminated in its load is first computed, again in terms of the normalized complex frequency y, and is given by

$$Z_{11}(y) = 1.771 \frac{y^2 + 0.642y + 0.284}{y^3 + 0.642y^2 + 1.278y + 0.497}. \tag{4.177}$$

Let

$$\hat{\omega} = \frac{\omega}{10^8}$$

be the normalized radian frequency. Then substituting $y = j\hat{\omega}$ in (4.177) and computing its real part give

$$\text{Re } Z_{11}(j\hat{\omega}) = \frac{0.25\hat{\omega}^2 + 0.25}{\hat{\omega}^6 - 2.144\hat{\omega}^4 + 0.995\hat{\omega}^2 + 0.247}. \quad (4.178)$$

From this, we compute

$$|1 + Z_{11}(j\hat{\omega})|^2 = \frac{\hat{\omega}^6 + \hat{\omega}^4 + \hat{\omega}^2 + 1}{\hat{\omega}^6 - 2.144\hat{\omega}^4 + 0.995\hat{\omega}^2 + 0.247}. \quad (4.179)$$

Substituting (4.178) and (4.179) in the first equation of (4.153) yields

$$G(\hat{\omega}^2) = \frac{\hat{\omega}^2 + 1}{\hat{\omega}^6 + \hat{\omega}^4 + \hat{\omega}^2 + 1} = \frac{1}{1 + \hat{\omega}^4}, \quad (4.180)$$

confirming our design.

EXAMPLE 4.8. In the above example, let ω_c be decreased from 10^8 rad/s to $\frac{1}{2} \times 10^8$ rad/s, everything else being the same. Then $\sigma_0 = 10^8$, $A(\sigma_0) = \frac{1}{3}$ and

$$y_0 = \sigma_0/\omega_c = 2, \quad (4.181a)$$

$$A(\sigma_0)q(1/y_0) = 1.957/3 = 0.652 < 1. \quad (4.181b)$$

Thus, Case 1 applies, indicating that $K_2 = 1$ is attainable by the insertion of an open RHS zero in $\rho(s)$, which according to (4.165) is located at

$$\sigma_1 = 10^8 \frac{1 - 0.652}{1 + 0.652} = 0.211 \times 10^8. \quad (4.182)$$

Using (4.137) in conjunction with (4.158), the bounded-real reflection coefficient becomes

$$\rho(y) = \frac{(y - 0.422)y^2}{(y + 0.422)(y^2 + 1.414y + 1)}. \quad (4.183)$$

Finally, by (4.138) the equalizer back-end impedance is obtained in terms of the normalized complex frequency $y = 2s/10^8$:

$$\begin{aligned}
Z_{22}(y) &= \frac{F(y)}{A(y) - \rho(y)} - z_l(y) \\
&= \frac{200 \dfrac{y^2 - 4}{(1+y)^2}}{\dfrac{y-1}{y+1} - \dfrac{(y-0.422)y^2}{(y+0.422)(y^2+1.414y+1)}} - 100 \frac{y+4}{y+1} \\
&= 775.19 \frac{(y^2-4)(y^3+1.836y^2+1.597y+0.422)}{(y+1)(y^3+0.708y^2-4.554y-1.6356)} - 100 \frac{y+4}{y+1} \\
&= 775.19 \frac{(y+2)(y^3+1.836y^2+1.597y+0.422)}{(y+1)(y^2+2.708y+0.818)} - 100 \frac{y+4}{y+1} \\
&= 775.19 \frac{y^4+3.708y^3+4.404y^2+2.113y+0.422}{(y+1)(y^2+2.708y+0.818)} \\
&= 775.19 \frac{y^3+2.708y^2+1.697y+0.422}{y^2+2.708y+0.818},
\end{aligned} \qquad (4.184)$$

whose realization requires an ideal transformer, as expected, since we have introduced an open RHS zero in $\rho(y)$.

8.2. Chebyshev transducer power-gain characteristic

Consider the same problem discussed in the preceding section except that we wish to achieve the nth-order low-pass Chebyshev transducer power gain, as given in (4.46). Now we follow the eight steps outlined earlier.

Steps 1–3 are the same as in §8.1, and are omitted here.

Step 4. The minimum-phase factorization of (4.48a) is given in (4.51), and we write

$$\hat{\rho}(y) = \frac{\hat{p}(y)}{p(y)}, \qquad (4.185)$$

where $\hat{p}(y)$ and $p(y)$ denote the numerator and denominator polynomials of (4.51), respectively.

Step 5. Like the Butterworth case, no series expansions are required in Step 6, and thus are omitted here.

Step 6. The basic constraint can easily be deduced from (4.159) and is given by

$$\pm \eta(\sigma_0)\hat{p}(y_0) = A(\sigma_0)p(y_0), \quad (4.186a)$$

where

$$y_0 = \sigma_0/\omega_c. \quad (4.186b)$$

We remark that since the coefficients of $\hat{p}(y)$ have not yet been determined at this point, depending on K_n through (4.48b), $\hat{p}(y_0)$ is a function of K_n. For $K_n = 0$, $\hat{p}(y_0) = p(y_0)$ and (4.186a) is satisfied with $\pm \eta(\sigma_0) = A(\sigma_0)$, which shows that (4.186a) possesses a solution in the range $0 \leq K_n \leq 1$. Our objective is to maximize K_n in the range $0 \leq K_n \leq 1$ so that it satisfies (4.186a). For this reason, we first determine the condition under which $K_n = 1$ can be achieved. To this end, let $K_n = 1$. Then the numerator polynomial of (4.48a) becomes $\epsilon^2 C_n^2(-jy)$, and the corresponding Hurwitz factorization $\hat{p}(y)$, denoted by $\hat{p}_m(y)$, is given by (Problem 4.32)

$$\hat{p}_m(y) = j^n 2^{1-n} C_n(-jy), \quad (4.187)$$

which is a real polynomial. For example, from (3.51) we have

$$\hat{p}_m(y) = y^3 + 0.75y \quad \text{for } n = 3,$$
$$\hat{p}_m(y) = y^4 + y^2 + 0.125 \quad \text{for } n = 4.$$

For an adequate discussion of (4.186a), two cases are considered.

Case 1. $A(\sigma_0)p(y_0) < \hat{p}_m(y_0)$. Then from (4.186a) we conclude that $K_n = 1$ is not attainable without the insertion of open RHS zeros in $\rho(y)$. Fortunately, it is never necessary to insert anything but a single positive-real-axis zero. For if we let

$$\pm \eta(s) = \frac{s - \sigma_1}{s + \sigma_1}, \quad (4.188)$$

where σ_1 is to be determined from the constraint (4.186a), we obtain

$$\sigma_1 = \sigma_0 \frac{\hat{p}_m(y_0) - A(\sigma_0)p(y_0)}{\hat{p}_m(y_0) + A(\sigma_0)p(y_0)} > 0. \quad (4.189)$$

THE PASSIVE LOAD   283

Case 2. $A(\sigma_0)p(y_0) \geqq \hat{p}_m(y_0)$. From (4.186a), we see that $\eta(\sigma_0) <$ 1 increases $\hat{p}(y_0)$ over its value for $\pm \eta(\sigma_0) = 1$, and consequently $\hat{a}$ of (4.53b) increases (Problem 4.33) or equivalently $\hat{e}$ decreases. According to (4.48b) this decrease in $\hat{e}$ decreases K_n. Thus, the best we can do is to have $\pm \eta(s) = 1$. Under this condition, (4.186a) becomes

$$\hat{p}(y_0) = A(\sigma_0)p(y_0), \qquad (4.190)$$

where $\hat{p}(y_0)$ is a polynomial in sinh $\hat{a}$ (Problem 4.35). This equation can then be solved to yield a real positive sinh $\hat{a}$. That there exists such a solution will now be demonstrated.

For $K_n = 0$, $\hat{p}(y_0) = p(y_0)$ and we have $\hat{p}(y_0) > A(\sigma_0)p(y_0)$. For $K_n = 1$, $\hat{p}(y_0) = \hat{p}_m(y_0)$ and we have $\hat{p}(y_0) \leqq A(\sigma_0)p(y_0)$. A plot of $\hat{p}(y_0)$ versus K_n shows that, for a given n and a prescribed ϵ, $\hat{p}(y_0)$ decreases monotonically from $p(y_0)$ to $\hat{p}_m(y_0)$ as K_n is increased from 0 to 1. On the same coordinates, let us plot the constant line $A(\sigma_0)p(y_0)$, which lies between the lines $p(y_0)$ and $\hat{p}_m(y_0)$. The intersection of the curve $\hat{p}(y_0)$ and the constant line $A(\sigma_0)p(y_0)$ yields the desired value for K_n. In other words, the solution of (4.190) is always physical, and gives the maximum attainable K_n.

Steps 7 and 8 are the same as before, and are omitted here.

EXAMPLE 4.9. Design a lossless matching network that equalizes the Darlington type-C load with $R_1 = 100\,\Omega$, $R_2 = 300\,\Omega$ and $C = 200/3$ pF to a generator of internal resistance of $1\,\Omega$, and that achieves the second-order Chebyshev transducer power gain whose peak-to-peak ripple in the passband must not exceed 1 dB. The cutoff frequency of the equalizer is $50/\pi$ MHz.

From Example 4.2, the passband tolerance of 1 dB corresponds to a ripple factor $\epsilon = 0.50885$. The Hurwitz polynomial formed by the open LHS zeros of $1 + \epsilon^2 C_2^2(-js) = 0$ can be obtained from (3.79) and is given by

$$p(s) = s^2 + 1.098s + 1.102. \qquad (4.191)$$

For $n = 2$, (4.187) becomes

$$\hat{p}_m(y) = y^2 + 0.5, \qquad (4.192)$$

where $y = s/10^8$. Then we have

$$A(\sigma_0)p(y_0) = (1/3) \times 3.2 = 1.067 < 1.5 = \hat{p}_m(y_0), \quad (4.193)$$

where, as before, $\sigma_0 = 10^8$ and $y_0 = 1$. Thus, Case 1 applies, which implies that $K_2 = 1$ cannot be achieved without the insertion of the open RHS zeros in $\rho(s)$. From (4.189), the desired zero is located at

$$\sigma_1 = 10^8 \frac{1.5 - 1.067}{1.5 + 1.067} = 0.169 \times 10^8. \quad (4.194)$$

We next compute the bounded-real reflection coefficient, which yields

$$\rho(y) = \eta(y)\frac{\hat{p}_m(y)}{p(y)} = \frac{(y - 0.169)(y^2 + 0.5)}{(y + 0.169)(y^2 + 1.098y + 1.102)}. \quad (4.195)$$

Finally, the equalizer back-end driving-point impedance is computed by the formula

$$\begin{aligned}
Z_{22}(y) &= \frac{F(y)}{A(y) - \rho(y)} - z_l(y) \\
&= \frac{200\dfrac{y^2 - 1}{(y + 0.5)^2}}{\dfrac{y - 0.5}{y + 0.5} - \dfrac{(y - 0.169)(y^2 + 0.5)}{(y + 0.169)(y^2 + 1.098y + 1.102)}} - 100\frac{y + 2}{y + 0.5} \\
&= \frac{100(2y^4 + 4.098y^3 + 3.562y^2 + 1.545y + 0.27)}{(y + 0.5)(0.436y^2 + 0.676y + 0.051)} \\
&= \frac{100(2y^3 + 3.098y^2 + 2.012y + 0.54)}{0.436y^2 + 0.676y + 0.051} \\
&= 458.7y + \frac{1}{0.00245y + 0.00306\dfrac{y + 0.0938}{y + 0.3040}}, \quad (4.196)
\end{aligned}$$

whose realization requires an ideal transformer. This is the price we paid in attaining additional flexibility by introducing an open RHS zero in $\rho(s)$.

EXAMPLE 4.10. Consider the same problem as in Example 4.9 except that now we decrease the resistance R_1 from 100 Ω to 5 Ω, everything else being the same. Then from (4.157), (4.155b) and (4.186b), we have $\sigma_0 = 3.905 \times 10^8$, $A(\sigma_0) = 0.773$, $y_0 = 3.905$ and

$$A(\sigma_0)p(y_0) = 0.773 \times 20.639 = 15.954 > 15.749 = \hat{p}_m(y_0). \quad (4.197)$$

Thus, Case 2 applies, and $K_2 = 1$ cannot be achieved. Using (4.190), we can compute the maximum obtainable K_2 by the equation

$$\hat{p}(y_0) = 15.954. \quad (4.198)$$

The left-hand side polynomial can be obtained from (3.79), and is given by the relation

$$\begin{aligned}\hat{p}(s) &= (s + \sinh \hat{a} \sin \pi/4 + j \cosh \hat{a} \cos \pi/4) \\ &\quad \times (s + \sinh \hat{a} \sin 3\pi/4 + j \cosh \hat{a} \cos 3\pi/4) \\ &= s^2 + 1.414 s \sinh \hat{a} + \sinh^2 \hat{a} + 0.5, \quad (4.199)\end{aligned}$$

and (4.198) becomes a polynomial in $\sinh \hat{a}$:

$$\sinh^2 \hat{a} + 5.522 \sinh \hat{a} - 0.205 = 0, \quad (4.200a)$$

which yields

$$\sinh \hat{a} = 0.0367 \text{ and } -5.559. \quad (4.200b)$$

Thus, let $\sinh \hat{a} = 0.0367$ or $\hat{a} = 0.0367$, and from (4.53b), we have

$$\hat{\epsilon} = 1/(\sinh 2\hat{a}) = 1/0.0734. \quad (4.201)$$

Finally, from (4.48b), we can compute the maximum obtainable K_2, which is

$$K_2 = 1 - (\epsilon/\hat{\epsilon})^2 = 1 - (0.5088 \times 0.0734)^2 = 0.9986. \quad (4.202)$$

From (4.191) and (4.199), the bounded-real reflection coefficient is obtained as

$$\hat{\rho}(y) = \frac{\hat{p}(y)}{p(y)} = \frac{y^2 + 0.052y + 0.501}{y^2 + 1.098y + 1.102}. \quad (4.203)$$

Using (4.138) in conjunction with (4.155), the equalizer back-end driving-point impedance can now be computed, and is given by

$$Z_{22}(y) = \frac{F(y)}{A(y) - \hat{\rho}(y)} - z_l(y)$$

$$= \frac{\dfrac{40y^2 - 610}{(2y+1)^2}}{\dfrac{2y-1}{2y+1} - \dfrac{y^2 + 0.052y + 0.501}{y^2 + 1.098y + 1.102}} - \frac{10y + 305}{2y + 1}$$

$$= 433.839 \frac{y^3 + 5.003y^2 + 5.39y + 4.303}{(y + 4.453)(2y + 1)} - \frac{10y + 305}{2y + 1}$$

$$= 216.92 \frac{y^2 + 4.489y + 2.344}{y + 4.453}, \qquad (4.204a)$$

which can then be realized as an LC ladder terminating in a resistor by a continued fraction expansion:

$$Z_{22}(y) = 216.92y + \cfrac{1}{1.97 \times 10^{-3} y + \cfrac{1}{412.2}}. \qquad (4.204b)$$

The realization is shown in Fig. 4.22. After the frequency is

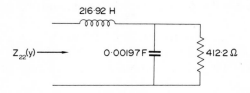

FIG. 4.22. A ladder realization of the impedance (4.204).

denormalized by a factor of 10^8, the equalizer together with its resistive generator and load is presented in Fig. 4.23, in which the ideal transformer performs the usual function of changing the resistance level at the generator end.

THE PASSIVE LOAD   287

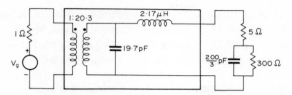

FIG. 4.23. An equalizer together with its terminations possessing the second-order Chebyshev transducer power-gain characteristic.

8.3. Elliptic transducer power-gain characteristic

We now consider and discuss the gain-bandwidth limitations imposed by the Darlington type-C load of Fig. 4.19 on the design of a lossless equalizer that achieves the nth-order low-pass elliptic transducer power-gain characteristic, as shown in (4.67). For illustrative purposes, we again follow the eight steps outlined in §7.

Steps 1–3 are the same as in §8.1, and are omitted here.

Step 4. The minimum-phase factorization of (4.68a) is given in (4.70) and we write

$$\hat{\rho}(y) = \lambda \frac{\hat{r}(y)}{r(y)}, \qquad (4.205)$$

where $\hat{r}(y)$ and $r(y)$ denote the numerator and denominator polynomials of (4.70), respectively, as in (3.249), and λ is given in (3.250).

Step 5. Like the previous two cases, no series expansions are required in Step 6, and thus are omitted here.

Step 6. The basic constraint can easily be deduced from (4.159) and is given by

$$\pm \lambda \eta(\sigma_0)\hat{r}(y_0) = A(\sigma_0)r(y_0), \qquad (4.206)$$

where $y_0 = \sigma_0/\omega_c$.

As in the Chebyshev response, since the coefficients of $\hat{r}(y)$ have not yet been determined at this point, depending on H_n through (4.68b), $\hat{r}(y_0)$ is a function of H_n. Our objective is to maximize H_n in the range

$$0 \leq H_n \leq 1, \qquad (4.207)$$

so that (4.206) is satisfied. For this reason, we first determine the condition for which $H_n = 1$ can be attained. Thus, let $H_n = 1$. Then (4.68a) reduces to (4.69) and we write $\hat{r}(y)$ as $\hat{r}_m(y)$. Since the zeros of $F_n(-jy)$ are restricted to the real-frequency axis, we have

$$\hat{r}_m(y) = \zeta \text{ [the numerator polynomial of } F_n(-jy)], \quad (4.208)$$

ζ being a constant. For an adequate discussion, two cases are distinguished.

Case 1. $A(\sigma_0)r(y_0) < \lambda \hat{r}_m(y_0)$. Then, like the Chebyshev response, $H_n = 1$ is not attainable without the insertion of open RHS zeros in $\rho(y)$. Fortunately, it is never necessary to insert anything but a single positive-real-axis zero. For if we let

$$\pm \eta(s) = \frac{s - \sigma_1}{s + \sigma_1} \quad (4.209)$$

in (4.206) and solve for σ_1, we obtain

$$\sigma_1 = \sigma_0 \frac{\lambda \hat{r}_m(y_0) - A(\sigma_0)r(y_0)}{\lambda \hat{r}_m(y_0) + A(\sigma_0)r(y_0)} > 0. \quad (4.210)$$

Case 2. $A(\sigma_0)r(y_0) \geqq \lambda \hat{r}_m(y_0)$. As in Case 2 of the Chebyshev response, the best choice is $\pm \eta(s) = 1$. Under this situation, (4.206) becomes

$$\lambda \hat{r}(y_0) = A(\sigma_0)r(y_0). \quad (4.211)$$

To show that there exists an H_n in the range (4.207) satisfying (4.211), we observe that for $H_n = 0$, $\hat{r}(y_0) = r(y_0)$, giving $\lambda \hat{r}(y_0) > A(\sigma_0)r(y_0)$; and for $H_n = 1$, $\hat{r}(y_0) = \hat{r}_m(y_0)$, yielding $\lambda \hat{r}(y_0) \leqq A(\sigma_0)r(y_0)$. Thus, the plots of $A(\sigma_0)r(y_0)$ and $\lambda \hat{r}(y_0)$ versus H_n must have the same form as in Case 2 of the Chebyshev response. Their intersection gives the desired value of H_n.

Steps 7 and 8 are the same as before, and are omitted here.

We now illustrate the above results by the following two examples.

EXAMPLE 4.11. Consider the same problem discussed in Example 4.4 for the load impedance of the network of Fig. 4.21.

THE PASSIVE LOAD

From Examples 4.4, 4.7 and 3.8 of Chapter 3, we have

$$\sigma_0 = 10^8, \tag{4.212a}$$

$$y_0 = 1, \tag{4.212b}$$

$$A(s) = \frac{2 \times 10^{-8} s - 1}{2 \times 10^{-8} s + 1}, \tag{4.212c}$$

$$r(y) = y^3 + 1.21995 y^2 + 1.48376 y + 0.91865, \tag{4.212d}$$

$$F_3(-js) = -j3.11629 \frac{s(s^2 + 0.81206)}{1 + 0.41432 s^2}, \tag{4.212e}$$

and from (3.250), $\lambda = 1$ since n is odd. From these, we obtain

$$A(\sigma_0) = \tfrac{1}{3}, \tag{4.213a}$$

$$r_m(s) = s(s^2 + 0.81206), \tag{4.213b}$$

$$r(y_0) = 4.62236, \tag{4.213c}$$

giving

$$A(\sigma_0) r(y_0) = 1.54079 < 1.81206 = \lambda \hat{r}_m(y_0). \tag{4.214}$$

Thus, Case 1 applies, and we can attain $H_3 = 1$ by inserting an open RHS zero in $\rho(s)$. From (4.210), the zero is located at

$$\sigma_1 = 10^8 \frac{1.81206 - 1.54079}{1.81206 + 1.54079} = 0.08091 \times 10^8. \tag{4.215}$$

This leads to a bounded-real reflection coefficient

$$\begin{aligned}\rho(y) &= \eta(y)\hat{\rho}(y) \\ &= \frac{y(y - 0.08091)(y^2 + 0.81206)}{(y + 0.08091)(y^3 + 1.21995 y^2 + 1.48376 y + 0.91865)} \\ &= \frac{y(y^3 - 0.08091 y^2 + 0.81206 y - 0.06570)}{y^4 + 1.30086 y^3 + 1.58247 y^2 + 1.03870 y + 0.07433}. \end{aligned} \tag{4.216}$$

Finally, the equalizer back-end impedance is computed by the formula

$$Z_{22}(y) = \frac{F(y)}{A(y) - \rho(y)} - z_l(y)$$

$$= 200 \frac{2y^5 + 1.21995y^4 - 0.36901y^3 - 0.56782y^2 - 2.13448y - 0.14866}{0.76354y^4 + 0.32087y^3 - 0.18572y^2 - 0.82434y - 0.07430}$$

$$= 200 \frac{2y^4 + 3.21995y^3 + 2.85094y^2 + 2.28312y + 0.14864}{0.76354y^3 + 1.08441y^2 + 0.89869y + 0.07435}$$

$$= 523.876y + 99.39754 \frac{y^3 + 1.30954y^2 + 5.50337y + 0.39170}{y^3 + 1.42024y^2 + 1.17700y + 0.09738}.$$

(4.217)

The first term, of course, corresponds to an inductance, and the second term can be realized as the input impedance of a lossless two-port network terminated in a resistor.

EXAMPLE 4.12. In Example 4.11, suppose that we decrease the resistance R_1 from 100 Ω to 5 Ω, everything else being the same. Then from Example 4.10, we have $\sigma_0 = 3.905 \times 10^8$, $A(\sigma_0) = 0.773$, $y_0 = 3.905$ and

$$A(\sigma_0)r(y_0) = 65.59927 > 62.71854 = \lambda \hat{r}_m(y_0), \quad (4.218)$$

which is obtained from (4.212d) and (4.213b). Thus, Case 2 applies, and we must solve the equation

$$\hat{r}(3.905) = 65.59927 \quad (4.219)$$

for H_3 in the range (4.207). A solution was shown to exist.

8.4. Equalizer back-end impedance

In the present section, we derive explicit formula for the equalizer back-end impedance $Z_{22}(s)$ that simplifies its computation and improves considerably its numerical accuracy.

To avoid introducing additional symbols, we shall express $Z_{22}(s)$ in terms of the bounded-real reflection coefficient $\rho(s)$. To this end,

THE PASSIVE LOAD

we substitute the functions of (4.155) in (4.138), giving

$$Z_{22}(s) = \frac{R_1 R_2 Cs[1+\rho(s)] - (R_1 + R_2)[1-\rho(s)]}{R_2 Cs[1-\rho(s)] - [1+\rho(s)]}. \quad (4.220)$$

We remark that the terms $1+\rho(s)$ and $1-\rho(s)$ in the formula can be replaced by the sum and the difference of the denominator and the numerator polynomials of $\rho(s)$, respectively. Also, the numerator and denominator polynomials of (4.220) always possess the common factor $(s-\sigma_0)$, which can be used as a check for the correctness of the computation. To see this, we recall that the basic constraint imposed on $\rho(s)$ by the load is $A(\sigma_0) = \rho(\sigma_0)$. Using this in conjunction with (4.155b) and (4.157), it is straightforward to verify that both the numerator and denominator of (4.220) vanish at $s = \sigma_0$.

Thus, for the problem considered in Example 4.7, we have

$$Z_{22}(s) = \frac{200y(2y^2 + 1.544y + 1.008) - 400(1.284y + 0.992)}{2y(1.284y + 0.992) - (2y^2 + 1.544y + 1.008)}$$

$$= 200 \frac{2y^2 + 3.544y + 1.983}{0.568y + 1.008}, \quad (4.221a)$$

confirming (4.175). We remark that, in addition to the common factor $(2y+1)$ that was avoided in the computation, there was another common factor $(y-1)$ in $Z_{22}(s)$ before cancellation.

As another example, we shall compute (4.204a) by means of (4.220). From (4.203), we have

$$Z_{22}(s) = \frac{10y(2y^2 + 1.150y + 1.603) - 305(1.046y + 0.601)}{2y(1.046y + 0.601) - (2y^2 + 1.150y + 1.603)}$$

$$= \frac{20y^2 + 89.523y + 46.241}{0.092y + 1.603}, \quad (4.221b)$$

confirming (4.204a). As expected, before cancellation the numerator and denominator of $Z_{22}(s)$ possess the common factor $(y-3.905)$.

Finally, we compute (4.217) by means of (4.220). From (4.216), we have

$$Z_{22}(s) = \frac{200y(2y^4 + 1.21995y^3 + 2.39453y^2 + 0.97300y + 0.07433)}{2y(1.38177y^3 + 0.77041y^2 + 1.10440y + 0.07433) -}$$
$$\frac{- 400(1.38177y^3 + 0.77041y^2 + 1.10440y + 0.07433)}{(2y^4 + 1.21995y^3 + 2.39453y^2 + 0.97300y + 0.07433)}$$
$$= 200 \frac{2y^4 + 3.21995y^3 + 2.85094y^2 + 2.28312y + 0.14865}{0.76354y^3 + 1.08441y^2 + 0.89868y + 0.07434},$$

(4.221c)

confirming (4.217). Again, before cancellation the numerator and denominator polynomials of (4.221c) possess the common factor $(y - 1)$.

9. Constant transducer power gain

At this point, it is natural to ask as to whether or not it is possible to equalize a given load to a resistive generator to achieve a transducer power gain that is constant over the entire sinusoidal frequency spectrum. Clearly, this is not possible for any load. However, if the zeros of transmission of the load all belong to Class I, then a lumped lossless matching network always exists to accomplish this. The proof requires an appreciation of the problem of interpolation with positive-real functions. For this reason, in the present section, we shall first prove the assertion for a special class of loads that have only one pair of complex zeros or one real zero of transmission, and then state and prove the general situation. Although it is possible to obtain the general constraints first, and treat the restricted class as a special case, the restricted class is sufficiently important to be considered separately, and its formulas are much the simplest.

THEOREM 4.1. *Assume that a load impedance has only one pair of Class I complex zeros or one Class I real zero of transmission. Then it is always possible to match this load to a resistive generator by a lossless equalizer to achieve a constant transducer power gain over the entire sinusoidal frequency spectrum.*

THE PASSIVE LOAD

Proof. Let $z_l(s)$ be the given load, and let

$$G(\omega^2) = \zeta > 0. \tag{4.222}$$

Substituting (4.222) in (4.9) yields

$$\rho(s)\rho(-s) = 1 - \zeta, \tag{4.223}$$

whose general solution is given by

$$\rho(s) = \pm \eta(s)\gamma \tag{4.224a}$$

with

$$\gamma = (1 - \zeta)^{1/2}, \tag{4.224b}$$

$\eta(s)$ being an arbitrary real all-pass function. Our objective is to show that a value of γ in the range $0 \leq \gamma < 1$ can always be found subject to the constraints of Class I zeros. In order to maximize ζ, we shall look for minimum γ that can be attained. To this end, we consider two cases.

Case 1. σ_0 is a real and positive zero of transmission. The corresponding coefficient constraint, as stated in §3, becomes $A_0 = \rho_0$, which implies that

$$A(\sigma_0) = \rho(\sigma_0). \tag{4.225}$$

Substituting (4.224) in (4.225) yields

$$\gamma = \left| \frac{A(\sigma_0)}{\eta(\sigma_0)} \right|. \tag{4.226}$$

Since $|A(\sigma_0)| < 1$ and $|\eta(\sigma_0)| \leq 1$, a value of γ in the range $0 \leq \gamma < 1$ can always be found by the proper choice of $\eta(s)$. The minimum value of γ, denoted by $\gamma_{\min}$, is obtained by letting $\eta(s) = 1$, which gives $\gamma_{\min} = |A(\sigma_0)|$.

Case 2. $s_0 = \sigma_0 + j\omega_0$ and $\bar{s}_0 = \sigma_0 - j\omega_0$ are a pair of Class I zeros of transmission with $\sigma_0 > 0$ and $\omega_0 \neq 0$. Because of reality of $A(s)$ and $\rho(s)$, which implies $A(\bar{s}) = \bar{A}(s)$ and $\rho(\bar{s}) = \bar{\rho}(s)$, the restriction at $\bar{s}_0$ is automatically satisfied if it is satisfied at s_0. Without loss of generality, we shall only consider the restriction at s_0. For an adequate discussion, we further divide this case into two subcases.

Subcase 1. $A(s_0)$ is real. Following Case 1, the minimum γ is attained by letting $\eta(s) = 1$, which results in

$$\gamma_{\min} = |A(s_0)|. \tag{4.227}$$

Subcase 2. $A(s_0)$ is complex. For the sake of definiteness, we choose the plus sign in (4.224a), incorporating the possible minus sign in $\eta(s)$. Define

$$z(s) = \frac{\gamma + \rho(s)}{\gamma - \rho(s)}. \tag{4.228}$$

Substituting (4.224a) in (4.228) yields

$$z(s) = \frac{1 + \eta(s)}{1 - \eta(s)}, \tag{4.229}$$

which is the familiar bilinear transformation possessing the properties that [see (1.106) of Chapter 1]

$$\operatorname{Re} z(s) \gtreqless 0 \quad \text{if and only if} \quad |\eta(s)| \lesseqgtr 1. \tag{4.230}$$

Since $|\eta(s)| \leq 1$, $\operatorname{Re} s \geq 0$, being bounded-real, $z(s)$ is positive-real. As a matter of fact, $z(s)$ is a reactance function (Problem 4.39). For a rational positive-real function, it is necessary and sufficient that

$$|\arg z(s)| \leq |\arg s| \quad \text{for} \quad |\arg s| \leq \pi/2, \tag{4.231}$$

as stated in Corollary 1.3 of Chapter 1. At $s = s_0$, the coefficient constraint is $A(s_0) = \rho(s_0)$, which from (4.228) is equivalent to the condition

$$z(s_0) = \frac{\gamma + A(s_0)}{\gamma - A(s_0)}. \tag{4.232}$$

Let

$$z(s_0) = r_0 + jx_0, \tag{4.233a}$$

$$A(s_0) = a_1 + ja_2. \tag{4.233b}$$

THE PASSIVE LOAD

Substituting (4.233) in (4.232), we get

$$r_0 = \frac{\gamma^2 - a_1^2 - a_2^2}{(\gamma - a_1)^2 + a_2^2}, \qquad (4.234a)$$

$$x_0 = \frac{2\gamma a_2}{(\gamma - a_1)^2 + a_2^2}. \qquad (4.234b)$$

Since $z(s)$ is known to be positive-real, (4.231) applies for $s = s_0$, which is equivalent to the inequality

$$\frac{r_0}{\sigma_0} \geq \left|\frac{x_0}{\omega_0}\right|. \qquad (4.235)$$

Combining (4.234) and (4.235) gives the inequality

$$\gamma^2 - 2\sigma_0|a_2|\gamma|\omega_0| - |A(s_0)|^2 \geq 0. \qquad (4.236)$$

Now we show that we can always find a value of γ in the range $0 \leq \gamma \leq 1$ satisfying the above inequality. The minimum of such values satisfying (4.236) occurs when the equality is attained, which yields the value

$$\gamma_{\min} = \sigma_0|a_2/\omega_0| + [(\sigma_0 a_2/\omega_0)^2 + |A(s_0)|^2]^{1/2}. \qquad (4.237)$$

To show that $\gamma_{\min} \leq 1$, we observe that the function

$$\frac{1 + A(s)}{1 - A(s)}, \qquad (4.238)$$

having the same functional form as (4.229), is positive-real. Following the same procedure outlined in (4.232)–(4.235), we conclude that (4.236) is valid with $\gamma = 1$, which in conjunction with (4.237) yields precisely the condition $\gamma_{\min} \leq 1$. In the proof of Theorem 4.2, we shall show that $0 < \gamma_{\min} < 1$, Q.E.D.

Several points are worth mentioning that might go unnoticed. First of all, in Case 1 and Subcase 1 of Case 2, the equalizer turns out to be an ideal transformer, since $\rho(s)$ and $Z_{22}(s)$ are constants. To show that $Z_{22}(s)$ is indeed a constant, we consider the even part of $Z_{22}(s)$:

$$R_{22}(s) = \tfrac{1}{2}[Z_{22}(s) + Z_{22}(-s)]$$
$$= \frac{r_l(s)[1 - \rho(s)\rho(-s)]}{[A(s) - \rho(s)][A(-s) - \rho(-s)]}. \qquad (4.239)$$

Under the described situation, we have $\rho(s)\rho(-s) = \gamma^2 < 1$. Thus, the closed RHS zeros of $R_{22}(s)$ can only be those of $r_l(s)/[A(s) - \rho(s)]$. Notice that the poles of $A(-s)$ cannot be zeros of $R_{22}(s)$ since they are also poles of $r_l(s)$. But the closed RHS zeros of $r_l(s)$, being the same as the zeros of transmission of $z_l(s)$, are also those of $A(s) - \rho(s)$. This means that $R_{22}(s)$, being an even function, is devoid of zeros in the entire complex plane. Thus, $R_{22}(s)$ must be a constant, so does $Z_{22}(s)$ since from (4.10) $Z_{22}(s)$ has neither poles nor zeros on the $j\omega$-axis.

Secondly, in Subcase 2 of Case 2, if we choose $\gamma = \gamma_{\min}$, then (4.235) is satisfied with the equality sign, and $z(s)$ represents either a pure inductor or a pure capacitor, i.e. $z(s) = Ls$ or $1/Cs$, L and C being nonnegative. For $z(s) = Ls$, the value of the inductance can be deduced from (4.234) and (4.235) and is given by

$$L = \frac{r_0}{\sigma_0} = \left|\frac{x_0}{\omega_0}\right| = \frac{\gamma_{\min}^2 - |A(s_0)|^2}{\sigma_0[(\gamma_{\min} - a_1)^2 + a_2^2]}$$
$$= \frac{2\gamma_{\min}|a_2|}{|\omega_0|[(\gamma_{\min} - a_1)^2 + a_2^2]}, \qquad (4.240a)$$

and the corresponding bounded-real reflection coefficient is obtained from (4.228) as

$$\rho(s) = \gamma_{\min}\frac{Ls - 1}{Ls + 1}. \qquad (4.240b)$$

For $z(s) = 1/Cs$, the values are given by

$$C = \frac{\sigma_0}{|s_0|^2 r_0} = \frac{|\omega_0|}{|s_0|^2|x_0|} = \frac{|\omega_0|[(\gamma_{\min} - a_1)^2 + a_2^2]}{2\gamma_{\min}|a_2||s_0|^2}, \qquad (4.241a)$$

$$\rho(s) = \gamma_{\min}\frac{1 - Cs}{1 + Cs}. \qquad (4.241b)$$

THE PASSIVE LOAD

Again, from (4.239) we conclude that the closed RHS zeros of the even part of the back-end impedance $Z_{22}(s)$ are exactly those of

$$\frac{1-\rho(s)\rho(-s)}{\rho(-s)}, \qquad (4.242)$$

since, as before, the closed RHS zeros of $r_l(s)$ are also the zeros of $A(s)-\rho(s)$, and the poles of $A(-s)$ are also those of $r_l(s)$. Note that $z_l(s)$ is minimum reactance and its even part $r_l(s)$ is devoid of zeros on the entire $j\omega$-axis (Problem 4.41). Substituting (4.224a) in (4.242) yields

$$\frac{1-\gamma^2}{\gamma}\eta(s), \qquad (4.243)$$

meaning that the closed RHS zeros of $R_{22}(s)$ are the same as those of the chosen $\eta(s)$. For $\gamma = \gamma_{\min}$, the corresponding $\eta(s)$ is simply $\rho(s)/\gamma_{\min}$ of (4.240b) or (4.241b), as the case may be, indicating that $R_{22}(s)$ has only a single real zero in the open RHS. This information is extremely important for the realization of the equalizer.

Finally, we mention that in Case 2 we assume that $\omega_0 \neq 0$. However, if we permit $\omega_0 \to 0$, then in the limit, $s_0 \to \sigma_0$, and $z_l(s)$ has a double-order zero of transmission at σ_0. The corresponding formulas can then be obtained from those of Subcase 2 by the limiting process. For this purpose, we first expand $A(s)$ by the Taylor series expansion about the point σ_0:

$$A(s) = A(\sigma_0) + \delta(\sigma_0)(s-\sigma_0) + \cdots, \qquad (4.244a)$$

where

$$\delta(\sigma_0) = \frac{dA(s)}{ds}\bigg|_{s=\sigma_0}. \qquad (4.244b)$$

In (4.244a), set $s = s_0$. As $s_0 \to \sigma_0$, $A(s_0)$ approaches to $A(\sigma_0) = a_1$, where $A(s_0) = a_1 + ja_2$, and (4.244a) becomes

$$\lim_{s_0 \to \sigma_0} \frac{a_2}{\omega_0} = \delta(\sigma_0). \qquad (4.245)$$

Substituting (4.245) in (4.237), (4.240a) and (4.241a) gives the desired

BROADBAND MATCHING NETWORKS

formulas (Problem 4.43):

$$\gamma_{\min} = \sigma_0|\delta(\sigma_0)| + [\sigma_0^2\delta^2(\sigma_0) + A^2(\sigma_0)]^{1/2}, \quad (4.246a)$$

$$L = \frac{\gamma_{\min}^2 - A^2(\sigma_0)}{\sigma_0[\gamma_{\min} - A(\sigma_0)]^2} = \frac{2\gamma_{\min}\delta(\sigma_0)}{[\gamma_{\min} - A(\sigma_0)]^2}, \quad \delta(\sigma_0) > 0, \quad (4.246b)$$

$$C = \frac{[\gamma_{\min} - A(\sigma_0)]^2}{2\gamma_{\min}\sigma_0^2|\delta(\sigma_0)|}, \quad \delta(\sigma_0) < 0. \quad (4.246c)$$

We shall illustrate the above results by the following examples.

EXAMPLE 4.13. Suppose that we wish to design an equalizer that matches the load impedance

$$z_l(s) = \frac{2s+1}{s+2} \quad (4.247)$$

to a resistive generator to achieve a maximum truly-flat transducer power gain over the entire sinusoidal frequency spectrum.

We first compute the even part of $z_l(s)$, $r_l(s)/z_l(s)$ and $A(s)$ which are given by

$$r_l(s) = \frac{2(s^2-1)}{(s^2-4)}, \quad (4.248a)$$

$$\frac{r_l(s)}{z_l(s)} = \frac{2(s^2-1)}{(s-2)(2s+1)}, \quad (4.248b)$$

$$A(s) = \frac{s-2}{s+2}. \quad (4.248c)$$

Thus, $z_l(s)$ has a simple zero of transmission in the open RHS at $s_0 = 1$, and Case 1 applies. From (4.226), the minimum γ is obtained as

$$\gamma_{\min} = |A(1)| = \tfrac{1}{3}. \quad (4.249)$$

Thus, from (4.224a) in conjunction with (4.225), we have

$$\rho(s) = -\tfrac{1}{3}, \quad (4.250)$$

and by (4.224b) the maximum attainable constant transducer power gain is determined by the relation

$$\zeta = 1 - \gamma_{\min}^2 = 1 - \tfrac{1}{9} = \tfrac{8}{9}. \tag{4.251}$$

Substituting (4.248) and $\rho(s) = -\tfrac{1}{3}$ in (4.138) yields the equalizer back-end impedance

$$Z_{22}(s) = \frac{2r_l(s)A(s)}{A(s) - \rho(s)} - z_l(s) = \frac{3(s+1)}{s+2} - \frac{2s+1}{s+2} = 1, \tag{4.252}$$

a result that was expected. This value can also be obtained from (4.8) in conjunction with (4.4) by setting $s = 0$. If the internal resistance of the generator is r_g, the matching network is simply an ideal transformer with turns ratio $\sqrt{r_g}:1$, as depicted in Fig. 4.24.

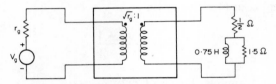

FIG. 4.24. An equalizer together with its terminations possessing a maximum truly-flat transducer power-gain characteristic over the entire sinusoidal frequency spectrum.

As a check, we compute the transducer power gain of the network of Fig. 4.24, giving

$$\begin{aligned} G(\omega^2) &= \frac{4r_l(j\omega)}{|1 + z_l(j\omega)|^2} \\ &= \frac{8(1+\omega^2)}{4+\omega^2} \cdot \frac{4+\omega^2}{9(1+\omega^2)} \\ &= \tfrac{8}{9} = \zeta. \end{aligned} \tag{4.253}$$

EXAMPLE 4.14. Match the load impedance

$$z_l(s) = \frac{s^2 + 2s + 1}{s^2 + s + 1} \tag{4.254}$$

to a resistive generator to achieve the largest flat transducer power gain over the entire sinusoidal frequency spectrum.

From (4.254), we find that

$$r_l(s) = \frac{(s^2 + \sqrt{2}s + 1)(s^2 - \sqrt{2}s + 1)}{(s^2 + s + 1)(s^2 - s + 1)} = \frac{s^4 + 1}{(s^4 + s^2 + 1)}, \quad (4.255a)$$

$$A(s) = \frac{s^2 - s + 1}{s^2 + s + 1}. \quad (4.255b)$$

Thus, $z_l(s)$ has a pair of Class I zeros of transmission of order 1 at

$$s_0, \bar{s}_0 = \frac{1}{\sqrt{2}} \pm \frac{1}{\sqrt{2}} = \sigma_0 \pm j\omega_0. \quad (4.256)$$

Substituting s_0 in (4.255b) gives

$$A(s_0) = \frac{\sqrt{2} - 1}{\sqrt{2} + 1}, \quad (4.257)$$

and Subcase 1 applies. Using (4.227), we get

$$\gamma_{min} = |A(s_0)| = \frac{\sqrt{2} - 1}{\sqrt{2} + 1}, \quad (4.258)$$

which yields

$$\zeta = 1 - \gamma_{min}^2 = 0.9706. \quad (4.259)$$

The bounded-real reflection coefficient is chosen as $\rho(s) = \gamma_{min}$. Since the equalizer back-end impedance is known to be a constant, as indicated in (4.239), it can be obtained from (4.8) in conjunction with (4.4) by setting $s = 0$:

$$\frac{\sqrt{2} - 1}{\sqrt{2} + 1} = \frac{Z_{22}(0) - 1}{Z_{22}(0) + 1},$$

which yields $Z_{22}(s) = \sqrt{2}$. If the internal resistance of the generator is r_g, the equalizer is an ideal transformer of turns ratio $(r_g^2/2)^{1/4}$.

EXAMPLE 4.15. It is required to equalize the load impedance

$$z_l(s) = \frac{s^2 + 9s + 8}{s^2 + 2s + 2} \qquad (4.260)$$

to a resistive generator to achieve the largest flat transducer power gain over the entire sinusoidal frequency spectrum.

The needed functions are computed as follows:

$$r_l(s) = \frac{(s^2 - 4)^2}{s^4 + 4} = \frac{(s+2)^2(s-2)^2}{s^4 + 4}, \qquad (4.261a)$$

$$A(s) = \frac{s^2 - 2s + 2}{s^2 + 2s + 2}, \qquad (4.261b)$$

$$\delta(s) = \frac{dA(s)}{ds} = \frac{4(s^2 - 2)}{(s^2 + 2s + 2)^2}. \qquad (4.261c)$$

Thus, $z_l(s)$ has a Class I zero of transmission of order 2 at $s_0 = \sigma_0 = 2$. At this zero, we have

$$A(\sigma_0) = 1/5, \qquad (4.262a)$$

$$\delta(\sigma_0) = 2/25. \qquad (4.262b)$$

Using (4.246a) and (4.246b), we get

$$\gamma_{\min} = 2 \times (2/25) + (16/625 + 1/25)^{1/2} = (4 + \sqrt{41})/25, \qquad (4.263a)$$

$$L = \frac{4(4 + \sqrt{41})/625}{[(4 + \sqrt{41})/25 - 1/5]^2} = (5 + \sqrt{41})/8. \qquad (4.263b)$$

Substituting (4.263) in (4.240b) yields

$$\rho(s) = \frac{4 + \sqrt{41}}{25} \cdot \frac{(5 + \sqrt{41})s - 8}{(5 + \sqrt{41})s + 8}. \qquad (4.264)$$

The irrational terms have been retained in order to facilitate the computation of $Z_{22}(s)$ by means of (4.138), in which the numerator and denominator of the first term possess a common factor $(s-2)^2$ as required:

$$Z_{22}(s) = \frac{F(s)}{A(s)-\rho(s)} - z_l(s)$$

$$= \frac{2\dfrac{(s^2-4)^2}{(s^2+2s+2)^2}}{\dfrac{s^2-2s+2}{s^2+2s+2} - \dfrac{(4+\sqrt{41})[(5+\sqrt{41})s-8]}{25[(5+\sqrt{41})s+8]}} - \frac{s^2+9s+8}{s^2+2s+2}$$

$$= \frac{12.5(s+2)^2[(5+\sqrt{41})s+8]}{(s^2+2s+2)[4(4+\sqrt{41})s+29+\sqrt{41}]} - \frac{s^2+9s+8}{s^2+2s+2}$$

$$= \frac{100.927s + 58.388}{41.613s + 35.403}, \qquad (4.265)$$

which can then be realized by a single Darlington type-C section terminated in a 1-Ω resistor (Youla, 1961).

We shall now proceed to state and prove the general assertion that if a load has an arbitrary number of Class I zeros of transmission, it is still possible to equalize such a load to a resistive generator to achieve a flat transducer power gain over the entire frequency spectrum. However, before we do this, we need a result on the existence of an all-pass function that interpolates to prescribed values at specified points, which can be deduced from the existence of an interpolating positive-real function (Youla and Saito, 1967). We shall state this result as a lemma, its proof being omitted here.

LEMMA 4.1. *Given n pairs of complex numbers* $\{(s_i, z_i); i = 1, 2, \ldots, n\}$, *in which the s_i are distinct and possess strict positive-real parts*:

$$\text{Re } s_i > 0, \qquad i = 1, 2, \ldots, n, \qquad (4.266a)$$

$$s_i \neq s_j, \qquad i \neq j. \qquad (4.266b)$$

Then the necessary and sufficient condition for the existence of a real all-pass function $\eta(s)$ of order m interpolating to z_i at s_i, i.e.

THE PASSIVE LOAD

$$\eta(s_i) = z_i, \quad i = 1, 2, \ldots, n, \quad (4.267)$$

is that the $n \times n$ hermitian matrix

$$\boldsymbol{D} = [d_{ij}], \quad (4.268a)$$

where

$$d_{ij} = \frac{1 - \bar{z}_i z_j}{\bar{s}_i + s_j}, \quad (4.268b)$$

be nonnegative-definite. If $\boldsymbol{D}$ is singular, then $\eta(s)$ is unique with $m = \operatorname{rank} \boldsymbol{D}$.

We remark that in the lemma, it is implicitly assumed that any complex s_i is accompanied by its conjugate mate $\bar{s}_i$ and that if (s_i, z_i) is in the set with complex s_i, $(\bar{s}_i, \bar{z}_i)$ must also be in the set. Also if s_i is real, so is z_i.

THEOREM 4.2. *Assume that a load impedance possesses only simple Class I zeros of transmission. Then it is always possible to match this load to a resistive generator by a lossless equalizer to achieve a constant transducer power gain over the entire sinusoidal frequency spectrum.*

Proof. The equations (4.222)–(4.224) are still valid for the general situation. Let s_i $(i = 1, 2, \ldots, n)$ be the simple Class I zeros of transmission of the load $z_l(s)$. According to (4.13), the coefficient constraints become $\rho_0 = A_0$ at each of the zeros of transmission, which is equivalent to

$$\rho(s_i) = A(s_i), \quad i = 1, 2, \ldots, n. \quad (4.269)$$

Substituting (4.224) in (4.269) gives

$$\pm \eta(s_i)\gamma = A(s_i), \quad i = 1, 2, \ldots, n. \quad (4.270)$$

In other words, at the zeros of transmission s_i, the all-pass function $\eta(s)$ must assume the preassigned values

$$z_i \equiv \eta(s_i) = A(s_i)/\gamma, \quad i = 1, 2, \ldots, n, \quad (4.271)$$

in which we choose the plus sign in (4.270), incorporating the possible minus sign in $\eta(s)$. Appealing to Lemma 4.1, $\eta(s)$ exists if and only if the hermitian matrix D of (4.268) is nonnegative definite.

Since by assumption all the zeros of transmission belong to Class I, $z_l(s)$ is a minimum reactance function, being devoid of poles on the entire $j\omega$-axis, infinity included; and also *all* the zeros of its even part $r_l(s)$ lie *off* the $j\omega$-axis. This implies that the number of poles of $z_l(s)$ is equal to the number of zeros of transmission, counting multiplicities. Thus, the all-pass function $A(s)$ of (4.6) defined by $z_l(s)$ is of order n, the number of poles of $A(s)$. But $A(s)$ is known *a priori* to be a physical all-pass function. Invoking Lemma 4.1, the $n \times n$ hermitian matrix

$$B = [b_{ij}], \qquad (4.272a)$$

where

$$b_{ij} = \frac{1 - \bar{A}(s_i)A(s_j)}{\bar{s}_i + s_j}, \qquad (4.272b)$$

is positive-definite (Problem 4.44).

Consider the elements d_{ij} of the matrix D as shown in (4.268), which can be written as

$$\gamma^2 d_{ij} = \frac{\gamma^2 - \bar{A}(s_i)A(s_j)}{\bar{s}_i + s_j} = b_{ij} - \frac{\zeta}{\bar{s}_i + s_j},$$

after appealing to (4.224b), (4.271) and (4.272b). This results in the decomposition of $\gamma^2 D$ into the difference of two matrices as

$$\gamma^2 D = B - \zeta H, \qquad (4.27a)$$

where the ith row and jth column element h_{ij} of H is defined by

$$h_{ij} = \frac{1}{\bar{s}_i + s_j}, \qquad i, j = 1, 2, \ldots, n. \qquad (4.274b)$$

Knowing that ζ must be bounded in the range $0 < \zeta < 1$, we conclude that D is nonnegative-definite if and only if $B - \zeta H$ is.

Thus, the maximum attainable ζ is the largest value of λ in $B - \lambda H$ such that $B - \lambda H$ is nonnegative-definite, which is equivalent to determining the smallest root $\lambda_{\min}$ of the equation $\det(B - \lambda H) = 0$ in the range $0 < \lambda_{\min} < 1$. Since by Problem 4.47 all the roots of $\det(B - \lambda H) = 0$ are real and since B is positive-definite, we have $\lambda_{\min} > 0$. To show that $\lambda_{\min} < 1$, we assume otherwise, i.e. $\lambda_{\min} \geqq 1$. Then for $B - \lambda_{\min} H$ to be nonnegative-definite, its diagonal elements must be nonnegative:

$$b_{ii} - \frac{\lambda_{\min}}{\bar{s}_i + s_i} = -\frac{(\lambda_{\min} - 1) + |A(s_i)|^2}{2 \operatorname{Re} s_i} \geqq 0, \qquad (4.275)$$

which is clearly impossible. Thus, we have $0 < \zeta < 1$, as expected. This completes the proof of the theorem, Q.E.D.

For illustrative purposes, we now deduce the condition (4.237) from the general constraints (4.274) by assuming that $z_i(s)$ possesses only a pair of Class I complex zeros of transmission at s_0, $\bar{s}_0 = \sigma_0 \pm j\omega_0$, $\omega_0 \neq 0$. Using this in conjunction with (4.272) and (4.233b), we obtain the pencil of matrices

$$B - \lambda H = \begin{bmatrix} \dfrac{1 - \lambda - |A(s_0)|^2}{2\sigma_0} & \dfrac{1 - \lambda - \bar{A}^2(s_0)}{2\bar{s}_0} \\ \dfrac{1 - \lambda - A^2(s_0)}{2s_0} & \dfrac{1 - \lambda - |A(s_0)|^2}{2\sigma_0} \end{bmatrix}, \qquad (4.276)$$

whose characteristic equation $\det(B - \lambda H) = 0$ is given by

$$(1 - \lambda)^2 - 2[2\sigma_0^2 a_2^2 / \omega_0^2 + |A(s_0)|^2](1 - \lambda) + |A(s_0)|^4 = 0. \qquad (4.277)$$

To obtain $\lambda_{\min}$, it is equivalent to determining the largest $(1 - \lambda)$ in (4.277) which corresponds to the solution

$$1 - \lambda_{\min} = \frac{2\sigma_0^2 a_2^2}{\omega_0^2} + |A(s_0)|^2 + \frac{2\sigma_0|a_2|}{|\omega_0|} \left[\frac{\sigma_0^2 a_2^2}{\omega_0^2} + |A(s_0)|^2\right]^{1/2}. \qquad (4.278)$$

Now observe that since the maximum obtainable constant gain ζ is

λ_{min}, from (4.224b), we have

$$\gamma_{min} = (1 - \lambda_{min})^{1/2} = \frac{\sigma_0 |a_2|}{|\omega_0|} + \left[\frac{\sigma_0^2 a_2^2}{\omega_0^2} + |A(s_0)|^2\right]^{1/2}, \quad (4.279)$$

confirming our result derived in (4.237).

10. Conclusions

In this chapter, we presented Youla's theory of broadband matching in detail. The objective is to design an optimum, lumped, reciprocal lossless equalizer to match an arbitrary passive, lumped load to a resistive generator and to achieve a preassigned transducer power-gain characteristic over the entire sinusoidal frequency spectrum. For a prescribed gain characteristic, we first stated a set of basic constraints in terms of the coefficients of the Laurent series expansions of the bounded-real reflection coefficient and two other functions determined solely by the load impedance, and then showed that these constraints are both necessary and sufficient for the physical realizability of the lossless equalizer. However, if these constraints cannot be satisfied, we may either alter the gain characteristic or introduce an all-pass function in the bounded-real reflection coefficient, which may lead to a solution to the constraint equations, but in the latter case, the resulting equalizer will require ideal transformers. Based on the discussions and proofs for the basic coefficient constraints, a simple procedure for the design of an optimum equalizer was outlined in eight steps. For practical and illustrative purposes, we considered both the Bode's parallel RC load and Darlington type-C load in great detail, each again being treated for the Butterworth, Chebyshev and elliptic transducer power-gain characteristics in its full generality, from which specific numerical examples were worked out.

Finally, we proved an assertion that if a load possesses only Class I simple zeros of transmission, then it is always possible to match this load to a resistive generator to achieve a transducer power-gain characteristic that is truly flat over the entire sinusoidal frequency spectrum.

THE PASSIVE LOAD 307

Problems

4.1. Consider the transducer power-gain characteristic

$$G(\omega^2) = \frac{K_n}{1 + \epsilon^2(\omega/\omega_c)^{2n}}, \qquad (4.280)$$

ϵ being a real constant. Derive an inequality similar to (4.28) for the parallel RC load.

4.2. Repeat the problem stated in Example 4.1 for $n = 4$.

4.3. Repeat the problem given in Example 4.2 for a 2-dB passband tolerance.

4.4. For $n = 1$, the minimum-phase solution $\hat{\rho}(s)$ of (4.17a) becomes

$$\hat{\rho}(s) = \frac{s + (1 - K_1)^{1/2}\omega_c}{s + \omega_c}. \qquad (4.281)$$

Using this, derive the equalizer back-end driving-point impedance $Z_{22}(s)$ and realize this impedance as a lossless two-port network terminated in a 1-Ω resistor.

4.5. Equating coefficients of like powers of s on the two sides of the equation representing the Laurent series expansion of a known function, derive the Laurent series expansions of the functions given in (4.24).

4.6. For a given function $F(s)$, define $f(s) = F(1/s)$. The Laurent series expansion of $F(s)$ about infinity is the same as that of $f(s)$ about the origin. Using this technique, obtain the expansions given in (4.24).

4.7. assume that $2(\sin \pi/2n)/RC\omega_c \geq 1$. Show that $K_n = 1$ is a solution of the inequality (4.28). In particular, we can set all $\lambda_i = 0$.

4.8. For Bode's parallel RC load, it can be shown by integrating around the basic contour that the input reflection coefficient $S_{11}(s)$ of the lossless equalizer is restricted along the $j\omega$-axis by the integral constraint

$$\int_0^\infty \ln \frac{1}{|S_{11}(j\omega)|} d\omega \leq \pi/RC. \qquad (4.282)$$

Using this integral constraint, derive the inequality (4.30) under the same assumption that $2(\sin \pi/2n)/RC\omega_c \leq 1$. [*Hint*. Invoke the relation $G(\omega^2) = 1 - |S_{11}(j\omega)|^2$.]

4.9. Consider the ideal brick-wall type of low-pass response for $G(\omega^2)$, and derive (4.33) directly from (4.282).

4.10. It is required to equalize the parallel combination of a 60-Ω resistor and a 150-pF capacitor to a resistive generator of internal resistance 100 Ω, and to achieve the fourth-order low-pass Butterworth transducer power gain with maximal attainable dc gain. The 3-dB bandwidth is 10^8 rad/s. Realize the desired lossless equalizer.

4.11. Consider the same problem as in Problem 4.10 except that now we wish to achieve the third-order Chebyshev transducer power gain having a maximum attainable K_3. The passband tolerance is 1.5 dB and the cutoff frequency is $50/\pi$ MHz. Design a lossless equalizer with the desired properties.

4.12. Derive the inequality (4.35b).

4.13. Show that K_n in (4.36) is maximized if we choose $d = 1$. Can we choose $d = -1$ to maximize K_n? Justify your statement.

4.14. For Bode's parallel RC load with Butterworth type of transducer power gain, we have indicated that the ideal limit (4.33) is approached from the above by K_n as n approaches to infinity. Give an explanation why this happens. [*Hint.* For finite n, ω_c represents 3-dB bandwidth; whereas, for the ideal brick-wall type of response, it is the bandwidth.]

4.15. In Problem 4.14, can the same thing be said for the Chebyshev type of transducer power gain?

4.16. Using formula (4.220), compute the equalizer back-end impedance $Z_{22}(s)$ of (4.184) and (4.196).

4.17. Design a lossless matching network to equalize a load composed of a series connection of an R-ohm resistor and a C-farad capacitor to a resistive generator and to achieve the third-order Butterworth transducer power gain of low-pass type. The radian cutoff frequency is ω_c. Derive the equalizer back-end bounded-real reflection coefficient and the corresponding impedance.

4.18. Repeat the problem given in Example 4.11 for the steepness $1/k = 1.3$, everything else being the same.

4.19. Repeat the problem given in Example 4.11 for $n = 4$.

4.20. Prove that the quantities ϵ, n and a in (4.53a) of the Chebyshev response function are related by the equation

$$a = \frac{1}{n} \ln [(1 + 1/\epsilon^2)^{1/2} + 1/\epsilon]. \quad (4.283)$$

4.21. Repeat Problem 4.17 for the load composed of a series connection of an R-ohm resistor and an L-henry inductor.

4.22. From (4.56), derive the inequality (4.57). Also show that K_n has a solution in the range $0 \leq K_n \leq 1$ only if the inequality (4.27) is satisfied.

4.23. It is desired to design a lossless matching network to equalize the load

$$z_l(s) = s + \frac{1}{s+1} \quad (4.284)$$

to a resistive generator and to achieve the third-order Butterworth transducer power gain having a maximum attainable dc gain K_3. The normalized radian cutoff frequency is $\omega_c = 1$. Show that
  (i) $K_3 = 1$ can always be achieved without the insertion of the open RHS zeros in $\rho(s)$,
  (ii) the corresponding equalizer back-end impedance $Z_{22}(s) = z_l(s)$,
  (iii) the scattering matrix of the resulting equalizer realized as a lossless ladder terminated in a 1-Ω resistor is given by

THE PASSIVE LOAD 309

$$S(s) = \frac{1}{(s+1)(s^2+s+1)} \begin{bmatrix} -s^3 & 1 \\ 1 & -s^3 \end{bmatrix}, \quad (4.285)$$

normalizing to the reference impedances 1 and $z_l(s)$.

4.24. Show that, like the Butterworth case, the non-minimum-phase factorization of (4.48a) cannot result in an increase of the maximum attainable constant K_n in relation to the minimum-phase solution.

4.25. Repeat Example 4.4 for the steepness $1/k = 1.3$, everything else being the same.

4.26. Referring to the Darlington type-C load of Fig. 4.19, let $R_1 = 100\,\Omega$, $R_2 = 200\,\Omega$ and $C = 50\,\mathrm{pF}$. Design a third-order maximally-flat equalizer for this load. The cutoff frequency is 30 MHz.

4.27. In the equation (4.59), prove that

$$\cosh_{\epsilon \to 0} \left(\sinh^{-1} \frac{1}{\epsilon} \right) \to \frac{1}{\epsilon}. \quad (4.286)$$

4.28. Repeat Problem 4.26 for a third-order equiripple equalizer.

4.29. Justify the assertion that if $\mathrm{Re}\, f(j\omega) \geq 0$ for all ω, then $\mathrm{Re}\, 1/f(j\omega) \geq 0$ for all ω.

4.30. Design an equalizer to match the load as indicated in (4.284) to a resistive generator and to achieve the third-order Chebyshev transducer power gain having a maximum attainable constant K_3. The passband tolerance is 1 dB and the normalized radian cutoff frequency $\omega_c = 1$.

4.31. Consider the same specifications as given in Example 4.4 except that the edge of the stopband starts at $60/\pi$ MHz and the ripple in the passband must not exceed 1 dB. Design this elliptic equalizer.

4.32. Show that the Hurwitz factorization of $C_n^2(-jy)$ is as given in (4.187).

4.33. Justify the statement that an increase of $\hat{a}$ in (4.53b) also increases $\hat{p}(y_0)$ of (4.186a).

4.34. Repeat the problem stated in Example 4.3 for $n = 3$.

4.35. Confirm the statement that the equation (4.190) is a polynomial of order n in $\sinh \hat{a}$.

4.36. Show that for $K_n = 1$ in (4.158), the numerator of $\hat{\rho}(s)$ becomes y^n, and (4.160a) reduces to

$$\pm \eta(\sigma_0) y_0^n = A(\sigma_0) q(y_0). \quad (4.287)$$

4.37. Consider the same problem as stated in Example 4.6 except now that we wish to achieve the fourth-order Chebyshev transducer power gain with a passband tolerance of 1 dB. Plot the transducer power gain of your realization as a function of ω.

4.38. Using the second-order Chebyshev transducer power gain with passband tolerance of 1 dB, show that the load given in Example 4.7 can always be matched to a resistive generator for any ω_c.

4.39. Prove that $z(s)$ defined in (4.229) is a reactance function, where $\eta(s)$ is an arbitrary real all-pass function.

4.40. Repeat the problem stated in Example 4.8 for $n = 3$.

4.41. Let $z_l(s)$ be a minimum reactance function, being devoid of poles on the entire $j\omega$-axis, whose even part $r_l(s)$ has no zeros on the entire $j\omega$-axis, infinity included. Show that the closed RHS zeros of the even part $R_{22}(s)$ of the equalizer back-end impedance, as shown in (4.239), are exactly those of (4.242).

4.42. Repeat the problem stated in Example 4.9 for $n = 3$.

4.43. Justify (4.246b) and (4.246c) for $\delta(\sigma_0) > 0$ and $\delta(\sigma_0) < 0$, respectively.

4.44. Using Lemma 4.1, prove that if there exists a real all-pass function of order n interpolating to z_i at s_i ($i = 1, 2, \ldots, n$), then the associated $n \times n$ hermitian matrix D defined in (4.268) is nonsingular.

4.45. Repeat the problem stated in Example 4.10 for $n = 3$.

4.46. Equalize the load impedance

$$z_l(s) = \frac{s^2 + s + 2}{2s^2 + s + 1} \tag{4.288}$$

to a resistive generator and to achieve a truly-flat transducer power gain over the entire real-frequency axis. Obtain the maximum attainable constant transducer power gain ζ, and compute the equalizer back-end impedance.

4.47. If A and B are hermitian matrices of order n and if B is positive-definite, show that all the roots of the equation $\det(A - \lambda B) = 0$ are real. [*Hint*. There exists a unitary matrix U such that U^*BU is diagonal.]

4.48. Repeat Problem 4.46 for the load impedance

$$z_l(s) = \frac{5s^2 + 3s + 4}{s^2 + 2s + 2}. \tag{4.289}$$

4.49. Repeat (i) and (ii) of Problem 4.23 for the load

$$z_l(s) = 3s + \frac{1}{s+1}. \tag{4.290}$$

[*Hint*. We must insert open RHS zeros in $\rho(s)$.]

4.50. Repeat the problem stated in Example 4.4 for a passband ripple of 0.43 dB, everything else being the same.

4.51. Show that, like the Butterworth and Chebyshev cases, the non-minimum-phase factorization of (4.68a) cannot result in an increase of the maximum attainable constant H_n in relation to the minimum-phase solution.

4.52. Referring to (4.34), let

$$\rho(s) = (1 - K_n)^{1/2} \frac{\hat{q}(x)}{q(y)} \tag{4.291}$$

be a general factorization of (4.17a). Show that this factorization can always be expressed as the product of an all-pass function $\bar{\eta}(s)$ and the minimum-phase factorization $\hat{\rho}(s)$ (4.20):

$$\rho(s) = \bar{\eta}(s)\hat{\rho}(s). \tag{4.292}$$

Using this result, justify the statement that, in order to maximize K_n, the bounded-real reflection coefficient should be devoid of zeros in the open RHS. [*Hint.* Use (4.28) and (4.29).]

4.53. Repeat Problem 4.52 for the Chebyshev response described in §4.2.

References

1. Bode, H. W. (1945) *Network Analysis and Feedback Amplifier Design*, Princeton, N.J.: Van Nostrand.
2. Carlin, H. J. and Crepeau, P. J. (1961) Theoretical limitations on the broadband matching of arbitrary impedances. *IRE Trans. Circuit Theory*, vol. CT-8, no. 2, p. 165.
3. Carlin, H. J. and La Rosa, R. (1952) Broadband reflectionless matching with minimum insertion loss. *Proc. Symp. Modern Network Synthesis*, Polytechnic Inst. of Brooklyn, New York, vol. 1, pp. 161–178.
4. Carlin, H. J. and Shen, R. C. (1972) Gain-bandwidth theory for optimizing transmission through a prescribed lossless two-port. *IEEE Trans. Circuit Theory*, vol. CT-19, no. 1, pp. 98–100.
5. Chen, W. K. (1975) On the minimum-phase reflection coefficient in broadband equalizers. *Int. J. Electronics*, vol. 39, no. 3, pp. 357–360.
6. Chen, W. K. (1975) Equalization of Darlington type-C load to give Chebyshev or elliptic transducer power-gain characteristics. *Int. J. Electronics*, vol. 39, no. 6, pp. 667–680.
7. Chen, W. K. (1975) Synthesis of Chebyshev and elliptic impedance-matching networks for the Darlington type-C load. *Proc. 18th Midwest Symp. Circuits and Systems*, Concordia University, Montreal, Canada, pp. 327–331, August 11–12.
8. Chen, W. K. (1976) On the design of the broadband elliptic impedance-matching networks. *J. Franklin Inst.*, vol. 301, no. 6.
9. Fano, R. M. (1950) Theoretical limitations on the broadband matching of arbitrary impedances. *J. Franklin Inst.*, vol. 249, nos. 1 and 2, pp. 57–83 and 139–154.
10. Fielder, D. C. (1958) Numerical determination of cascaded *LC* network elements from return loss coefficients. *IRE Trans. Circuit Theory*, vol. CT-5, no. 4, pp. 356–359.
11. Fielder, D. C. (1961) Broad-band matching between load and source systems. *IRE Trans. Circuit Theory*, vol. CT-8, no. 2, pp. 138–153.
12. Green, E. (1954) Synthesis of ladder networks to give Butterworth or Chebÿshev response in the passband. *Proc. IEE* (London), vol. 101, Pt. IV, no. 2, pp. 192–203.
13. Koo, R. L. and Sobral, M., Jr. (1975) On the choice of reflection coefficient zeros of coupling networks. *J. Franklin Inst.*, vol. 300, no. 3, pp. 197–202.
14. Ku, W. H. (1964) A broad-banding theory for varactor parametric amplifiers. *IEEE Trans. Circuit Theory*, vol. CT-11, no. 1, pp. 50–66.
15. Ku, W. H. (1970) Some results in the theory of optimum broad-band matching. *IEEE Trans. Circuit Theory*, vol. CT-17, no. 3, pp. 420–423.
16. Levy, R. (1964) Explicit formulas for Chebÿshev impedance-matching networks, filters and interstages. *Proc. IEE* (London), vol. 111, no. 6, pp. 1099–1106.
17. Matthaei, G. L. (1956) Synthesis of Tchebycheff impedance-matching networks, filters and interstages. *IRE Trans. Circuit Theory*, vol. CT-3, no. 3, pp. 163–172.

18. Plotkin, S. and Nahi, N. E. (1962) On limitations of broad-band impedance matching without transformers. *IRE Trans. Circuit Theory*, vol. CT-9, no. 2, pp. 125–132.
19. Scanlan, J. O. and Lim, J. T. (1964) Phase response and ripple in minimum-phase broadband equalizers. *IEEE Trans. Circuit Theory*, vol. CT-11, no. 4, pp. 507–508.
20. Shvarts, N. Z. and Uvbarkh, V. I. (1968) New relationships for the synthesis of Chebyshev bandpass matching ladder circuits with nonresonant sections. *Radio Engrg.*, vol. 23, no. 10, pp. 68–71.
21. Van Valkenburg, M. E. (1960) *Introduction to Modern Network Synthesis*, New York: Wiley.
22. Weinberg, L. (1962) *Network Analysis and Synthesis*, New York: McGraw-Hill.
23. Wohlers, M. R. (1965) On gain-bandwidth limitations for physically realizable systems. *IEEE Trans. Circuit Theory*, vol. CT-12, no. 3, pp. 329–333.
24. Youla, D. C. (1961) A new theory of cascade synthesis. *IRE Trans. Circuit Theory*, vol. CT-8, no. 3, pp. 244–260.
25. Youla, D. C. (1964) A new theory of broad-band matching. *IEEE Trans. Circuit Theory*, vol. CT-11, no. 1, pp. 30–50.
26. Youla, D. C. and Saito, M. (1967) Interpolation with positive-real functions. *J. Franklin Inst.*, vol. 284, no. 2, pp. 77–108.
27. Zysman, G. I. and Carlin, H. J. (1965) Restrictions on linear phase, low-pass networks with prescribed parasitic loads. *IEEE Trans. Circuit Theory*, vol. CT-12, no. 3, pp. 387–392.
28. Chien, T. M. (1974) A theory of broadband matching of a frequency-dependent generator and load. *J. Franklin Inst.*, vol. 298, no. 3, pp. 181–221.

CHAPTER 5

Theory of Broadband Matching: The Active Load

IN THE preceding chapter, we were concerned with the problem of matching a given strictly passive load impedance to a resistive generator to achieve a preassigned transducer power-gain characteristic. The central problem was to ascertain the restrictions imposed upon the transducer power-gain characteristic by the passive load impedance. The restrictions were stated in terms of the coefficients of the Laurent series expansions of various quantities defined by the load impedance, which were then shown to be both necessary and sufficient for the existence of a lossless two-port coupling network called an equalizer. Since we admit only passive networks in the study, it is clear that at any sinusoidal frequency the maximal attainable gain cannot exceed unity. Thus, no amplification can be achieved. On the other hand, if we admit a load impedance $z_2(s)$ which is active

$$\text{Re } z_2(j\omega) < 0$$

over a frequency band of interest, we may attain the desired amplification. This is especially significant in view of the continuing development of new one-port devices such as the tunnel diode. In the present chapter, we shall show that with suitable manipulations of the scattering parameters, the theory of broadband matching can be applied to the design of negative-resistance amplifiers.

We shall first define a special class of active impedances and a general configuration for negative-resistance amplifiers. We then derive some useful relations among the scattering parameters of networks with active and its associated passive loads. The application of Youla's theory of broadband matching to this special class of active impedances together with three basic configurations for the amplifiers comprises the bulk of treatment of the chapter. Although it is possible to extend the results to general active impedances, we restrict ourselves to the special class because most practical devices such as the tunnel diode can be approximated by impedances belonging to this class. Also the theory is much the simplest, and the developments can be used as a guide when other amplifier design problems are encountered.

1. Special class of active impedances

The special class of active load impedances considered here is the class of impedances $z_l(s)$ that are active over a frequency band of interest and such that the function defined by the relation

$$z_3(s) = -z_l(-s) \tag{5.1}$$

is a strictly passive impedance function. A pure negative resistance, for example, belongs to this class. The complete equivalent network for a tunnel diode, including both the effects of series inductance and loss, is shown in Fig. 5.1(a). Some typical values of the parameters for high-quality tunnel diodes are given by

$$R_d = 1\,\Omega,$$
$$L_d = 0.4\,\text{nH},$$
$$C = 10\,\text{pF},$$
$$R = 50\,\Omega.$$

If the frequency is not too high, the loss represented by the resistance R_d can be ignored, and the approximation of Fig. 5.1(b) is adequate and leads to considerable simplification. The driving-point impedance of the simplified network becomes

$$z_l(s) = sL_d + \frac{R}{RCs - 1},$$

THE ACTIVE LOAD

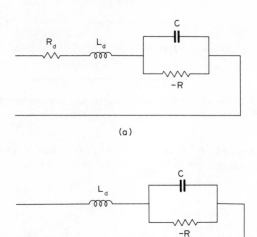

FIG. 5.1. (a) The complete equivalent network for a tunnel diode including both the effects of series inductance and loss. (b) The simplified model for a tunnel diode with the loss being ignored.

whose associated impedance

$$z_3(s) = -z_l(-s) = sL_d + \frac{R}{RCs + 1}$$

is clearly strictly passive. Thus, it also belongs to the special class. As a matter of fact, any active impedance which is formed by a lossless two-port network terminated at the output port by a negative resistor with resistance $-R\ \Omega$, as depicted in Fig. 5.2, belongs to the special class. To see this, we compute the driving-point impedance $z_l(s)$ of Fig. 5.2 in terms of the open-circuit and short-circuit immittance parameters $z_{ij}(s)$ and $y_{ij}(s)$ of the lossless two-port and the load resistor, which gives (Problem 5.1)

$$z_l(s) = z_{11}(s)\frac{1/y_{22}(s) - R}{z_{22}(s) - R}. \tag{5.2}$$

316 BROADBAND MATCHING NETWORKS

FIG. 5.2. An active impedance formed by a lossless two-port network terminated in a negative resistor.

Since the two-port is lossless, $z_{11}(s)$, $z_{22}(s)$ and $y_{22}(s)$, being the reactance functions, are odd functions of s, and hence satisfy the property

$$f(-s) = -f(s).$$

Using this property in conjunction with (5.2) yields

$$-z_l(-s) = z_{11}(s) \frac{1/y_{22}(s) + R}{z_{22}(s) + R}, \qquad (5.3)$$

which is recognized to be the driving-point impedance of the same two-port terminated at the output port in a positive resistor of resistance R Ω. Thus, $z_3(s) = -z_l(-s)$ is a strictly passive impedance.

2. General configuration of the negative-resistance amplifiers

The most general configuration of a negative-resistance amplifier in which an active one-port impedance is embedded in a two-port network connected between a source and a load can be represented by a three-port network terminated in a strictly passive impedance $z_1(s)$ in series with a voltage source at port 1, a strictly passive load impedance $z_2(s)$ at port 2, and an active impedance $z_l(s)$ at port 3, as depicted in Fig. 5.3. The three-port network is chosen to be lossless since, intuitively, one would expect that a lossless three-port would yield a higher realizable gain for the amplifier than a lossy one. Also, for simplicity, throughout the remainder of the

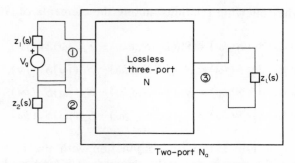

FIG. 5.3. The general configuration of a negative-resistance amplifier.

chapter, we assume that the active impedance $z_l(s)$ belongs to the special class. Let

$$S(s) = [S_{ij}], \qquad (5.4a)$$

$$S_a(s) = [S_{ija}] \qquad (5.4b)$$

be the scattering matrices of the three-port N and two-port N_a, respectively, normalizing to the reference impedances $z_1(s)$, $z_2(s)$ and $z_3(s) = -z_l(-s)$, and $z_1(s)$ and $z_2(s)$. The scattering matrix $S_a(s)$ can be readily expressed in terms of the elements of $S(s)$ by means of the interconnection formula (2.167) derived in Chapter 2, which results in

$$S_a(s) = \begin{bmatrix} S_{11} & S_{12} \\ S_{21} & S_{22} \end{bmatrix} + \begin{bmatrix} S_{13} \\ S_{23} \end{bmatrix} \left(\frac{1}{\rho} - S_{33}\right)^{-1} [S_{31} \quad S_{32}]$$

$$= \frac{1}{S_{33}} \begin{bmatrix} S_{11}S_{33} - S_{13}S_{31} & S_{12}S_{33} - S_{13}S_{32} \\ S_{21}S_{33} - S_{23}S_{31} & S_{22}S_{33} - S_{23}S_{32} \end{bmatrix}, \qquad (5.5)$$

where $1/\rho$, which is zero, is the reciprocal of the reflection coefficient of the active one-port impedance $z_l(s)$ normalized with respect to $z_3(-s)$. Since the three-port N is lossless, $S(j\omega)$ is unitary, which implies that

$$S^*(j\omega) = S^{-1}(j\omega), \qquad (5.6)$$

yielding the following relations among the elements of $S(j\omega)$:

$$\bar{S}_{22}(j\omega) \det S(j\omega) = S_{11}(j\omega)S_{33}(j\omega) - S_{13}(j\omega)S_{31}(j\omega), \quad (5.7a)$$

$$\bar{S}_{21}(j\omega) \det S(j\omega) = S_{13}(j\omega)S_{32}(j\omega) - S_{12}(j\omega)S_{33}(j\omega), \quad (5.7b)$$

$$\bar{S}_{12}(j\omega) \det S(j\omega) = S_{23}(j\omega)S_{31}(j\omega) - S_{21}(j\omega)S_{33}(j\omega), \quad (5.7c)$$

$$\bar{S}_{11}(j\omega) \det S(j\omega) = S_{22}(j\omega)S_{33}(j\omega) - S_{23}(j\omega)S_{32}(j\omega). \quad (5.7d)$$

Substituting these in (5.5) in conjunction with the fact that det $S(j\omega) = \pm 1$, the magnitudes of the elements of $S_a(j\omega)$ can be further simplified and are given by

$$|S_{11a}(j\omega)| = \left|\frac{S_{22}(j\omega)}{S_{33}(j\omega)}\right|, \quad (5.8a)$$

$$|S_{12a}(j\omega)| = \left|\frac{S_{21}(j\omega)}{S_{33}(j\omega)}\right|, \quad (5.8b)$$

$$|S_{21a}(j\omega)| = \left|\frac{S_{12}(j\omega)}{S_{33}(j\omega)}\right|, \quad (5.8c)$$

$$|S_{22a}(j\omega)| = \left|\frac{S_{11}(j\omega)}{S_{33}(j\omega)}\right|. \quad (5.8d)$$

These formulas are important because they can be applied directly to the determination of the gain-bandwidth limitation and optimum synthesis procedures for a negative-resistance amplifier. Clearly, the optimum amplifier having transducer power gain $|S_{21a}(j\omega)|^2$ as given in (5.8c) should have a maximum $|S_{12}(j\omega)|$ and a minimum $|S_{33}(j\omega)|$ over the frequency band of interest. Since $S(j\omega)$ is unitary, $|S_{12}(j\omega)| \leq 1$ and the transducer power gain $G(\omega^2)$ of the optimum amplifier is bounded by

$$G(\omega^2) \leq \frac{1}{|S_{33}(j\omega)|} \quad (5.9)$$

over the frequency band of interest.

3. Nonreciprocal amplifiers

Having succeeded in expressing the transducer power gain of the amplifier in terms of the scattering parameters of the lossless three-port N, we now proceed to discuss the specific configurations of the three-port N. One simple arrangement that will allow $|S_{12}(j\omega)|$ to be designed independently of $|S_{33}(j\omega)|$ is presented in Fig. 5.4 where the original three-port N is in the form of an interconnection of an ideal three-port circulator N_c and two two-ports N_α and N_β. The three-port circulator provides the needed isolation among the desired parameters, while the frequency shaping is achieved with two lossless equalizers placed between the circulator and the active impedance and the circulator and the load impedance. For simplicity, throughout the remainder of this section, we further stipulate that $z_1(s) = R_1$ and $z_2(s) = R_2$ are positive resistances.

We now derive an expression that relates the scattering parameters of the lossless three-port N in terms of those of the

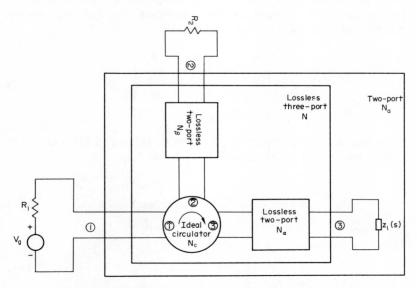

FIG. 5.4. A specific configuration of the nonreciprocal amplifiers.

component multiports. For this purpose, let

$$S_\alpha(s) = [S_{ij\alpha}], \qquad (5.10a)$$

$$S_\beta(s) = [S_{ij\beta}] \qquad (5.10b)$$

be the scattering matrices of the two-ports N_α and N_β normalizing to the reference impedances $\hat{R}_3$ and $z_3(s)$, and $\hat{R}_2$ and R_2, respectively, where $z_3(s)$ is defined in (5.1) and $\hat{R}_2$ and $\hat{R}_3$ are arbitrary real numbers. With proper normalization, the scattering matrix $S_c(s)$ of the ideal circulator N_c with respect to the reference resistances R_1, R_2 and $\hat{r}_3$ can be expressed in the form (Problem 2.20)

$$S_c(s) = \begin{bmatrix} 0 & 1 & 0 \\ 0 & 0 & 1 \\ 1 & 0 & 0 \end{bmatrix}. \qquad (5.11)$$

The three-port N can be viewed as an interconnection of a three-port N_c whose partitioned scattering matrix is shown in (5.11) and a four-port formed by the two-ports N_α and N_β, whose partitioned scattering matrix is given by

$$\begin{bmatrix} S_{11\beta} & 0 & S_{12\beta} & 0 \\ 0 & S_{11\alpha} & 0 & S_{12\alpha} \\ S_{21\beta} & 0 & S_{22\beta} & 0 \\ 0 & S_{21\alpha} & 0 & S_{22\alpha} \end{bmatrix}. \qquad (5.12)$$

Applying the interconnection formulas (2.165) to (5.11) and (5.12) yields the desired result (Problems 5.2 and 5.3):

$$S(s) = \begin{bmatrix} S_{11\alpha}S_{11\beta} & S_{12\beta} & S_{12\alpha}S_{11\beta} \\ S_{11\alpha}S_{21\beta} & S_{22\beta} & S_{12\alpha}S_{21\beta} \\ S_{21\alpha} & 0 & S_{22\alpha} \end{bmatrix}. \qquad (5.13)$$

Thus, we have

$$S_{12}(s) = S_{12\beta}(s), \qquad (5.14a)$$

$$S_{33}(s) = S_{22\alpha}(s), \qquad (5.14b)$$

THE ACTIVE LOAD 321

and from (5.8c) the transducer power gain of the amplifier becomes

$$G(\omega^2) = |S_{21a}(j\omega)|^2 = \left|\frac{S_{12}(j\omega)}{S_{33}(j\omega)}\right|^2 = \left|\frac{S_{12\beta}(j\omega)}{S_{22\alpha}(j\omega)}\right|^2. \quad (5.15)$$

We conclude from the above observations that the problem of designing an optimum negative-resistance amplifier for the special class of active impedances $z_l(s)$ is equivalent to that of designing two lossless two-port networks N_α and N_β. The lossless two-port N_α is required to match a strictly passive load impedance $z_3(s) = -z_l(-s)$ to a constant resistance $\hat{R}_3$ to achieve a minimum magnitude of the reflection coefficient $S_{22\alpha}(j\omega)$ at the output port of N_α over a given band of frequencies, and N_β is a lossless two-port that matches a positive resistance R_2 to another positive resistance $\hat{R}_2$ to achieve a maximum transducer power gain over the same given band of frequencies. To facilitate our discussion, we shall consider the design of the lossless two-ports N_α and N_β in the following sections separately.

3.1. Design considerations for N_α

As indicated in Fig. 5.5, let $\hat{Z}_{11\alpha}(s)$ and $Z_{11\alpha}(s)$ be the input impedances when the output port is terminated in $z_l(s)$ and $z_3(s)$, respectively. Also let $Z_{22\alpha}(s)$ be the output impedance when the input port is terminated in $\hat{R}_3$. Following (2.11), define

$$\rho_1^I(s) = \frac{\hat{Z}_{11\alpha}(s) - \hat{R}_3}{\hat{Z}_{11\alpha}(s) + \hat{R}_3}, \quad (5.16a)$$

$$\rho_2^I(s) = \frac{Z_{22\alpha}(s) - z_l(-s)}{Z_{22\alpha}(s) + z_l(s)}. \quad (5.16b)$$

They are the current-based reflection coefficients of the input and output ports with respect to the reference impedances $\hat{R}_3$ and $z_l(s)$, respectively. From (2.112) and (5.16b) we have

$$|S_{22\alpha}(j\omega)| = \left|\frac{Z_{22\alpha}(j\omega) - z_3(-j\omega)}{Z_{22\alpha}(j\omega) + z_3(j\omega)}\right| = \left|\frac{Z_{22\alpha}(j\omega) + z_l(j\omega)}{Z_{22\alpha}(j\omega) - z_l(-j\omega)}\right|$$

$$= \frac{1}{|\rho_2^I(j\omega)|}. \quad (5.17)$$

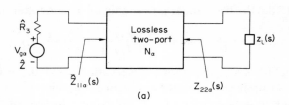

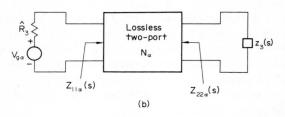

FIG. 5.5. (a) A general active impedance matching problem. (b) The associated passive impedance matching problem.

Note that $S_{22\alpha}(s)$ is the reflection coefficient at the output port normalizing to the strictly passive impedance $z_3(s) = -z_l(-s)$. This shows that in order to minimize the magnitude of the passive output reflection coefficient $S_{22\alpha}(j\omega)$, it is equivalent to maximizing the magnitude of the active output reflection coefficient $\rho_2^I(j\omega)$. Since N_α is lossless and terminated in $z_l(s)$, it is simple to show that $|\rho_1^I(j\omega)| = |\rho_2^I(j\omega)|$ (Problem 5.4). Summarizing these results, we obtain the desired formula for the two-port N_α:

$$|\rho_1^I(j\omega)| = |\rho_2^I(j\omega)| = \frac{1}{|S_{11\alpha}(j\omega)|} = \frac{1}{|S_{22\alpha}(j\omega)|}. \qquad (5.18)$$

From (5.18) and in conjunction with the relation

$$|S_{22\alpha}(j\omega)| = |S_{11\alpha}(j\omega)| = \left|\frac{Z_{11\alpha}(j\omega) - \hat{R}_3}{Z_{11\alpha}(j\omega) + \hat{R}_3}\right|, \qquad (5.19)$$

we recognize that in order to minimize $|S_{22\alpha}(j\omega)|$, $Z_{11\alpha}(j\omega)$ should be approximately equal to the resistance $\hat{R}_3$ over the frequency band of interest. Similarly, from (5.16a), $|\rho_1^I(j\omega)|$ is maximized over the desired frequency band if $\hat{Z}_{11\alpha}(j\omega)$ is approximately equal to $-\hat{R}_3$. In other words, in designing an optimum N_α, it is equivalent to synthesizing a lossless coupling network that presents to the source resistor an impedance that is approximately equal to the source resistance over the frequency band of interest when the output port is terminated in a strictly passive impedance $z_3(s) = -z_l(-s)$, and that is approximately equal to the negative of the source resistance over the frequency band of interest when the output port is terminated in the active load impedance $z_l(s)$. These are various ways in stating the same thing. Thus, we have successfully converted an amplifier problem to a broadband matching problem. However, there is one thing that we have to consider whenever active impedances are involved; namely, the stability of the overall amplifier. The natural frequencies of the network of Fig. 5.5(a) are the zeros of

$$Z_{22\alpha}(s) + z_l(s).$$

For the network to be stable, we require that all the zeros of the above impedance be restricted to the open LHS. Since from (4.10) and (4.11)

$$S_{22\alpha}(s) = \hat{\eta}(s)A(s)\frac{Z_{22\alpha}(s) + z_l(s)}{Z_{22\alpha}(s) + z_3(s)} \equiv \hat{\eta}(s)\hat{S}_{22\alpha}(s), \quad (5.20)$$

$\hat{\eta}(s)$ being an all-pass function, and since the denominator of (5.20) is strictly passive, we conclude that the closed RHS zeros of $Z_{22\alpha}(s) + z_l(s)$ are also zeros of $\hat{S}_{22\alpha}(s)$. We remark that the open RHS zeros of $A(s)$ cancel with the open RHS poles of $z_3(-s)$, as required, and thus will not appear in $\hat{S}_{22\alpha}(s)$. Therefore, for the amplifier to be stable, it is necessary that $\hat{S}_{22\alpha}(s)$ be devoid of zeros in the closed RHS.

3.2. Design considerations for N_β

Having chosen the magnitude function of the transmission coefficient $S_{12\beta}(j\omega)$ from (5.15), the reflection coefficients of the lossless two-port N_β can be determined from the known relation

$$|S_{11\beta}(j\omega)|^2 = |S_{22\beta}(j\omega)|^2 = 1 - |S_{12\beta}(j\omega)|^2. \quad (5.21)$$

Since, as indicated in Fig. 5.6, the lossless equalizer N_β is required to match only a resistive load to a resistive source, the problem, as stated in § 6 of Chapter 4, always possesses a solution provided that $S_{22\beta}(s)$ is a bounded-real function since the coefficient conditions are always satisfied.

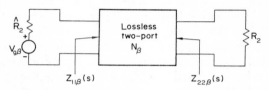

FIG. 5.6. The problem of matching a resistive load to a resistive source.

3.3. Design considerations for N_c

The scattering matrix $S_c(s)$ of the ideal circulator N_c in Fig. 5.4 is normalized to the reference impedances R_1, $\hat{R}_2$ and $\hat{R}_3$. From (2.204a) and (2.193b) it is straightforward to show that the admittance matrix $Y_c(s)$ of the ideal circulator is given by

$$\begin{aligned} Y_c(s) &= r^{-1/2}[U_3 + S_c(s)]^{-1}[U_3 - S_c(s)]r^{-1/2} \\ &= r^{-1/2}[U_3 - S_c(s)][U_3 + S_c(s)]^{-1}r^{-1/2} \\ &= \begin{bmatrix} 0 & -R_1^{-1/2}\hat{R}_2^{-1/2} & R_1^{-1/2}\hat{R}_3^{-1/2} \\ R_1^{-1/2}\hat{R}_2^{-1/2} & 0 & -\hat{R}_2^{-1/2}\hat{R}_3^{-1/2} \\ -R_1^{-1/2}\hat{R}_3^{-1/2} & \hat{R}_2^{-1/2}\hat{R}_3^{-1/2} & 0 \end{bmatrix}, \end{aligned} \quad (5.22a)$$

where

$$r^{\pm 1/2} = \begin{bmatrix} R_1^{\pm 1/2} & 0 & 0 \\ 0 & \hat{R}_2^{\pm 1/2} & 0 \\ 0 & 0 & \hat{R}_3^{\pm 1/2} \end{bmatrix}. \quad (5.22b)$$

We remark that since $U_3 - S_c(s)$ is singular or equivalently $Y_c(s)$ is singular, the ideal circulator having scattering matrix (5.11) does not possess the impedance matrix.

Alternatively, we can choose the scattering matrix

$$S_c(s) = \begin{bmatrix} 0 & 1 & 0 \\ 0 & 0 & -1 \\ 1 & 0 & 0 \end{bmatrix}, \qquad (5.22c)$$

as given in (2.194), instead of (5.11) for the ideal circulator. With the exception of a few sign changes in (5.13), all the results derived above for the nonreciprocal amplifier remain valid, *mutatis mutandis*. Under this situation, the impedance matrix $Z_c(s)$ of the ideal circulator becomes

$$\begin{aligned} Z_c(s) &= r^{1/2}[U_3 - S_c(s)]^{-1}[U_3 + S_c(s)]r^{1/2} \\ &= r^{1/2}[U_3 + S_c(s)][U_3 - S_c(s)]^{-1}r^{1/2} \\ &= \begin{bmatrix} 0 & R_1^{1/2}\hat{R}_2^{1/2} & -R_1^{1/2}\hat{R}_3^{1/2} \\ -R_1^{1/2}\hat{R}_2^{1/2} & 0 & -\hat{R}_2^{1/2}\hat{R}_3^{1/2} \\ R_1^{1/2}\hat{R}_3^{1/2} & \hat{R}_2^{1/2}\hat{R}_3^{1/2} & 0 \end{bmatrix}. \end{aligned} \qquad (5.22d)$$

As expected, since $U_3 + S_c(s)$ is singular, the circulator having scattering matrix (5.22c) does not possess the admittance matrix. Notice that $Z_c(s)$ does not represent the inverse of $Y_c(s)$. To avoid confusion, throughout the remainder of the chapter, when we speak of $S_c(s)$, we mean the matrix given in (5.11).

Now we show that the driving-point impedance looking into one of the ports of the circulator when the other ports are terminated in their reference impedances is equal to the reference impedance of that port. To this end and without loss of generality, we consider the admittance matrix $Y_c(s)$ and compute the driving-point impedance of port $\hat{1}$ when port $\hat{2}$ and port $\hat{3}$ are terminated in their reference impedances $\hat{R}_2$ and $\hat{R}_3$, respectively. The other possibilities are left as exercises (Problems 5.8 and 5.9). Let port $\hat{1}$ be excited by a current generator I_g, and let V_{1c}, V_{2c} and V_{3c} be the port voltages of the circulator. Then

$$\begin{bmatrix} I_g \\ 0 \\ 0 \end{bmatrix} = \begin{bmatrix} 0 & -R_1^{-1/2}\hat{R}_2^{-1/2} & R_1^{-1/2}\hat{R}_3^{-1/2} \\ R_1^{-1/2}\hat{R}_2^{-1/2} & \hat{R}_2^{-1} & -\hat{R}_2^{-1/2}\hat{R}_3^{-1/2} \\ -R_1^{-1/2}\hat{R}_3^{-1/2} & \hat{R}_2^{-1/2}\hat{R}_3^{-1/2} & \hat{R}_3^{-1} \end{bmatrix} \begin{bmatrix} V_{1c} \\ V_{2c} \\ V_{3c} \end{bmatrix}.$$

(5.22e)

Solving for V_{1c} in terms of I_g gives the driving-point impedance at port $\hat{1}$, which is R_1. The significance of the above assertion is that it provides the needed resistive terminations for the equalizers N_α and N_β, as depicted schematically in Fig. 5.4. This will be illustrated in the following examples.

We remark that methods for synthesizing the immittance matrix of a circulator are available (see, for example, Carlin and Giordano, 1964), and we will not discuss this aspect of the problem any further, since it would take us far afield into another subject that has been adequately covered in other books.

3.4. Illustrative examples

In this section, we shall work out several detailed examples on the design of an optimum negative-resistance amplifier based on the theory of broadband matching discussed in Chapter 4. We first give a detailed numerical example, and then present a fairly complete and systematic account of the gain-bandwidth and stability limitations of an amplifier when a tunnel diode, represented by its simplified model after the effects of series inductance and loss having been ignored, is employed. Formulas for amplifiers having Butterworth and Chebyshev characteristics of arbitrary order will be derived together with their design curves, each being presented in a separate section.

EXAMPLE 5.1. Consider an active device whose simplified linear model is given in Fig. 5.7. Suppose that we wish to use this device to design an optimum nonreciprocal negative-resistance amplifier and to achieve the fifth-order Chebyshev transducer power-gain

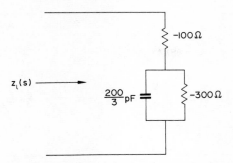

FIG. 5.7. A simplified linear network model of an active device.

characteristic, and suppose that the amplifier is of low-pass type whose cutoff frequency is $50/\pi$ MHz and whose passband tolerance must be within 1-dB ripple. The generator internal resistance and the load impedance are given by $R_1 = 50\ \Omega$ and $R_2 = 200\ \Omega$, as indicated in Fig. 5.8. The active impedance clearly belongs to the special class.

For computational purposes, the network of Fig. 5.8 is first magnitude-scaled down by a factor 10^{-2}, and frequency-scaled down by 10^{-8}. This results in the normalized quantities $R_1 = \frac{1}{2}\ \Omega$, $R_2 = 2\ \Omega$,

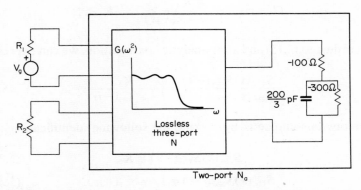

FIG. 5.8. The design of an optimum nonreciprocal negative-resistance amplifier to achieve the fifth-order Chebyshev transducer power-gain characteristic.

328 BROADBAND MATCHING NETWORKS

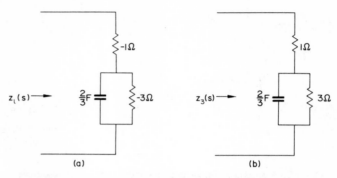

FIG. 5.9. (a) The normalized active impedance of Fig. 5.7. (b) The associated passive impedance of (a).

and the load $z_l(s)$ as shown in Fig. 5.9(a). From (3.62), the 1-dB peak-to-peak ripple in the passband corresponds to a ripple factor

$$\epsilon = (10^{0.1} - 1)^{1/2} = 0.50885.$$

The fifth-order low-pass Chebyshev transducer power gain is shown in (4.46), which after analytic continuation is given by

$$G(-s^2) = \frac{K_5}{1 + \epsilon^2 C_5^2(-js)}, \qquad K_5 \geqq 1. \tag{5.23}$$

According to (5.15) and after analytic continuation, we can express

$$\frac{S_{12\beta}(s)S_{12\beta}(-s)}{S_{22\alpha}(s)S_{22\alpha}(-s)} = \frac{K_5}{1 + \epsilon^2 C_5^2(-js)}. \tag{5.24}$$

One obvious choice is by making the following identifications:

$$S_{12\beta}(s)S_{12\beta}(-s) = K_5, \tag{5.25a}$$

$$S_{22\alpha}(s)S_{22\alpha}(-s) = 1 + \epsilon^2 C_5^2(-js). \tag{5.25b}$$

However, this is completely unacceptable since $|S_{12\beta}(j\omega)| \leqq 1$ and

$|S_{22\alpha}(j\omega)| \leq 1$. Another possibility is that by dividing the numerator and denominator of (5.24) by the quantity $K_5 + \epsilon^2 C_5^2(-js)$, we can then make the identifications. This gives

$$S_{12\beta}(s)S_{12\beta}(-s) = \frac{K_5}{K_5 + \epsilon^2 C_5^2(-js)}, \quad (5.26a)$$

$$S_{22\alpha}(s)S_{22\alpha}(-s) = \frac{1 + \epsilon^2 C_5^2(-js)}{K_5 + \epsilon^2 C_5^2(-js)}. \quad (5.26b)$$

This choice not only satisfies the requirement that $|S_{12\beta}(j\omega)| \leq 1$ and $|S_{22\alpha}(j\omega)| \leq 1$, but also realizes the desired gain characteristic.

A. *Realization of N_α*

The starting point is (5.26b), which can be expressed as

$$S_{22\alpha}(s)S_{22\alpha}(-s) = \frac{1}{\alpha^2} \cdot \frac{1 + \epsilon^2 C_5^2(-js)}{1 + \hat{\epsilon}^2 C_5^2(-js)}, \quad (5.27a)$$

where

$$\alpha^2 = K_5, \quad (5.27b)$$

$$\hat{\epsilon} = \frac{\epsilon}{\alpha}, \quad (5.27c)$$

$$C_5^2(-js) = -(16s^5 + 20s^3 + 5s)^2. \quad (5.27d)$$

Since $\epsilon = 0.50885$, the zeros of (5.27a) can be computed by the formula (3.79) in a straightforward manner. The strictly Hurwitz polynominal $q(s)$ formed by the open LHS zeros is given by

$$q(s) = s^5 + 0.937s^4 + 1.689s^3 + 0.974s^2 + 0.580s + 0.123. \quad (5.28)$$

Recall that our objective is to minimize $|S_{22\alpha}(j\omega)|$ over the passband in order to maximize the gain. This is equivalent to maximizing K_5 in (5.26b) or minimizing $\hat{\epsilon}$ in (5.27c). To this end, let $\hat{q}(s)$ be the strictly Hurwitz polynomial formed by the open LHS poles of (5.27a) and write

$$\hat{q}(s) = s^5 + b_4 s^4 + b_3 s^3 + b_2 s^2 + b_1 s + b_0. \quad (5.29)$$

Let $\hat{S}_{22\alpha}(s)$ be the minimum-phase factorization of (5.27a). Then we have

$$\hat{S}_{22\alpha}(s) = \pm \frac{s^5 + 0.937s^4 + 1.689s^3 + 0.974s^2 + 0.580s + 0.123}{s^5 + b_4 s^4 + b_3 s^3 + b_2 s^2 + b_1 s + b_0}. \quad (5.30)$$

We now follow the eight steps outlined in §7 of Chapter 4 to complete the realization. The needed functions are computed next and are given by

$$z_l(s) = -1 + \frac{3}{2s-1} = \frac{-2s+4}{2s-1}, \quad (5.31a)$$

$$z_3(s) = -z_l(-s) = \frac{2s+4}{2s+1}, \quad (5.31b)$$

$$r_3(s) = \tfrac{1}{2}[z_3(s) + z_3(-s)] = \frac{4s^2-4}{4s^2-1}, \quad (5.31c)$$

$$A(s) = \frac{2s-1}{2s+1}, \quad (5.31d)$$

$A(s)$ being an all-pass function whose zero is the open RHS pole of $z_3(-s)$,

$$F(s) = 2r_3(s)A(s) = \frac{8(s^2-1)}{(2s+1)^2}, \quad (5.32a)$$

$$\frac{r_3(s)}{z_3(s)} = \frac{4(s+1)(s-1)}{(2s-1)(2s+4)}, \quad (5.32b)$$

indicating that $s = 1$ is a Class I zero of transmission of order 1 of the impedance $z_3(s)$. Expanding the functions $\hat{S}_{22\alpha}(s)$ and $A(s)$ by Laurent series about the point $s = 1$ yields

$$\hat{S}_{22\alpha}(s) = S_0 + S_1(s-1) + S_2(s-1)^2 + \cdots, \quad (5.33a)$$

$$A(s) = A_0 + A_1(s-1) + A_2(s-1)^2 + \cdots. \quad (5.33b)$$

The coefficient constraint for a simple Class I zero of transmission is

$$S_0 = A_0, \quad (5.34)$$

THE ACTIVE LOAD 331

which is equivalent to choosing the plus sign in (5.30) with

$$\hat{S}_{22\alpha}(1) = A(1). \qquad (5.35)$$

Thus, from (5.29), (5.30) and (5.31d) we obtain

$$\hat{q}(1) = 15.909. \qquad (5.36)$$

Using formula (3.79), the polynomial $\hat{q}(s)$ can be expressed as

$$\hat{q}(s) = (s + \sinh \hat{a})(s + 0.309 \sinh \hat{a} + j0.951 \cosh \hat{a})$$
$$\times (s + 0.309 \sinh \hat{a} - j0.951 \cosh \hat{a})$$
$$\times (s + 0.809 \sinh \hat{a} + j0.588 \cosh \hat{a})$$
$$\times (s + 0.809 \sinh \hat{a} - j0.588 \cosh \hat{a}), \qquad (5.37a)$$

where

$$\hat{a} = \frac{1}{5} \sinh^{-1} \frac{1}{\hat{\epsilon}}. \qquad (5.37b)$$

Substituting (5.37a) in (5.36) yields

$$x^5 + 3.2361x^4 + 6.4861x^3 + 8.1631x^2 + 6.4756x - 13.3465 = 0, \quad (5.38a)$$

where

$$x = \sinh \hat{a}. \qquad (5.38b)$$

The equation can be solved with the aid of a computer, and its roots are given by

$$0.7473, \quad -1.8638 \pm j1.129, \quad -0.1278 \pm j1.935. \qquad (5.39)$$

Thus, let $x = 0.7473$, and from (5.38b), (5.37b) and (5.27) we obtain

$$\hat{a} = 0.691, \qquad (5.40a)$$

$$\alpha = 8.046, \qquad (5.40b)$$

$$K_5 = 64.7. \qquad (5.40c)$$

332 BROADBAND MATCHING NETWORKS

Substituting these in (5.37a) and (5.30) gives

$$\hat{q}(s) = s^5 + 2.419s^4 + 4.174s^3 + 4.366s^2 + 2.956s + 0.988, \quad (5.41)$$

$$\hat{S}_{22\alpha}(s) = \frac{s^5 + 0.937s^4 + 1.689s^3 + 0.974s^2 + 0.58s + 0.123}{s^5 + 2.419s^4 + 4.174s^3 + 4.366s^2 + 2.956s + 0.988}. \quad (5.42)$$

Finally, from (4.138) the equalizer back-end impedance $Z_{22\alpha}(s)$ can be determined by

$$Z_{22\alpha}(s) = \frac{F(s)}{A(s) - \hat{S}_{22\alpha}(s)} - z_3(s)$$

$$= \frac{4s^5 + 10.712s^4 + 16.511s^3 + 17.263s^2 + 10.741s + 3.46}{0.964s^4 + 2.578s^3 + 3.51s^2 + 2.911s + 1.111}$$

$$= 4.156s + \cfrac{1}{0.502s + \cfrac{1}{4.343s + \cfrac{1}{0.342s + \cfrac{1}{1.163s + 3.114}}}}, \quad (5.43)$$

which can readily be realized as a lossless ladder terminated in a resistor, as shown in Fig. 5.10. Denormalizing the element values with regard to magnitude-scaling by a factor 100 and frequency-scaling by 10^8 gives the final design of the equalizer, which together with the active load is presented in Fig. 5.11.

For illustrative purposes, we shall compute the active reflection coefficients $\rho_1^I(s)$ and $\rho_2^I(s)$, as defined in (5.16), for the network of

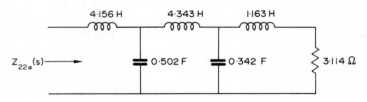

FIG. 5.10. A ladder realization of the impedance (5.43).

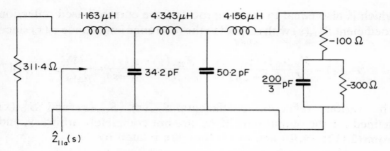

FIG. 5.11. A lossless equalizer that matches an active load to a passive resistive load.

Fig. 5.11. To simplify our computation, we consider the normalized network of Fig. 5.10 and the normalized active load of Fig. 5.9(a). The input impedance when the output port is terminated in $z_l(s)$ is obtained as

$$\hat{Z}_{11\alpha}(s) = \frac{7.197s^6 - 5.34s^5 + 33.12s^4 - 21.14s^3 + 31.95s^2 - 11.66s + 4}{6.16s^5 - 4.56s^4 + 12.96s^3 - 6.68s^2 + 5.37s - 1},$$

(5.44)

from which we compute

$$\rho_1^I(s) = \frac{\hat{Z}_{11\alpha}(s) - 3.114}{\hat{Z}_{11\alpha}(s) + 3.114}$$

$$= \frac{(s-1)(s^5 - 2.42s^4 + 4.16s^3 - 4.37s^2 + 2.96s - 0.99)}{(s+1)(s^5 + 0.93s^4 + 1.69s^3 + 0.97s^2 + 0.58s + 0.12)}. \quad (5.45)$$

Substituting (5.31a) and (5.43) in (5.16b) yields

$$\rho_2^I(s) = \frac{Z_{22\alpha}(s) - z_l(-s)}{Z_{22\alpha}(s) + z_l(s)}$$

$$= \frac{(2s-1)(s^5 + 2.42s^4 + 4.18s^3 + 4.37s^2 + 2.95s + 0.988)}{(2s+1)(s^5 + 0.94s^4 + 1.69s^3 + 0.97s^2 + 0.59s + 0.123)},$$

(5.46)

which is also equal to the reciprocal of the current-based reflection coefficient $S_{22\alpha}^I(s)$ with respect to the reference impedance $z_3(s)$ since

$$S_{22\alpha}^I(s) = \frac{Z_{22\alpha}(s) - z_3(-s)}{Z_{22\alpha}(s) + z_3(s)} = \frac{Z_{22\alpha}(s) + z_l(s)}{Z_{22\alpha}(s) - z_l(-s)} = \frac{1}{\rho_2^I(s)}. \quad (5.47)$$

The various reflection coefficients $S_{22\alpha}(s)$, $\hat{S}_{22\alpha}(s)$ and $S_{22\alpha}^I(s)$ defined for the output port of N_α are not completely arbitrary, and from (2.112), (4.10) and (4.11) they are related by

$$\begin{aligned}S_{22\alpha}(s) &= \frac{h_3(s)}{h_3(-s)} S_{22\alpha}^I(s) = \frac{(s-1)(2s-1)}{(s+1)(2s+1)} S_{22\alpha}^I(s) \\ &= \frac{(s-1)}{(s+1)} A(s) S_{22\alpha}^I(s) = \frac{(s-1)}{(s+1)} \hat{S}_{22\alpha}(s) \\ &= \frac{(s-1)}{(s+1)} \cdot \frac{q(s)}{\hat{q}(s)}, \end{aligned} \quad (5.48)$$

where $h_3(s)h_3(-s)$ is the para-hermitian part of $z_3(s)$. Thus, we have

$$|S_{22\alpha}(j\omega)| = |S_{22\alpha}^I(j\omega)| = |\hat{S}_{22\alpha}(j\omega)| = \frac{1}{|\rho_1^I(j\omega)|} = \frac{1}{|\rho_2^I(j\omega)|}. \quad (5.49)$$

We remark that (5.49) is valid in general (Problem 5.5).

B. *Realization of N_β*
From (5.26a) and (5.21), we obtain

$$S_{22\beta}(s) S_{22\beta}(-s) = 1 - \frac{K_5}{K_5 + \epsilon^2 C_5^2(-js)} = \frac{\epsilon^2 C_5^2(-js)}{K_5 + \epsilon^2 C_5^2(-js)}. \quad (5.50)$$

Using (5.27d) in conjunction with (5.41), the minimum-phase factorization of (5.50) can be chosen to be

$$\hat{S}_{22\beta}(s) = \pm \frac{s^5 + 1.25s^3 + 0.313s}{s^5 + 2.419s^4 + 4.174s^3 + 4.366s^2 + 2.956s + 0.988}. \quad (5.51)$$

THE ACTIVE LOAD

Suppose that we choose the positive sign for $\hat{S}_{22\beta}(s)$ in (5.51). From (4.138), the equalizer back-end impedance $Z_{22\beta}(s)$ is computed as

$$\frac{Z_{22\beta}(s)}{R_2} = \frac{2A(s)}{A(s) - \hat{S}_{22\beta}(s)} - 1 = \frac{2}{1 - \hat{S}_{22\beta}(s)} - 1$$

$$= \frac{2s^5 + 2.419s^4 + 5.424s^3 + 4.366s^2 + 3.268s + 0.988}{2.419s^4 + 2.924s^3 + 4.366s^2 + 2.644s + 0.988}$$

$$= 0.827s + \cfrac{1}{1.334s + \cfrac{1}{1.652s + \cfrac{1}{1.341s + \cfrac{1}{0.829s + 1}}}}, \quad (5.52)$$

which can easily be identified as a lossless ladder terminated in a resistor. After denormalization, we arrive at the final design of the equalizer N_β, as shown in Fig. 5.12.

Instead of choosing the positive sign in (5.51), suppose that we choose the negative sign. The corresponding equalizer back-end impedance $Z_{22}(s)/R_2$ turns out to be the reciprocal of that given in (5.52). After denormalization, the corresponding ladder realization together with the load impedance R_2 is presented in Fig. 5.13. Comparing these two realizations, we recognize that in choosing the negative sign the resulting network allows some shunt capacitance

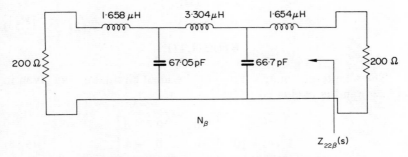

FIG. 5.12. A lossless equalizer that matches a resistive load to another.

336 BROADBAND MATCHING NETWORKS

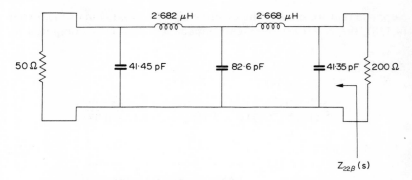

FIG. 5.13. A lossless equalizer that matches a resistive load to another.

at its ports to be absorbed into the equalizer N_β, while choosing the positive sign yields an equalizer of the form used for network N_α and allows some series inductance at its ports to be absorbed into N_α.

C. Realization of N_c

Recall that the scattering matrix $S_c(s)$, as given in (5.11), of the ideal circulator N_c is normalized with respect to the reference impedances R_1, $\hat{R}_2$ and $\hat{R}_3$. Up to this point, these quantities have not played any role in the realization. We shall specify

$$R_1 = 50 \, \Omega,$$
$$\hat{R}_2 = 200 \, \Omega, \qquad (5.53)$$
$$\hat{R}_3 = 311.4 \, \Omega.$$

The admittance matrix $Y_c(s)$ of the ideal circulator was given in (5.22a) and is evaluated to be

$$Y_c(s) = 10^{-3} \begin{bmatrix} 0 & -10 & 8 \\ 10 & 0 & -4 \\ -8 & 4 & 0 \end{bmatrix}, \qquad (5.54)$$

THE ACTIVE LOAD

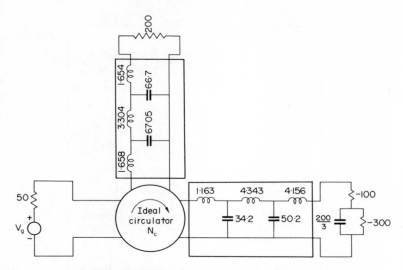

FIG. 5.14. A complete nonreciprocal negative-resistance amplifier employing the active impedance of Fig. 5.7.

which can be realized by using one gyrator and an ideal transformer bank (see, for example, Carlin and Giordano, 1964).

The complete nonreciprocal negative-resistance amplifier is presented in Fig. 5.14, in which all the inductances are in μH, all the capacitances in pF and all the resistances in Ω and where the circulator provides the required resistive terminations for the equalizers N_α and N_β. Substituting (5.28) and (4.40c) in (5.23) yields the transducer power gain of the amplifier:

$$G(\omega^2) = \frac{64.7}{256\epsilon^2 |q(j\omega)|^2}$$

$$= \frac{0.976}{(0.937\omega^4 - 0.974\omega^2 + 0.123)^2 + (\omega^5 - 1.689\omega^3 + 0.58\omega)^2},$$
(5.55)

which has a dc gain of 64.7 or 18.1 dB. The gain characteristic (5.55) can be plotted as a function of ω and is shown in Fig. 5.15.

338 BROADBAND MATCHING NETWORKS

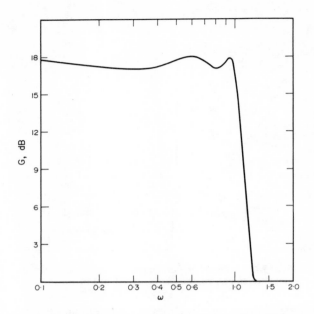

FIG. 5.15. The transducer power-gain characteristic of the nonreciprocal negative-resistance amplifier of Fig. 5.14.

3.4.1. The tunnel diode amplifier: maximally-flat transducer power gain

Consider a complete equivalent network for a tunnel diode as shown in Fig. 5.1(a). Suppose that the frequency is not too high, so that the effects of series inductance and loss can be ignored, which leads to considerable simplification. In this section, we use the simplified model of Fig. 5.16 to determine the gain-bandwidth restrictions imposed on the design of the lossless equalizer N_α that matches the active load

$$z_l(s) = \frac{-R}{1 - RCs} \tag{5.56}$$

to a resistive generator of internal resistance r_g, and that achieves

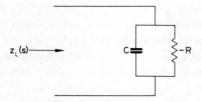

FIG. 5.16. A simplified model for a tunnel diode.

the nth-order low-pass Butterworth transducer power-gain characteristic

$$G(\omega^2) = \frac{K_n}{1+(\omega/\omega_c)^{2n}}, \qquad K_n \geqq 1, \tag{5.57}$$

where ω_c is the 3-dB radian bandwidth. For this load $z_l(s)$, we have the associated passive load

$$z_3(s) = \frac{R}{1+RCs}, \tag{5.58}$$

indicating that $z_l(s)$ belongs to the special class of active impedances.

Following (5.26) and after analytic continuation, we choose

$$S_{12\beta}(s)S_{12\beta}(-s) = \frac{K_n}{K_n + (-s^2/\omega_c^2)^n}, \tag{5.59a}$$

$$S_{22\alpha}(s)S_{22\alpha}(-s) = \frac{1+(-s^2/\omega_c^2)^n}{K_n + (-s^2/\omega_c^2)^n}, \tag{5.59b}$$

which can further be simplified to

$$S_{12\beta}(s)S_{12\beta}(-s) = \frac{1}{1+(-1)^n x^{2n}}, \tag{5.60a}$$

$$S_{22\alpha}(s)S_{22\alpha}(-s) = K_n^{-1}\frac{1+(-1)^n y^{2n}}{1+(-1)^n x^{2n}}, \tag{5.60b}$$

where

$$x = K_n^{-1/2n} s/\omega_c, \tag{5.61a}$$

$$y = s/\omega_c. \tag{5.61b}$$

A. Realization of N_α

The numerator and denominator polynomials of (5.60b) can be factored in terms of the roots of $(-1)^n s^{2n} + 1 = 0$. Let $q(s)$ be the nth-order Hurwitz polynomial formed by the open LHS zeros. Then the minimum-phase factorization of (5.60b) can be written as

$$\hat{S}_{22\alpha}(s) = \pm K_n^{-1/2} \frac{q(y)}{q(x)}. \quad (5.62)$$

We next compute the needed functions and their Laurent series expansions about the zero of transmission, which were done in § 4.1 of Chapter 4, and the results are given by

$$A(s) = \frac{s-\tau}{s+\tau} = 1 - 2\tau/s + 2\tau^2/s^2 + \cdots, \quad (5.63a)$$

$$F(s) = \frac{-2\tau/C}{(s+\tau)^2} = -2\tau/Cs^2 + 4\tau^2/Cs^3 + \cdots, \quad (5.63b)$$

$$\pm \hat{S}_{22\alpha}(s) = K_n^{-1/2}(y^n + a_{n-1}y^{n-1} + \cdots + a_0)(x^n + a_{n-1}x^{n-1} + \cdots + a_0)^{-1}$$
$$= K_n^{-1/2}(y^n + a_{n-1}y^{n-1} + \cdots)(x^{-n} - a_{n-1}x^{-n-1} + \cdots)$$
$$= 1 + a_{n-1}/y - a_{n-1}/x + \cdots$$
$$= 1 + S_1/s + \cdots, \quad (5.63c)$$

where a_m ($m = 1, 2, \ldots, n$) are defined in (4.18), $\tau = 1/RC$, and

$$S_1 = (1 - K_n^{1/2n})\omega_c / \sin(\pi/2n). \quad (5.64)$$

Since the impedance $z_3(s)$ of (5.58) has only a Class II zero of transmission of order 1, the basic constraints on the coefficients (5.63) become

$$A_0 = S_0, \quad (5.65a)$$

$$(A_1 - S_1)/F_2 \geq 0, \quad (5.65b)$$

where $A_0 = 1$, $A_1 = -2\tau$, $S_0 = \pm 1$ and $F_2 = -2\tau/C$. Thus, to satisfy (5.65a) we must choose the plus sign in (5.62). To satisfy the second constraint, we substitute the appropriate quantities in (5.65b), which yields the inequality

$$K_n \leq \left[1 + \frac{\sin(\pi/2n)}{\pi RCf_c}\right]^{2n}, \qquad (5.66)$$

where, as before, $\omega_c = 2\pi f_c$. In the limit as $n \to \infty$,

$$K_\infty \leq \exp(1/RCf_c) \qquad (5.67)$$

(Problem 5.7). Thus, the right-hand side of (5.66) represents the maximum attainable transducer power gain for the amplifier, which has been plotted in decibels as a function of n and RCf_c. The curves are presented in Fig. 5.17. From these curves, we note that the maximum power gain increases monotonically as n increases. In other words, the higher the order of the Butterworth, the larger is the gain for the same bandwidth. This result is in direct contrast with that given in § 4.1 of the preceding chapter, in which we showed that with the exception for $n = 1$, K_∞ is approached from the above rather than below.

B. *Realization of N_β*

From (5.21) and (5.60a), we obtain

$$S_{22\beta}(s)S_{22\beta}(-s) = \frac{(-1)^n x^{2n}}{1 + (-1)^n x^{2n}}, \qquad (5.68)$$

whose minimum-phase factorization is given by

$$\hat{S}_{22\beta}(s) = \pm \frac{x^n}{q(x)}. \qquad (5.69)$$

Since the coefficient conditions are always satisfied, the equalizer back-end impedance $Z_{22\beta}(s)$ can be determined by formula (4.138) which yields

$$\frac{Z_{22\beta}(x)}{R_2} = \frac{2}{1 - \hat{S}_{22\beta}(s)} - 1 = \frac{q(x) \pm x^n}{q(x) \mp x^n}. \qquad (5.70)$$

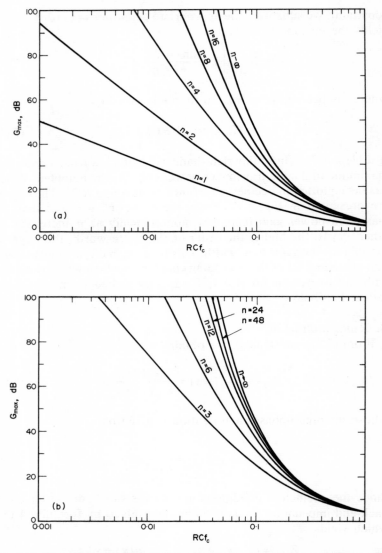

FIG. 5.17. The maximum attainable dc gain K_n for tunnel diode amplifiers with maximally-flat transducer power-gain characteristics.

THE ACTIVE LOAD 343

Expanding the function in a continued fraction gives a network realization which is recognized as a lossless ladder terminated in a resistor.

EXAMPLE 5.2. It is desired to design a nonreciprocal amplifier to have a maximally-flat low-pass characteristic for its transducer power gain. The amplifier is to be operated between 50-Ω and 200-Ω terminations for which the passband gain must be at least 30 dB for the bandwidth of $50/\pi$ MHz. For the tunnel diode, $-R = -100\ \Omega$ and $C = 50$ pF.

From the above specification, we have

$$R_1 = 50\ \Omega, \qquad (5.71\text{a})$$

$$R_2 = 200\ \Omega, \qquad (5.71\text{b})$$

$$-R = -100\ \Omega, \qquad (5.71\text{c})$$

$$C = 50\ \text{pF}, \qquad (5.71\text{d})$$

$$\omega_c = 10^8\ \text{rad/s}. \qquad (5.71\text{e})$$

Since the maximum deviation within the passband, which is about 3 dB down, occurs at the cutoff frequency, the value of the dc gain K_n must be, therefore, at least 33 dB. From (5.66) or the curves of Fig. 5.17, we see that for $n = 5$ the gain K_5 satisfies the requirement since

$$K_5 = (1 + 4 \sin 18°)^{10} = 2.2361^{10} = 3125.5 \text{ or } 34.95 \text{ dB}. \qquad (5.72)$$

Thus, from (5.62) the minimum-phase back-end reflection coefficient $\hat{S}_{22\alpha}(s)$ of N_α can be computed from (3.18), giving

$$\hat{S}_{22\alpha}(s) = \frac{y^5 + 3.2361 y^4 + 5.2361 y^3 + 5.2361 y^2 + 3.2361 y + 1}{y^5 + 7.2361 y^4 + 26.18 y^3 + 58.541 y^2 + 80.9 y + 55.9},$$
(5.73)

where y is defined in (5.61b). Finally, the equalizer back-end impedance $Z_{22\alpha}(s)$ can be determined from (4.138) and is given by (Problem 5.65)

$$Z_{22\alpha}(s) = \frac{\frac{-800}{(y+2)^2}}{\frac{y-2}{y+2} - \hat{S}_{22\alpha}(s)} - \frac{200}{y+2}$$

$$= 200 \frac{4y^5 + 28.944y^4 + 95.193y^3 + 184.275y^2 + 210.232y + 109.802}{9.528y^4 + 67.945y^3 + 213.151y^2 + 340.55y + 227.604}$$

$$= 200 \frac{4y^4 + 20.94y^3 + 53.31y^2 + 77.64y + 54.98}{9.528y^3 + 49.898y^2 + 113.36y + 113.8}$$

$$= 83.96y + \cfrac{1}{0.0083y + \cfrac{1}{52.52y + \cfrac{1}{0.00198y + 0.0104}}}, \quad (5.74)$$

which can be realized as a lossless ladder terminated in a resistor, whose resistance can also be determined directly from the formula given in Problem 5.10. After denormalization, the lossless equalizer N_α together with its load is presented in Fig. 5.18.

From (5.70), the back-end impedance $Z_{22\beta}(s)$ of N_β is given by

$$\left[\frac{Z_{22\beta}(s)}{200}\right]^{\pm 1} = \frac{2y^5 + 7.2361y^4 + 26.18y^3 + 58.541y^2 + 80.9y + 55.9}{7.2361y^4 + 26.18y^3 + 58.541y^2 + 80.9y + 55.9}$$

$$= 0.276y + \cfrac{1}{0.724y + \cfrac{1}{0.894y + \cfrac{1}{0.723y + \cfrac{1}{0.277y + 1}}}}. \quad (5.75)$$

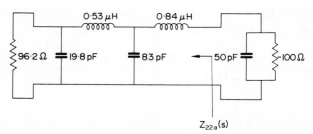

FIG. 5.18. A lossless equalizer that matches an active impedance to a passive resistance.

Denormalizing the element values to $\omega_c = 10^8$ gives the ladder realizations as shown in Fig. 5.19. Figure 5.19(a) corresponds to the choice of the plus sign in (5.69), while Fig. 5.19(b) corresponds to the choice of the minus sign.

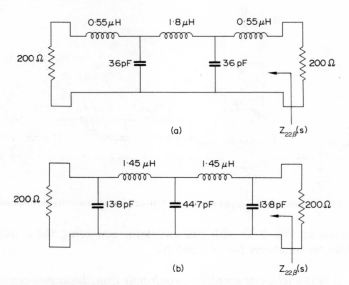

FIG. 5.19. (a) A ladder realization of the impedance (5.75) corresponding to the choice of the plus sign. (b) A ladder realization of the impedance (5.75) corresponding to the choice of the minus sign.

Finally, to compute the admittance matrix $Y_c(s)$ of the ideal circulator, we specify that $R_1 = 50 \, \Omega$, $\hat{R}_2 = 200 \, \Omega$ and $\hat{R}_3 = 96.2 \, \Omega$, which from (5.22a) yields

$$Y_c(s) = 10^{-2} \begin{bmatrix} 0 & -1 & 1.442 \\ 1 & 0 & -0.721 \\ -1.442 & 0.721 & 0 \end{bmatrix}. \quad (5.76)$$

$Y_c(s)$ can be realized by using one gyrator and an ideal transformer bank as mentioned in Example 5.1. The complete amplifier is

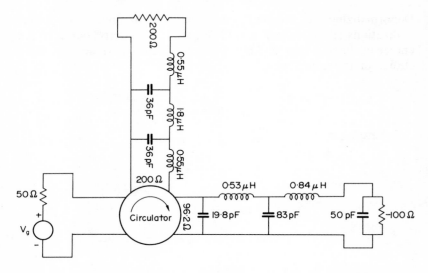

FIG. 5.20. A tunnel diode amplifier with the fifth-order Butterworth gain response.

presented in Fig. 5.20 with the circulator providing the required resistive terminations for N_α and N_β.

3.4.2. *The tunnel diode amplifier: equiripple transducer power gain*

Consider the same problem discussed in the preceding section except now we wish to achieve the nth-order low-pass Chebyshev transducer power-gain characteristic

$$G(\omega^2) = \frac{K_n}{1 + \epsilon^2 C_n^2(\omega/\omega_c)}, \qquad K_n \geq 1, \qquad (5.77)$$

where $\omega_c = 2\pi f_c$ and f_c is the bandwidth. Following (5.26), we choose

$$S_{12\beta}(s)S_{12\beta}(-s) = \frac{K_n}{K_n + \epsilon^2 C_n^2(-js/\omega_c)}, \qquad (5.78a)$$

$$S_{22\alpha}(s)S_{22\alpha}(-s) = \frac{1 + \epsilon^2 C_n^2(-js/\omega_c)}{K_n + \epsilon^2 C_n^2(-js/\omega_c)}. \qquad (5.78b)$$

A. Realization of N_α

Equation (5.78b) can be rewritten as

$$S_{22\alpha}(s)S_{22\alpha}(-s) = \alpha^{-2}\frac{1+\epsilon^2 C_n^2(-jy)}{1+\hat{\epsilon}^2 C_n^2(-jy)},\qquad(5.79a)$$

where

$$\alpha^2 = K_n,\qquad(5.79b)$$

$$\hat{\epsilon} = \epsilon/\alpha,\qquad(5.79c)$$

$$y = s/\omega_c,\qquad(5.79d)$$

and $C_n(x)$ is the nth-order Chebyshev polynomial as defined in (3.45). Like the Butterworth case, let $\hat{S}_{22\alpha}(s)$ be the minimum-phase factorization of (5.79a). Write

$$\hat{S}_{22\alpha}(s) = \pm \frac{y^n + b_{n-1}y^{n-1} + \cdots + b_1 y + b_0}{y^n + \hat{b}_{n-1}y^{n-1} + \cdots + \hat{b}_1 y + \hat{b}_0},\qquad(5.80)$$

where $b_i > 0$ and $\hat{b}_i > 0$ ($i = 0, 1, \ldots, n-1$), as computed by (3.79). Expanding (5.80) by Laurent series about the zero of transmission as in (5.63c) yields

$$\pm \hat{S}_{22\alpha}(s) = 1 + (b_{n-1} - \hat{b}_{n-1})\omega_c/s + \cdots.\qquad(5.81)$$

From the coefficient conditions of (5.65), we see that we must choose the plus sign on the left-hand side of (5.81), and that, in addition, the inequality

$$\frac{2}{RC\omega_c} \geq \hat{b}_{n-1} - b_{n-1}\qquad(5.82)$$

must be satisfied. Substituting (4.52a) and (4.53a) in (5.82) and solving for K_n give the maximum attainable gain within the passband for the bandwidth ω_c:

$$K_n \leq \epsilon^2 \sinh^2\left\{n \sinh^{-1}\left[\sinh\left(\frac{1}{n}\sinh^{-1}\frac{1}{\epsilon}\right) + \frac{\sin(\pi/2n)}{\pi RC f_c}\right]\right\}.\qquad(5.83)$$

348 BROADBAND MATCHING NETWORKS

In the limit as $n \to \infty$ and $\epsilon \to 0$, we have

$$\lim_{\epsilon \to 0}\lim_{n \to \infty} K_n \leq \lim_{\epsilon \to 0} \epsilon^2 \sinh^2\left[\sinh^{-1}\left(\frac{1}{\epsilon}\right) + \frac{1}{2RCf_c}\right]$$

$$= \lim_{\epsilon \to 0} \epsilon^2 \left[\sinh\left(\frac{1}{2RCf_c}\right)\cosh\left(\sinh^{-1}\frac{1}{\epsilon}\right) + \frac{1}{\epsilon}\cosh\left(\frac{1}{2RCf_c}\right)\right]^2$$

$$= \left[\sinh\left(\frac{1}{2RCf_c}\right) + \cosh\left(\frac{1}{2RCf_c}\right)\right]^2 = \exp\left(\frac{1}{RCf_c}\right). \quad (5.84)$$

The right-hand side of (5.83) represents the maximum attainable K_n, which has been plotted in dB as a function of RCf_c and n for various values of ripple width. The curves are presented in Fig. 5.21. An inspection of these curves shows that for a given bandwidth and a passband tolerance, K_n is raised as n is increased. To evaluate the relative performance of the equiripple and maximally-flat amplifiers, we can compare the curves in Fig. 5.17 with those in Fig. 5.21(d), the reason being that they are both plotted for a passband tolerance of 3 dB. It is found that for the same network complexity, the equiripple amplifiers yield higher K_n than that of the maximally-flat amplifiers. The explanation lies in the facts that the gain of the maximally-flat characteristic stays more uniformly at a higher level within the passband and that maximum deviation occurs at the cutoff frequency while the gain of the equiripple characteristic varies within the 3-dB bandwidth throughout the passband. Thus, to maintain the same maximum gain, the maximally-flat amplifier uses up more of the available area under the gain curve than that of the equiripple case.

We remark that the right-hand side of (5.67) and (5.84) also represents the maximum attainable constant gain for the ideal brick-wall type of low-pass response for the transducer power gain with a preassigned bandwidth ω_c. To see this, we use (5.15) in conjunction with (4.282), giving the inequality

$$\int_0^\infty \ln G(\omega^2)d\omega = 2\int_0^\infty \ln\left|\frac{S_{12\beta}(j\omega)}{S_{22\alpha}(j\omega)}\right|d\omega \leq 2\int_0^\infty \ln\left|\frac{1}{S_{22\alpha}(j\omega)}\right|d\omega$$

$$= 2\int_0^\infty \ln\left|\frac{1}{S_{11\alpha}(j\omega)}\right|d\omega \leq \frac{2\pi}{RC}, \quad (5.85)$$

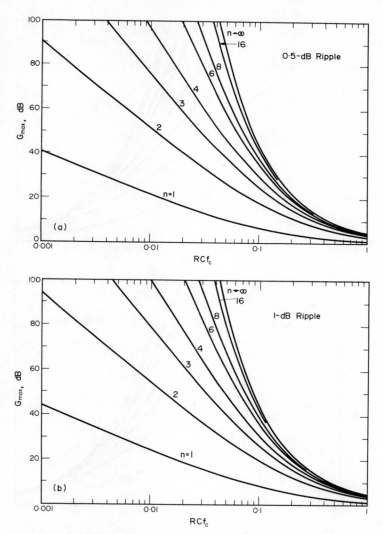

FIG. 5.21. (a) The maximum attainable K_n for tunnel diode amplifiers with equiripple transducer power-gain characteristics. Passband ripple: $\frac{1}{2}$ dB. (b) The maximum attainable K_n for tunnel diode amplifiers with equiripple transducer power-gain characteristics. Passband ripple: 1 dB.

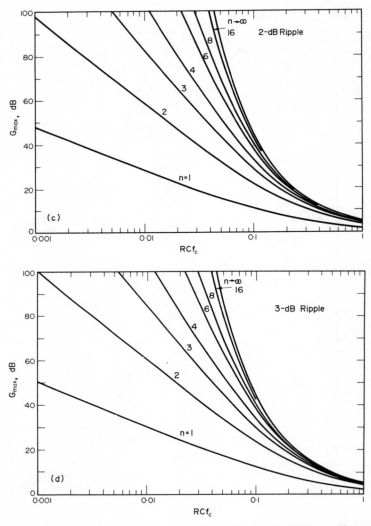

FIG. 5.21 (*cont.*). (c) The maximum attainable K_n for tunnel diode amplifiers with equiripple transducer power-gain characteristics. Passband ripple: 2 dB. (d) The maximum attainable K_n for tunnel diode amplifiers with equiripple transducer power-gain characteristics. Passband ripple: 3 dB.

THE ACTIVE LOAD 351

from which we obtain the desired result (5.67) or (5.84) by letting $G(\omega^2)$ to be a constant over the passband and zero outside.

B. *Realization of N_β*

From (5.21) and (5.78a), we obtain

$$S_{22\beta}(s)S_{22\beta}(-s) = \frac{\hat{\epsilon}^2 C_n^2(-jy)}{1 + \hat{\epsilon}^2 C_n^2(-jy)}, \quad (5.86)$$

whose minimum-phase factorization is given by

$$\hat{S}_{22\beta}(s) = \pm \frac{\hat{p}_m(y)}{\hat{q}(y)}, \quad (5.87a)$$

where $\hat{p}_m(y)$ is defined in (4.187) and

$$\hat{q}(y) = y^n + \hat{b}_{n-1}y^{n-1} + \cdots + \hat{b}_1 y + \hat{b}_0 = \sum_{m=0}^{n} \hat{b}_m y^m, \quad (5.87b)$$

$\hat{b}_n = 1$. Since the coefficient constraints are always satisfied, the equalizer back-end impedance $Z_{22\beta}(s)$ can be determined by the formula (4.138) which yields

$$\frac{Z_{22\beta}(s)}{R_2} = \frac{2}{1 - \hat{S}_{22\beta}(s)} - 1 = \frac{\hat{q}(y) \pm \hat{p}_m(y)}{\hat{q}(y) \mp \hat{p}_m(y)}. \quad (5.88)$$

As before, expanding the function in a continued fraction gives a network realization which is recognized as a lossless ladder terminated in a resistor.

EXAMPLE 5.3. Consider the same problem as in Example 5.2 except now that we wish to have an equiripple characteristic for its transducer power gain. In order to compare the present result with that of the maximally-flat case, we further stipulate that the passband tolerance must be within 3 dB.

From the curves of Fig. 5.21(d), we see that for $n = 3$ the maximum passband gain is about 35 dB, which can also be computed

352　BROADBAND MATCHING NETWORKS

from the formula (5.83), as follows:

$$K_3 \leqq (0.9976)^2 \sinh^2\{3\sinh^{-1}[\sinh(\tfrac{1}{3}\sinh^{-1}1.0024)+4\sin 30°]\}$$
$$= (0.9952)(55.48)^2 = 3063 \text{ or } 34.86 \text{ dB}, \tag{5.89}$$

where the 3-dB tolerance in the passband corresponds to $\epsilon = 0.9976$. Thus, the minimum gain in the passband is 31.86 dB, 1.86 dB above the minimum requirement. Using (3.79) in conjunction with (4.53), (5.79) and (5.80) yields

$$\hat{S}_{22\alpha}(s) = \frac{y^3 + 0.597y^2 + 0.928y + 0.251}{y^3 + 4.58y^2 + 11.242y + 13.736}, \tag{5.90}$$

where $y = s/10^8$. The equalizer back-end impedance $Z_{22\alpha}(s)$ is determined by

$$Z_{22\alpha}(s) = \frac{\dfrac{-800}{(y+2)^2}}{\dfrac{y-2}{y+2} - \dfrac{y^3 + 0.597y^2 + 0.928y + 0.251}{y^3 + 4.58y^2 + 11.242y + 13.736}} - \frac{200}{y+2}$$

$$= 200\,\frac{3.983y^2 + 10.314y + 13.485}{0.017y^3 + 0.0397y^2 + 10.856y + 27.974}$$

$$= \cfrac{1}{21.3 \times 10^{-14}s + \cfrac{1}{7.38 \times 10^{-7}s + \cfrac{1}{40 \times 10^{-12}s + \cfrac{1}{96.41}}}}, \tag{5.91}$$

which can be identified as a lossless ladder terminated in a resistor as shown in Fig. 5.22.

From (5.88) and in conjunction with (4.187), we obtain the back-end impedance of the lossless equalizer N_β as

$$Z_{22\beta}(s) = 200\,\frac{(y^3 + 4.58y^2 + 11.242y + 13.736) + (y^3 + 0.75y)}{(y^3 + 4.58y^2 + 11.242y + 13.736) - (y^3 + 0.75y)}$$

$$= 200\,\frac{2y^3 + 4.58y^2 + 11.992y + 13.736}{4.58y^2 + 10.492y + 13.736}$$

$$= 87.34y + \cfrac{1}{0.00382y + \cfrac{1}{87.28y + 200}} \tag{5.92a}$$

THE ACTIVE LOAD

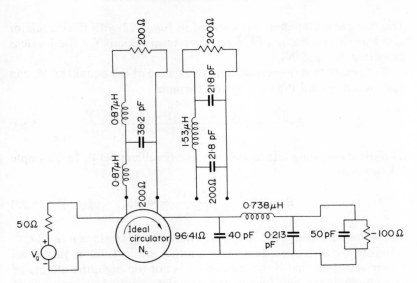

FIG. 5.22. A tunnel diode amplifier with the third-order Chebyshev gain response.

for the choice of the plus sign in (5.87a), and for the negative sign we have

$$Z_{22\beta}(s) = \cfrac{1}{0.00218y + \cfrac{1}{152.8y + \cfrac{1}{0.00218y + \cfrac{1}{200}}}}, \quad (5.92b)$$

which after denormalization can be realized as two lossless ladders terminated in a resistor of resistance 200 Ω, as shown in Fig. 5.22.

Finally, to compute the admittance matrix $Y_c(s)$ of the ideal circulator, we specify that $R_1 = 50$ Ω, $\hat{R}_2 = 200$ Ω, and $\hat{R}_3 = 96.41$ Ω, which from (5.22a) gives

$$Y_c(s) = 10^{-2} \begin{bmatrix} 0 & -1 & 1.44 \\ 1 & 0 & -0.72 \\ -1.44 & 0.72 & 0 \end{bmatrix}. \quad (5.93)$$

The complete amplifier is presented in Fig. 5.22 with the circulator again providing the required resistive terminations for the lossless equalizer N_α and N_β.

We remark that the terminating resistance of the equalizer N_α can also be computed directly by the formula

$$\hat{R}_3 = R \frac{\{[K_n + \epsilon^2 C_n^2(0)]^{1/2} - [1 + \epsilon^2 C_n^2(0)]^{1/2}\}^2}{K_n - 1}, \qquad (5.94)$$

its derivation being left as an exercise (Problem 5.14). In Example 5.3, we have

$$\hat{R}_3 = 100 \frac{(55.344 - 1)^2}{3063 - 1} = 96.45, \qquad (5.95)$$

consistent with that given in (5.91), where $C_3(0) = 0$ since n is odd.

Before we turn our attention to another subject, let us justify an earlier assertion that if the frequency is not too high, the effects of series inductance and loss in the complete equivalent network of Fig. 5.1(a) for a tunnel diode can be ignored. To see this, we compute the driving-point impedance $Z_d(s)$ of the network in Fig. 5.1(a). On the $j\omega$-axis, the real part of $Z_d(j\omega)$ ceases to be negative when the *resistive cutoff frequency*

$$f_r = (R/R_d - 1)^{1/2}/(2\pi RC) \qquad (5.96)$$

is reached or surpassed, and the corresponding *self-resonant frequency* is determined by (Problem 5.17)

$$f_s = (R^2 C/L_d - 1)^{1/2}/(2\pi RC). \qquad (5.97)$$

The tunnel diode used in Examples 5.2 and 5.3 has the following parameters:

$$\begin{aligned} -R &= -100\,\Omega, & C &= 50\,\text{pF}, \\ R_d &= 1\,\Omega, & L_d &= 0.4\,\text{nH}. \end{aligned} \qquad (5.98)$$

The resistive cutoff frequency is given by

$$f_r = (100 - 1)^{1/2} \times 10^8/\pi = 316.7\,\text{MHz}, \qquad (5.99\text{a})$$

and the diode self-resonant frequency is obtained as

$$f_s = (5000/4 - 1)^{1/2} \times 10^8/\pi = 1124.9 \text{ MHz}. \quad (5.99b)$$

The required bandwidth in the examples is 15.92 MHz, which is entirely consistent with the assumption that the effects of R_d and L_d are negligible. Also we remark that, in many instances, the equalizer N_α allows some series inductance at its ports to be absorbed into N_α. Thus the effects of the series inductance L_d in the equivalent network for a tunnel diode can be included in the design. For example, in Fig. 5.20, in order to absorb the inductance $L_d = 0.4$ nH of the tunnel diode, the inductor with inductance 0.84 μH should be replaced by one with inductance 0.8396 μH.

3.5. Extension and stability

So far we are dealing exclusively with the case where the load impedance $z_2(s)$ at the output port is purely resistive. The development can easily be extended to any strictly passive load $z_2(s)$ by making slight changes in the normalization of the scattering matrix. Refer first to the schematic of the general single-stage nonreciprocal negative-resistance amplifier of Fig. 5.4, where the load resistance R_2 is now replaced by a strictly passive impedance $z_2(s)$. Retaining the reference impedances R_1 and $z_3(s)$ at port 1 and port 3, the scattering matrix $S(s)$ of the lossless three-port network N is now normalized to the strictly passive impedances $R_1, z_2(s)$ and $z_3(s)$. With the exception of a few subsections on the lossless two-port N_β, all the results derived above for the nonreciprocal amplifier remain valid. For the two-port N_β, the problem is equivalent to matching a strictly passive load $z_2(s)$ to a resistive load $\hat{R}_2$ to achieve a prescribed transducer power-gain characteristic. According to the results presented in Chapter 4, the problem is solvable if and only if the back-end bounded-real reflection coefficient $S_{22\beta}(s)$ satisfies the coefficient constraints. We shall not pursue this line any further since the topic has been treated extensively in the preceding chapter.

We now turn our attention to the stability of the complete amplifier. As mentioned in §3.1, whenever an active impedance is involved, stability must be tested for the ports of the lossless three-port N. For illustrative purposes, we shall carry out the test for each of the ports.

For stability at port 3, the impedance $Z_{33}(s) = Z_{22\alpha}(s)$ looking into port 3 with the other two ports terminating in their reference impedances R_1 and $z_2(s)$ must not be equal to $-z_l(s)$ for all s in the closed RHS. This requires that the zeros of the function $Z_{22\alpha}(s) + z_l(s)$ be restricted to the open LHS, which according to (5.20) are also the zeros of $\hat{S}_{22\alpha}(s)$. But, by the very philosophy of optimum design, $\hat{S}_{22\alpha}(s)$ is chosen to be a minimum-phase function, and thus is devoid of zeros in the open RHS. According to (5.62) and (5.80), $S_{22\alpha}(j\omega) \neq 0$ for all ω for either Butterworth or Chebyshev characteristics. Thus, we conclude that at port 3 it is stable for either Butterworth or Chebyshev transducer power-gain characteristics. In addition, for gain characteristics other than these two, it is always at least marginally stable, meaning that it is at most unstable on the $j\omega$-axis.

The impedance $Z_{11}(s)$ looking into port 1 when port 2 is terminated in $z_2(s)$ and port 3 in $z_l(s)$ is given by (Problem 5.18)

$$Z_{11}(s) = R_1 \frac{1 + S_{11a}(s)}{1 - S_{11a}(s)}, \qquad (5.100)$$

where from (5.5)

$$S_{11a}(s) = S_{11}(s) - \frac{S_{13}(s)S_{31}(s)}{S_{33}(s)}. \qquad (5.101)$$

For stability, $Z_{11}(s) \neq -R_1$ for all s in the closed RHS, which is equivalent to requiring that $|S_{11a}(s)|$ be finite in the region. But by (5.101), $|S_{11a}(s)|$ is finite in the closed RHS if and only if $S_{33}(s) \neq 0$ in the region. Since from (5.14b) and (5.20), $S_{33}(s) = S_{22\alpha}(s) = \hat{\eta}(s)\hat{S}_{22\alpha}(s)$, invoking the same argument as above for port 3 leads to the same conclusion.

The impedance $Z_{22}(s)$ looking into port 2 when port 1 is terminated in R_1 and port 3 in $z_l(s)$ is given by (Problem 5.18)

$$Z_{22}(s) = z_2(s) \frac{\dfrac{h_2(s)z_2(-s)}{h_2(-s)z_2(s)} + S_{22a}(s)}{\dfrac{h_2(s)}{h_2(-s)} - S_{22a}(s)}, \qquad (5.102)$$

$h_2(s)h_2(-s)$ being the para-hermitian part of $z_2(s)$, where from (5.5)

$$S_{22a}(s) = S_{22}(s) - \frac{S_{23}(s)S_{32}(s)}{S_{33}(s)}. \quad (5.103)$$

Again, stability demands that $Z_{22}(s) \neq -z_2(s)$ for all s in the closed RHS. This together with the fact that $z_2(s)$ is strictly passive requires that $|S_{22a}(s)|$ be finite in the closed RHS. But by (5.103), $|S_{22a}(s)|$ is finite in the closed RHS if and only if $S_{33} \neq 0$ in the region. Following the same reasoning as above for port 1 and port 3, we again conclude that at port 2 it is stable for either the Butterworth or Chebyshev characteristics. For the transducer power-gain characteristics other than these two, it is always at least marginally stable.

4. Transmission-power amplifiers

Another extremely practical amplifier configuration is one in which the active impedance is equalized by a lossless two-port network, as shown in Fig. 5.23. The configuration can be considered

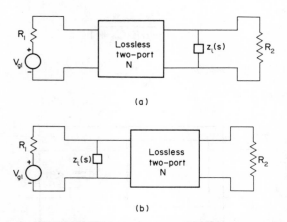

FIG. 5.23. (a) The general configuration of a transmission-power amplifier with the active impedance $z_i(s)$ in shunt with the load. (b) The general configuration of a transmission-power amplifier amplifier with the active impedance $z_i(s)$ in shunt with the generator.

as the degenerate types of the most general arrangement of Fig. 5.3, in which the output port 2 or the input port 1 is connected directly to the active impedance at the port 3. The active impedance that will be considered in the design in this section is represented by the simplified model of Fig. 5.16 for the tunnel diode. The arrangement is attractive in that it facilitates the design of the amplifier which requires the use of only one or two transformers with simple supply for diode bias. We shall first discuss its gain-bandwidth limitations, and then derive and present formulas for amplifiers having Butterworth and Chebyshev characteristics of arbitrary order.

4.1. Tunnel diode in shunt with the load

As indicated in Fig. 5.23, there are two possible arrangements for the tunnel diode. In the present section, we analyze the configuration where the tunnel diode is connected in shunt with the load, as shown in Fig. 5.24. The parallel combination of the load resistor R_2 and the tunnel diode is replaced by an equivalent network composed of the parallel connection of a capacitor C and a resistor $-\alpha R_2$, whose impedance is given by

$$z_2(s) = \frac{-\alpha R_2}{-\alpha R_2 C s + 1}, \qquad (5.104a)$$

where

$$\alpha = \frac{R}{R_2 - R}. \qquad (5.104b)$$

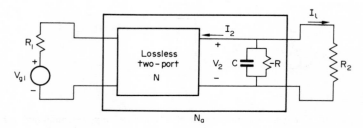

Fig. 5.24. The general configuration of a transmission-power amplifier with the tunnel diode in shunt with the load.

THE ACTIVE LOAD

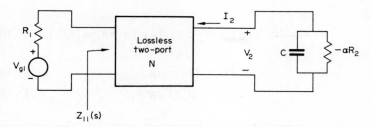

FIG. 5.25. An equivalent representation of the network of Fig. 5.24.

The modified equivalent network is depicted in Fig. 5.25. To describe the lossless two-port network N, let

$$\hat{S}(s) = [\hat{S}_{ij}] \tag{5.105}$$

be its scattering matrix normalized with respect to the strictly passive impedances R_1 and

$$\hat{z}_2(s) = \frac{|\alpha|R_2}{|\alpha|R_2 C s + 1}. \tag{5.106}$$

Depending upon the relative values of R and R_2, the resistance $-\alpha R_2$ may be either positive or negative, each case being considered separately in a section.

4.1.1. *Transducer power gain*: $R_2 > R$

In this case, α is positive and $z_2(s)$ and $\hat{z}_2(s)$ are related by

$$-\hat{z}_2(-s) = z_2(s). \tag{5.107}$$

From the familiar properties of the scattering parameters which arise when the output port of Fig. 5.25 is actually terminated in $-\hat{z}_2(-s)$, as discussed in § 3.5 of Chapter 2, the normalized reflected wave $b_2(s)$ at the output port is identically zero. Under this situation, we have

$$0 = \hat{S}_{21}(s)a_1(s) + \hat{S}_{22}(s)a_2(s) \tag{5.108}$$

and

$$a_2(s) = \tfrac{1}{2}\hat{h}_2^{-1}(s)[V_2(s) + \hat{z}_2(s)I_2(s)] = \hat{h}_2(-s)I_2(s), \quad (5.109)$$

where

$$V_2(s) = \hat{z}_2(-s)I_2(s), \quad (5.110a)$$

$$a_1(s) = \tfrac{1}{2}V_{g1}(s)/R_1^{1/2}, \quad (5.110b)$$

$\hat{h}_2(s)\hat{h}_2(-s)$ being the para-hermitian part of the reference impedance $\hat{z}_2(s)$. Eliminating the variables $a_1(s)$ and $a_2(s)$ from the above equations yields

$$I_2(s) = -\frac{V_{g1}(s)\hat{S}_{21}(s)}{2R_1^{1/2}\hat{h}_2(-s)\hat{S}_{22}(s)}, \quad (5.111)$$

where

$$\hat{h}_2(s) = \frac{(\alpha R_2)^{1/2}}{\alpha R_2 Cs + 1}. \quad (5.112)$$

Referring to the amplifier shown in Fig. 5.24, the transducer power gain is defined as

$$G(\omega^2) = \frac{|I_l(j\omega)|^2 R_2}{\dfrac{|V_{g1}(j\omega)|^2}{4R_1}}, \quad (5.113)$$

in which the current $I_l(s)$ can be expressed in terms of $I_2(s)$ by the relation

$$I_l(s) = -I_2(s) - (Cs - 1/R)V_2(s)$$

$$= -\frac{\alpha I_2(s)}{\alpha R_2 Cs - 1}. \quad (5.114)$$

Substituting these in (5.113) gives the desired transducer power gain

$$G(\omega^2) = \alpha\left[\frac{1}{|\hat{S}_{22}(j\omega)|^2} - 1\right], \quad \alpha > 0. \quad (5.115)$$

Thus, we have successfully converted the problem of designing an optimum transmission-power amplifier having a preassigned transducer power-gain characteristic over a frequency band of interest to that of equalizing a strictly passive impedance $\hat{z}_2(s)$ to a constant resistance R_1 to achieve a minimum magnitude of the back-end reflection coefficient $\hat{S}_{22}(j\omega)$ of N over the same frequency band of interest. For illustrative purposes, we shall discuss the gain-bandwidth limitations of amplifiers having Butterworth and Chebyshev gain characteristics of arbitrary order.

A. Maximally-flat low-pass amplifiers

Consider the nth-order low-pass Butterworth transducer power-gain characteristic

$$G(\omega^2) = \frac{K_n}{1+(\omega/\omega_c)^{2n}}, \qquad K_n \geq 1, \tag{5.116}$$

where ω_c is the 3-dB radian bandwidth. Substituting (5.116) in (5.115), solving for $|\hat{S}_{22}(j\omega)|^2$, and after analytic continuation yield

$$\hat{S}_{22}(s)\hat{S}_{22}(-s) = \hat{K}_n^{-1} \frac{1+(-1)^n y^{2n}}{1+(-1)^n x^{2n}}, \tag{5.117}$$

where

$$\hat{K}_n = 1 + K_n/\alpha, \tag{5.118a}$$

$$y = s/\omega_c, \tag{5.118b}$$

$$x = \hat{K}_n^{-1/2n} s/\omega_c. \tag{5.118c}$$

Observe that (5.117) has the same form as (5.60b), meaning that we can write an inequality similar to (5.66) for the coefficient constraints:

$$\hat{K}_n \leq \left[1 + \frac{\sin(\pi/2n)}{\pi\alpha R_2 C f_c}\right]^{2n}. \tag{5.119}$$

In addition, the curves plotted in Fig. 5.17 for the nonreciprocal amplifiers can still be used for obtaining the maximum $\hat{K}_n$ in (5.119) provided that the terms K_n and R in the figures are replaced by $\hat{K}_n$ and αR_2, respectively. Substituting (5.119) in (5.118a) gives the maximum attainable dc gain

$$K_n \leq \alpha\left[\left(1 + \frac{\sin(\pi/2n)}{\pi\alpha R_2 C f_c}\right)^{2n} - 1\right]. \quad (5.120)$$

In the limit as $n \to \infty$,

$$K_\infty \leq \alpha\left[\exp\left(\frac{1}{\alpha R_2 C f_c}\right) - 1\right], \quad (5.121)$$

(Problem 5.20). As before, the right-hand side of (5.121) also represents the maximum attainable constant gain for the ideal brick-wall type of low-pass response for the transducer power gain with a preassigned bandwidth ω_c (Problem 5.21).

We shall illustrate the above by the following example.

EXAMPLE 5.4. It is desired to design a transmission-power amplifier having a maximally-flat characteristic for its transducer power gain. The amplifier is to be operated between 90-Ω and 500-Ω terminations for which the passband gain must be at least 37 dB for the bandwidth of 40 MHz. For the tunnel diode, we use the following set of parameters:

$$\begin{aligned} -R &= -143\,\Omega, & C &= 7\,\text{pF}, \\ R_d &= 1\,\Omega, & L_d &= 5\,\text{nH}. \end{aligned} \quad (5.122)$$

From the above specification, we have

$R_1 = 90\,\Omega,$ $R_2 = 500\,\Omega,$

$f_c = 40\,\text{MHz},$ $\omega_c = 2.513 \times 10^8\,\text{rad/s},$

$f_r = 1.89\,\text{GHz},$ $f_s = 0.84\,\text{GHz},$

f_r and f_s being the resistive cutoff frequency and the self-resonant frequency of the tunnel diode, as computed from (5.96) and (5.97). Thus, a bandwidth of 0.04 GHz is consistent with the assumption that the effects R_d and L_d are negligible. Since the maximum deviation in the passband is 3 dB for the Butterworth characteristic,

THE ACTIVE LOAD

the required dc gain K_n must be at least 40 dB, which from (5.118a) is equivalent to requiring a value for $\hat{K}_n$ of 24,939 or 43.97 dB. From the curves given in Fig. 5.17, we see that for $n = 5$ the gain $\hat{K}_5$ is about 44 dB. For illustrative purposes, we compute $\hat{K}_5$ by means of (5.119), giving

$$\hat{K}_5 = \left(\frac{\sin 18°}{\pi 0.401 \times 500 \times 7 \times 10^{-12} \times 40 \times 10^6} + 1 \right)^{10} = (2.752)^{10}$$
$$= 24{,}916 \quad \text{or} \quad 43.96 \text{ dB},$$

in which

$$\alpha = R/(R_2 - R) = 143/(500 - 143) = 0.401.$$

From (5.117), the minimum-phase factorization can be expressed as

$$\tilde{S}_{22}(s) = \frac{y^5 + 3.236y^4 + 5.236y^3 + 5.236y^2 + 3.236y + 1}{(2.752)^5(x^5 + 3.236x^4 + 5.236x^3 + 5.236x^2 + 3.236x + 1)}$$
$$= \frac{y^5 + 3.236y^4 + 5.236y^3 + 5.236y^2 + 3.236y + 1}{y^5 + 8.906y^4 + 39.658y^3 + 109.144y^2 + 185.642y + 157.883}.$$

(5.123)

Using (4.93), the equalizer back-end impedance $Z_{22}(s)$ is computed as

$$Z_{22}(s) = 200.5 \frac{5.6699y^4 + 34.4223y^3 + 103.908y^2 + 182.402y + 156.883}{8.242y^3 + 50.038y^2 + 133.539y + 158.883}$$

$$= 0.55 \times 10^{-6} s + \cfrac{1}{13.6 \times 10^{-12} s + \cfrac{1}{0.37 \times 10^{-6} s + \cfrac{1}{3.3 \times 10^{-12} s + \cfrac{1}{197.8}}}},$$

(5.124)

which can then be identified as a lossless ladder terminated in a resistor as shown in Fig. 5.26. The complete amplifier together with its bias is depicted in Fig. 5.27, where the choke in the bias circuit provides high-frequency decoupling. To avoid instability, the bias

364 BROADBAND MATCHING NETWORKS

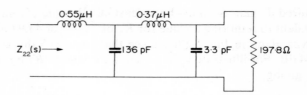

Fig. 5.26. A ladder realization of the impedance (5.124).

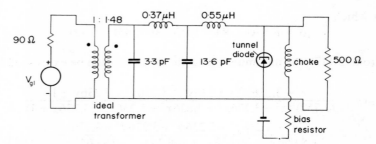

Fig. 5.27. A transmission-power amplifier with the fifth-order Butterworth gain response.

resistor must be such that its load line intersects the diode characteristic at only a single point, which can easily be accomplished by choosing the bias load resistance smaller than $R = 143\,\Omega$. We remark that in formula (4.93) the term corresponding to RC is $\alpha R_2 C \omega_c$ due to normalization and that the terminating resistance of the lossless ladder can be determined directly from the formula (5.266). To avoid the necessity of expanding $Z_{22}(s)$ in a continued fraction, explicit formulas for the element values have recently been derived by Chen (1975, 1976).

B. *Equiripple low-pass amplifiers*

Consider the nth-order low-pass Chebyshev transducer power-gain characteristic

$$G(\omega^2) = \frac{K_n}{1 + \epsilon^2 C_n^2(\omega/\omega_c)}, \qquad K_n \geqq 1. \tag{5.125}$$

THE ACTIVE LOAD

Proceeding as in the Butterworth case, we have

$$\hat{S}_{22}(s)\hat{S}_{22}(-s) = \hat{K}_n^{-1}\frac{1+\epsilon^2 C_n^2(-jy)}{1+\hat{\epsilon}^2 C_n^2(-jy)}, \quad (5.126)$$

where

$$\hat{\epsilon} = \hat{K}_n^{-1/2}\epsilon, \quad (5.127)$$

y and $\hat{K}_n$ being the same as in (5.118). Comparing (5.126) with (5.79a) shows that we can obtain a relation similar to (5.83) with (Problem 5.23)

$$\hat{K}_n \leq \epsilon^2 \sinh^2\left\{n \sinh^{-1}\left[\sinh\left(\frac{1}{n}\sinh^{-1}\frac{1}{\epsilon}\right) + \frac{\sin(\pi/2n)}{\pi\alpha R_2 Cf_c}\right]\right\}. \quad (5.128)$$

As before, we can again use the curves plotted in Fig. 5.21 for the nonreciprocal amplifiers for obtaining the maximum $\hat{K}_n$ in (5.128) provided that the terms K_n and R in the figures are replaced by $\hat{K}_n$ and αR_2, respectively. Substituting (5.128) in (5.118a) gives the bona fide gain-bandwidth limitation:

$$K_n \leq \alpha\left\{\epsilon^2 \sinh^2\left(n \sinh^{-1}\left[\sinh\left(\frac{1}{n}\sinh^{-1}\frac{1}{\epsilon}\right) + \frac{\sin(\pi/2n)}{\pi\alpha R_2 Cf_c}\right]\right) - 1\right\}.$$

$$(5.129)$$

In the limit as $n \to \infty$ and $\epsilon \to 0$, we have (Problem 5.24)

$$K_\infty \leq \alpha\left[\exp\left(\frac{1}{\alpha R_2 Cf_c}\right) - 1\right], \quad (5.130)$$

which is the same as (5.121) for the ideal brick-wall type of response, as expected.

We illustrate the Chebyshev case by the same problem given in Example 5.4.

EXAMPLE 5.5. Consider the same problem as in Example 5.4 except now that we wish to achieve an equiripple characteristic for its transducer power gain. For comparative purposes, we choose

$n = 5$ and let the passband tolerance be within 3 dB, which corresponds to having a ripple factor $\epsilon = 0.9976$, the reason being that both cases now have the same passband tolerance.

We first compute the maximum attainable K_5 from (5.129), yielding

$$K_5 = 0.401 \left\{ 0.9976^2 \sinh^2 \left(5 \sinh^{-1} \left[\sinh\left(\tfrac{1}{5} \sinh^{-1} 1.002\right) \right. \right. \right.$$
$$\left. \left. \left. + \frac{\sin 18°}{\pi 0.401 \times 500 \times 7 \times 10^{-12} \times 40 \times 10^6} \right] \right) - 1 \right\}$$
$$= 134{,}864 \quad \text{or} \quad 51.30 \qquad (5.131)$$

which from (5.118a) gives a value for $\hat{K}_n$ of 336,320 or 55.27 dB, a value consistent with that given in Fig. 5.21(d). From (5.127) and (4.53), we obtain

$$\hat{\epsilon} = 0.9976/(336{,}320)^{1/2} = 0.00172, \qquad (5.132a)$$

$$a = \frac{1}{5} \sinh^{-1} \frac{1}{0.9976} = 0.1766, \qquad (5.132b)$$

$$\hat{a} = \frac{1}{5} \sinh^{-1} \frac{1}{0.00172} = 1.4117 \qquad (5.132c)$$

Using these in conjunction with (3.79), the minimum-phase factorization of (5.126) is given by

$$\tilde{S}_{22}(s) = \frac{(y + 0.1775)(y + 0.0548 - j0.9659)(y + 0.0548 + j0.9659)}{(y + 1.9292)(y + 0.5961 - j2.0666)(y + 0.5961 + j2.0666)}$$
$$\times \frac{(y + 0.1436 - j0.5969)(y + 0.1436 + j0.5969)}{(y + 1.5607 - j1.2772)(y + 1.5607 + j1.2772)}$$
$$= \frac{y^5 + 0.5743y^4 + 1.4148y^3 + 0.5487y^2 + 0.4078y + 0.0626}{y^5 + 6.2428y^4 + 20.736y^3 + 43.239y^2 + 56.027y + 36.297}.$$
$$(5.133)$$

Using (4.93), the back-end impedance $Z_{22}(s)$ of the equalizer is computed as

$$Z_{22}(s) = 200.5 \frac{5.669y^4 + 19.322y^3 + 42.69y^2 + 55.619y + 36.235}{7.093y^3 + 24.169y^2 + 43.653y + 36.36}$$

$$= 0.638 \times 10^{-6}s$$

$$+ \cfrac{1}{18.04 \times 10^{-12}s + \cfrac{1}{0.581 \times 10^{-6}s + \cfrac{1}{5.86 \times 10^{-12}s + \cfrac{1}{199.81}}}} \quad (5.134)$$

which can be realized as a lossless ladder terminated in a resistor whose resistance can also be determined directly by the formula given in Problem 5.28. The complete amplifier together with its bias circuit is presented in Fig. 5.28. As expected, for a passband

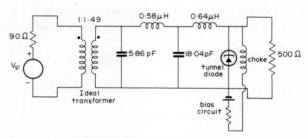

FIG. 5.28. A transmission-power amplifier with the fifth-order Chebyshev gain response.

tolerance of 3 dB and for the same network complexity, the amplifier having equiripple gain characteristic yields a higher K_n than that having the maximally-flat characteristic, which as explained in §3.4.2 is valid in general. In the present case, the fifth-order equiripple amplifier yields a dc gain of 51.30 dB, while corresponding maximally-flat amplifier gives a dc gain of 40 dB. From (5.130), these compare to the maximum allowable gain of 73.4 dB for the ideal brick-wall type of low-pass characteristic.

To avoid the necessity of expanding $Z_{22}(s)$ in a continued fraction, explicit formulas for the element values have been obtained by Chen (1975, 1976).

4.1.2. Transducer power gain: $R_2 < R$

In this case, α is negative and $z_2(s) = \hat{z}_2(s)$, a strictly passive impedance, which is also the normalizing impedance at port 2, as depicted in Fig. 5.29. From the familiar properties of the scattering

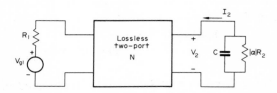

FIG. 5.29. An equivalent representation of the network of Fig. 5.24 with $R_2 < R$.

parameters which arise when the output port is terminated in its reference impedance $\hat{z}_2(s)$, the normalized incident wave $a_2(s)$ at the output port is identically zero. Under this situation, we have

$$b_2(s) = \hat{S}_{21}(s) a_1(s), \tag{5.135}$$

$$b_2(s) = \tfrac{1}{2} h_2^{-1}(-s)[V_2(s) - z_2(-s)I_2(s)] = -h_2(s)I_2(s), \tag{5.136}$$

where $h_2(s)h_2(-s)$ is the para-hermitian part of $z_2(s)$, and

$$V_2(s) = -z_2(s) I_2(s). \tag{5.137}$$

Combining these results together with (5.110b) gives

$$I_2(s) = -\frac{\hat{S}_{21}(s) V_{g1}(s)}{2 R_1^{1/2} h_2(s)}. \tag{5.138}$$

Since (5.113) and (5.114) are still valid under the present situation, this leads to the desired formula for the transducer power gain of the amplifier:

$$G(\omega^2) = |\alpha| \|\hat{S}_{21}(j\omega)|^2 = |\alpha|[1 - |\hat{S}_{22}(j\omega)|^2]. \tag{5.139}$$

The last equation follows from the fact that the two-port network N is lossless, whose scattering matrix $S(j\omega)$ must be unitary. Thus, to design an optimum amplifier having a preassigned transducer power-gain characteristic over a frequency band of interest, we must minimize the magnitude of the back-end reflection coefficient $\hat{S}_{22}(j\omega)$ of N over the same frequency band. But $|\hat{S}_{21}(j\omega)|$ and $|\hat{S}_{22}(j\omega)|$ are bounded by unity, which implies that the amplifier is not inherently high-gain. However, amplification in this case can be obtained by increasing the load resistance. Although some of the available power is wasted in the padding resistance, still more power is delivered to the load than before the padding. To this end, let sufficient resistance be placed in series with R_2 so that

$$R_2 + R_0 = R_2'' > R. \tag{5.140}$$

Defining

$$\alpha'' = \frac{R}{R_2'' - R} \tag{5.141}$$

and using (5.108)–(5.112) with α and R_2 being replaced by α'' and R_2'', respectively, the transducer power gain is determined as (Problem 5.25)

$$G(\omega^2) = \alpha'' \frac{R_2}{R_2''} \left[\frac{1}{|\hat{S}_{22}(j\omega)|^2} - 1 \right]. \tag{5.142}$$

Following the procedure used in deriving (5.120) and (5.129), the dc gain for a maximally-flat characteristic is bounded by

$$K_n \leq \alpha'' \frac{R_2}{R_2''} \left[\left(1 + \frac{\sin(\pi/2n)}{\pi \alpha'' R_2'' C f_c} \right)^{2n} - 1 \right], \tag{5.143}$$

and the constant term for an equiripple characteristic is given by

$$K_n \leq \alpha'' \frac{R_2}{R_2''} \left\{ \epsilon^2 \sinh^2 \left[n \sinh^{-1} \left(\sinh \left(\frac{1}{n} \sinh^{-1} \frac{1}{\epsilon} \right) \right. \right. \right.$$
$$\left. \left. \left. + \frac{\sin(\pi/2n)}{\pi \alpha'' R_2'' C f_c} \right) \right] - 1 \right\}. \tag{5.144}$$

370 BROADBAND MATCHING NETWORKS

We remark that (5.119) and (5.128) remain valid provided that we replace α and R_2 by α'' and R_2'', respectively, and define

$$\hat{K}_n = \frac{R_2'' K_n}{\alpha'' R_2} + 1. \tag{5.145}$$

This indicates that we can still use the curves plotted in Figs. 5.17 and 5.21 for estimating the values of $\hat{K}_n$.

As an illustration, consider the tunnel diode used in Examples 5.4 and 5.5. Let $R_2 = 100\,\Omega$ and $R_0 = 100\,\Omega$. Then $R_2'' = 200\,\Omega$ and

$$\alpha'' = R/(R_2'' - R) = 2.5088.$$

From Fig. 5.17, we have $\hat{K}_5 = 23$ dB for the fifth-order Butterworth response, which gives a value for K_5 of 24 dB; and from Fig. 5.21(d), the maximum $\hat{K}_5$ for the fifth-order Chebyshev response having a 3-dB passband tolerance is about 28 dB, which corresponds to a value for K_5 of about 29 dB.

Now suppose that the padding resistance is doubled, i.e. $R_0 = 200\,\Omega$. The Butterworth dc gain K_5 is increased to about 30.6 dB, which corresponds to a value for $\hat{K}_5$ of about 35.9 dB. The constant term K_5 for the Chebyshev response is increased to 39.7 dB corresponding to a value for $\hat{K}_5$ of about 44.9 dB. Thus, 6.6 dB additional gain is obtained for the Butterworth case and 10.7 dB for the Chebyshev response as a result of load resistance padding. For given values of R, C, f_c and R_2, the best choice of the padding resistance R_0 can be made graphically by plotting K_n or $\hat{K}_n$ as a function of RCf_c for various values of R/R_2''. Having chosen R/R_2'' graphically, the padding resistance R_0 can then be determined from (5.140).

4.2. Tunnel diode in shunt with the generator

As indicated in Fig. 5.23, there are two possible arrangements for the tunnel diode. In the present section, we analyze the configuration where the tunnel diode is connected in shunt with the generator, as shown in Fig. 5.30. The development is analogous to that used for the amplifier where the tunnel diode is in shunt with the load.

THE ACTIVE LOAD

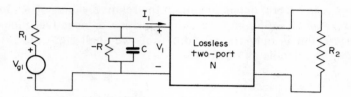

FIG. 5.30. The general configuration of a transmission-power amplifier with the tunnel diode in shunt with the generator.

The shunt combination of the resistive generator and the tunnel diode is replaced by an equivalent Thévenin source

$$\hat{V}_{g1}(s) = \frac{V_{g1}(s)z_1(s)}{R_1}, \tag{5.146}$$

where

$$z_1(s) = \frac{-\beta R_1}{-\beta R_1 Cs + 1}, \tag{5.147a}$$

$$\beta = \frac{R}{R_1 - R}, \tag{5.147b}$$

and an impedance $z_1(s)$, as depicted in Fig. 5.31. To describe the lossless two-port network N, let $\hat{S}(s)$ of (5.105) be its scattering matrix normalized with respect to the strictly passive impedances

$$\hat{z}_1(s) = \frac{|\beta|R_1}{|\beta|R_1 Cs + 1} \tag{5.148}$$

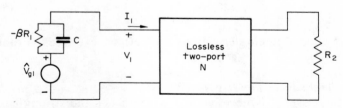

FIG. 5.31. An equivalent representation of the network of Fig. 5.30.

and R_2. As before, depending upon the relative values of R_1 and R, the resistance $-\beta R_1$ may be either positive or negative. Since the development is only trivially different, we shall state the results, leaving the details as obvious.

4.2.1. *Transducer power gain*: $R_1 > R$

Following the same procedure used in the foregoing, the transducer power gain of the amplifier is seen to be

$$G(\omega^2) = \beta \left[\frac{1}{|\hat{S}_{11}(j\omega)|^2} - 1 \right], \quad \beta > 0. \quad (5.149)$$

Proceeding as in the former case, the gain-bandwidth limitations are given by

$$K_n = \beta(\hat{K}_n - 1) \quad (5.150)$$

with

$$\hat{K}_n \leq \left[1 + \frac{\sin(\pi/2n)}{\pi \beta R_1 C f_c} \right]^{2n} \quad (5.151a)$$

for amplifiers having maximally-flat transducer power-gain characteristics, and

$$\hat{K}_n \leq \epsilon^2 \sinh^2 \left\{ n \sinh^{-1} \left[\sinh\left(\frac{1}{n} \sinh^{-1} \frac{1}{\epsilon}\right) + \frac{\sin(\pi/2n)}{\pi \beta R_1 C f_c} \right] \right\} \quad (5.151b)$$

for amplifiers having equiripple transducer power-gain characteristics. In the limit, as $n \to \infty$ or $n \to \infty$ and $\epsilon \to 0$, as the case may be, K_n is approached to

$$K_\infty \leq \beta \left[\exp\left(\frac{1}{\beta R_1 C f_c}\right) - 1 \right]. \quad (5.152)$$

The details of their derivations are left as exercises (Problems 5.35–5.37).

4.2.2. *Transducer power gain*: $R_1 < R$

In this case, β is negative and $z_1(s) = \hat{z}_1(s)$, the strictly passive normalizing impedance at port 1. Under this situation, the

THE ACTIVE LOAD

transducer power gain of the amplifier becomes (Problem 5.38)

$$G(\omega^2) = |\beta|[1 - |\hat{S}_{11}(j\omega)|^2], \qquad (5.153)$$

which, as expected from the former case, is not inherently high-gain. Again, amplification can be obtained by the use of padding resistance, which is now placed in series with R_1 so that

$$R_1 + R_0 = R_1'' > R. \qquad (5.154)$$

Defining

$$\beta'' = \frac{R}{R_1'' - R}, \qquad (5.155)$$

and following (5.142)–(5.145) yield the gain-bandwidth limitations:

$$K_n = \frac{\beta'' R_1}{R_1''}(\hat{K}_n - 1), \qquad (5.156)$$

where $\hat{K}_n$ is defined in (5.151) with the quantities R_1'' and β'' replacing R_1 and β, respectively, everything else being the same.

4.3. Stability

As for the stability of the transmission power-amplifier, its testing is the same as for the nonreciprocal amplifier. For simplicity, we shall only consider the situation where the tunnel diode is connected across the load, leaving the justification for the other configuration as an exercise (Problem 5.40).

Refer to the network of Fig. 5.25. The impedance $Z_{11}(s)$ looking into port 1 with port 2 terminating in $z_2(s)$, as depicted in the figure, is given by

$$Z_{11}(s) = R_1 \frac{1 + \hat{S}_{11a}(s)}{1 - \hat{S}_{11a}(s)}, \qquad (5.157)$$

$\hat{S}_{11a}(s)$ being the reflection coefficient of the resulting one-port network normalizing to R_1, which according to (5.108) is related to the passive scattering parameters $\hat{S}_{ij}(s)$ by (Problem 5.43)

$$\hat{S}_{11a}(s) = \hat{S}_{11}(s) - \frac{\hat{S}_{12}(s)\hat{S}_{21}(s)}{\hat{S}_{22}(s)} = \frac{\det \hat{\mathbf{S}}(s)}{\hat{S}_{22}(s)}. \quad (5.158)$$

For stability, $Z_{11}(s) \neq -R_1$ for all s in the closed RHS, requiring that $|\hat{S}_{11a}(s)|$ be finite in the region. But by (5.158), $|\hat{S}_{11a}(s)|$ is finite in the closed RHS if and only if $\hat{S}_{22}(s) \neq 0$ in the region. By the very philosophy of optimum design, $\hat{S}_{22}(s)$ is chosen to be a minimum-phase function, being devoid of zeros in the open RHS. Also, as in the case for the nonreciprocal amplifiers, $\hat{S}_{22}(j\omega) \neq 0$ for all ω for either Butterworth or Chebyshev characteristics with the possible exception of only one point at which $\hat{S}_{22}(j\omega)$ can equal to zero, which occurs only under the degenerate conditions where $R = 0$ and $\omega = 0$. Excluding this possibility, we can say that it is stable at port 1 for either Butterworth or Chebyshev response, and for gain characteristics other than these two, it is always at least marginally stable.

The impedance $Z_{22}(s)$ looking into port 2 with port 1 terminating in R_1 is related to the reflection coefficient $\hat{S}_{22}(s)$ by [see (4.11)]

$$\hat{S}_{22}(s) = \frac{\hat{h}_2(s)}{\hat{h}_2(-s)} \frac{Z_{22}(s) - \hat{z}_2(-s)}{Z_{22}(s) + \hat{z}_2(s)} \equiv \hat{\eta}(s)\tilde{s}_{22}(s), \quad (5.159)$$

$\hat{h}_2(s)\hat{h}_2(-s)$ being the para-hermitian part of $\hat{z}_2(s)$. For stability, $Z_{22}(s) \neq -z_2(s)$ for all s in the closed RHS. In the case $R_2 > R$, (5.107) applies, which in conjunction with (5.159) is equivalent to requiring $\tilde{s}_{22}(s) \neq 0$ for all s in the closed RHS. Invoking the same argument as above for port 1 leads to the same conclusion. As for the case where $R_2 < R$, $z_2(s) = \hat{z}_2(s)$ and $Z_{22}(s) + z_2(s)$ is strictly passive since, with $\hat{S}_{22}(s)$ satisfying the coefficient constraints, $Z_{22}(s)$ must be a positive-real function, as shown in §6 of Chapter 4. Thus, at port 2 we arrive at the same conclusion. This also follows from the fact that the device is reciprocal, meaning that if it is stable at one of the ports, it must also be stable at the other.

4.4. Sensitivity

In high-gain amplifiers, it is important to determine the percent of change in gain relative to the percent of deviation in the values of the critical network elements. In the present case, the most sensitive parameter is the negative resistance, which can be expected to undergo drifts of the order of 1 or 2 percent. The *sensitivity function* is defined as the ratio of fractional change in the transfer function $f(s)$ to the fractional change in a chosen parameter x under the situation when $\delta x = x - x_0$ approaches to zero, x_0 being the nominal value of the parameter x:

$$S_x^f(s) = \lim_{\delta x \to 0} \frac{\delta f(s,x)/f(s,x)}{\delta x / x} = \frac{x}{f(s,x)} \frac{\partial}{\partial x} f(s,x), \qquad (5.160)$$

in which the variable x was written explicitly to emphasize its functional dependency on x. In the present situation, the function of interest is the transducer power gain with $x = R$.

4.4.1. Tunnel diode in shunt with the load

Consider the network of Fig. 5.24. For sensitivity analysis, we need only consider the high-gain case where $R_2 > R$. Assume the amplifier has a low-pass characteristic. Then at dc, the reflection coefficient $\hat{S}_{22}(0)$ becomes

$$\hat{S}_{22}(0) = \frac{\hat{h}_2(0)}{\hat{h}_2(0)} \cdot \frac{R_{22} - \hat{z}_2(0)}{R_{22} + \hat{z}_2(0)} = \frac{R_{22} - \alpha R_2}{R_{22} + \alpha R_2}, \qquad (5.161)$$

where R_{22} is the dc resistance looking into port 2 of N with port 1 terminating in R_1. Substituting (5.161) in (5.115) yields the dc gain of the amplifier:

$$G(0) = \frac{4R^2 R_2 R_{22}}{(R_{22}R_2 - R_{22}R - R_2 R)^2}. \qquad (5.162)$$

Taking the partial derivative of $G(0)$ with respect to R and using (5.160) gives the dc sensitivity

$$S_R^G(0) = \frac{[R_2 R_{22} G(0)]^{1/2}}{R}, \qquad (5.163)$$

which shows that the sensitivity varies with the square root of the dc gain. Using (5.163), the amplifier sensitivities for the Butterworth and Chebyshev characteristics can easily be determined.

From Problem 5.27, the terminating resistance for the Butterworth response is given by

$$\hat{R}_2 = \alpha R_2 \frac{\hat{K}_n^{1/2} - 1}{\hat{K}_n^{1/2} + 1}, \tag{5.164}$$

where $\hat{K}_n$ is given in (5.119) and $\hat{R}_2$ is also the dc resistance R_{22}. Substituting (5.164) in (5.163) in conjunction with (5.116) and using the maximum $\hat{K}_n$ from (5.119) yield

$$S_R^G(0) = \frac{R_2}{R_2 - R} \left(\left[1 + \frac{\sin(\pi/2n)}{\pi \alpha R_2 C f_c} \right]^n - 1 \right). \tag{5.165}$$

In the Chebyshev case, the terminating resistance is given in Problem 5.28, which has two different expressions, depending upon whether n is odd or even:

$$R_{22} = \hat{R}_2 = \alpha R_2 \frac{\epsilon \sinh \gamma - 1}{\epsilon \sinh \gamma + 1} \tag{5.166a}$$

for n odd, and

$$R_{22} = \hat{R}_2 = \alpha R_2 \frac{\epsilon \cosh \gamma - (1 + \epsilon^2)^{1/2}}{\epsilon \cosh \gamma + (1 + \epsilon^2)^{1/2}} \tag{5.166b}$$

for n even, where

$$\gamma = n \sinh^{-1} \left[\sinh \left(\frac{1}{n} \sinh^{-1} \frac{1}{\epsilon} \right) + \frac{\sin(\pi/2n)}{\pi \alpha R_2 C f_c} \right]. \tag{5.166c}$$

Substituting (5.166) in (5.163) in conjunction with (5.125) gives

$$S_R^G(0) = \frac{R_2}{R_2 - R} (\epsilon \sinh \gamma - 1) \tag{5.167a}$$

for n odd, and

$$S_R^G(0) = \frac{R_2}{(R_2 - R)(1 + \epsilon^2)^{1/2}} [\epsilon \cosh \gamma - (1 + \epsilon^2)^{1/2}] \tag{5.167b}$$

THE ACTIVE LOAD 377

for n even. In the limit, as $n \to \infty$ in (5.165) or as $n \to \infty$ and $\epsilon \to 0$ in (5.167), we have (Problem 5.44)

$$S_R^G(0) = \frac{R_2}{R_2 - R}\left[\exp\left(\frac{1}{2\alpha R_2 C f_c}\right) - 1\right]. \qquad (5.168)$$

As an illustration, consider the amplifier designed in Example 5.4 for a fifth-order Butterworth transducer power gain with a 3-dB bandwidth of 40 MHz. Then from (5.165) we get

$$S_R^G(0) = \frac{500}{500 - 143}\left[\left(1 + \frac{\sin 18°}{\pi 0.401 \times 500 \times 7 \times 10^{-12} \times 40 \times 10^6}\right)^5 - 1\right]$$
$$= 1.4 \times [(2.752)^5 - 1] = 219.6 \quad \text{or} \quad 23.42 \text{ dB}, \qquad (5.169)$$

while $K_5 = 40$ dB. In Example 5.5, the amplifier was designed for a fifth-order Chebyshev gain characteristic with a 3-dB passband tolerance. From (5.167a) we obtain

$$S_R^G(0) = 1.4 \times (0.9976 \sinh 7.1 - 1)$$
$$= 845 \quad \text{or} \quad 29.27 \text{ dB}, \qquad (5.170\text{a})$$

where

$$\gamma = 5 \sinh^{-1}[\sinh(\tfrac{1}{5}\sinh^{-1} 1.002) + 1.752] = 7.1, \qquad (5.170\text{b})$$

while the dc gain $K_5 = 51.30$. From (5.168), the sensitivity for the ideal brick-wall type of low-pass characteristic is given by

$$S_R^G(0) = 1.4 \times (\exp 8.91 - 1) = 10{,}366 \quad \text{or} \quad 40.16 \text{ dB}. \qquad (5.171)$$

4.4.2. Tunnel diode in shunt with the generator

Consider the network of Fig. 5.30, operating in high-gain mode with $R_1 > R$ and having a low-pass power-gain characteristic. Proceeding as in the former case, the dc gain is obtained as (Problem 5.46)

$$G(0) = \frac{4R^2 R_1 R_{11}}{(R_{11}R_1 - R_{11}R - R_1 R)^2}, \qquad (5.172)$$

where R_{11} is the dc resistance looking into port 1 of N with port 2 terminating in R_2, giving the sensitivity

$$S_R^G(0) = \frac{[R_1 R_{11} G(0)]^{1/2}}{R}, \qquad (5.173)$$

whose derivation is left as an exercise (Problem 5.47). If the terms α and R_2 are replaced by β and R_1, respectively, in (5.165), (5.167) and (5.168), the resulting equations will give the sensitivities for the Butterworth, Chebyshev and ideal brick-wall type responses.

We remark that using the above derived explicit expressions for sensitivity, it is possible to carry out a design in which the sensitivity is less than a preassigned value. The analysis has also demonstrated that high gain and low sensitivity are not compatible. Thus, it will be necessary to sacrifice some gain in order to achieve a reasonable value of sensitivity. Since the sensitivity is of the order of the square root of the gain, a very well-regulated bias supply is essential in order to minimize drifts in the negative resistance.

5. Reciprocal amplifiers

As discussed in §2, the transducer power gain of the most general configuration of a negative-resistance amplifier, as shown in Fig. 5.3, is given by

$$G(\omega^2) = |S_{21a}(j\omega)|^2 = \frac{|S_{12}(j\omega)|^2}{|S_{33}(j\omega)|^2}. \qquad (5.174)$$

In § 3 we indicated that in order to allow $|S_{12}(j\omega)|$ to be designed independently of $|S_{33}(j\omega)|$, we use an ideal three-port circulator to attain the needed isolation. In the present section, we shall restrict the lossless three-port network N to be reciprocal, and determine its optimum gain-bandwidth relationship. Unlike the nonreciprocal amplifier, the present arrangement would allow power transmission in both directions with exactly the same gain characteristic.

THE ACTIVE LOAD

5.1. General gain-bandwidth limitations

Since N is lossless and reciprocal, its scattering matrix, as defined in (5.4a), is symmetric and, on the $j\omega$-axis, also unitary. Hence

$$S(s) = S'(s), \qquad (5.175)$$

$$S^*(j\omega)S(j\omega) = U_3 \qquad (5.176)$$

which yield

$$|S_{11}(j\omega)|^2 + |S_{12}(j\omega)|^2 + |S_{13}(j\omega)|^2 = 1, \qquad (5.177a)$$

$$|S_{12}(j\omega)|^2 + |S_{22}(j\omega)|^2 + |S_{23}(j\omega)|^2 = 1, \qquad (5.177b)$$

$$|S_{13}(j\omega)|^2 + |S_{23}(j\omega)|^2 + |S_{33}(j\omega)|^2 = 1. \qquad (5.177c)$$

Combining these equations gives

$$2|S_{12}(j\omega)|^2 = 1 + |S_{33}(j\omega)|^2 - |S_{11}(j\omega)|^2 - |S_{22}(j\omega)|^2. \qquad (5.178)$$

Rewriting (5.176) in the form of (5.6), we obtain, in addition to the four relations given in (5.7), an equality corresponding to the (3,3)-elements of (5.6), giving

$$\bar{S}_{33}(j\omega)\det S(j\omega) = S_{11}(j\omega)S_{22}(j\omega) - S_{12}^2(j\omega), \qquad (5.179)$$

in which $\det S(j\omega) = \pm 1$. Applying the well-known inequality

$$|x - y| \geq |x| - |y| \qquad (5.180)$$

to the magnitude of (5.179) results in

$$|S_{33}(j\omega)| = |\bar{S}_{33}(j\omega)| = |S_{12}^2(j\omega) - S_{11}(j\omega)S_{22}(j\omega)|$$
$$\geq |S_{12}(j\omega)|^2 - |S_{11}(j\omega)||S_{22}(j\omega)|, \qquad (5.181)$$

which together with the inequality

$$|S_{11}(j\omega)|^2 + |S_{22}(j\omega)|^2 \geq 2|S_{11}(j\omega)||S_{22}(j\omega)|, \qquad (5.182)$$

shows that

$$2|S_{12}(j\omega)|^2 \leq 2|S_{33}(j\omega)| + |S_{11}(j\omega)|^2 + |S_{22}(j\omega)|^2. \quad (5.183)$$

Adding (5.178) and (5.183) finally leads to

$$2|S_{12}(j\omega)| \leq 1 + |S_{33}(j\omega)|. \quad (5.184)$$

Thus, from (5.174) the transducer power gain of the amplifier is bounded by

$$G(\omega^2) \leq \frac{1}{4}\left[1 + \frac{1}{|S_{33}(j\omega)|}\right]^2. \quad (5.185)$$

We remark that the equality sign of (5.184) is attained only if $S_{12}^2(j\omega)$ and $S_{11}(j\omega)S_{22}(j\omega)$ are in phase and $|S_{11}(j\omega)| = |S_{22}(j\omega)|$ at that frequency. One network that attains this equality is shown in Fig. 5.32, whose scattering matrix normalizing to the resistances R_1, R_2 and R_3 is given by (Problem 5.74)

$$S(s) = \begin{bmatrix} -\frac{1}{2} & \frac{1}{2} & 2^{-1/2} \\ \frac{1}{2} & -\frac{1}{2} & 2^{-1/2} \\ 2^{-1/2} & 2^{-1/2} & 0 \end{bmatrix}. \quad (5.186)$$

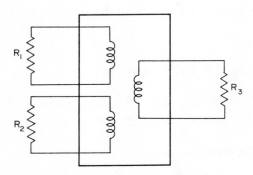

FIG. 5.32. An optimum coupling network.

However, such a network is of little practical value as a coupling network since it does not provide any equalization for the load, as required in the design. To avoid this difficulty, in the following section we shall consider the situation where a lossless two-port network is inserted between the active load and port 3 to provide the needed equalization. Before we do this, we derive the fundamental gain-bandwidth limitation for the reciprocal amplifiers. From (5.185) we recognize that there is no simple expression for $S_{33}(s)$ in terms of the given gain characteristic $G(\omega^2)$. For this reason, no simple and systematic expressions are possible for arbitrary orders of given maximally-flat or equiripple gain characteristics even for the simplified model of a tunnel diode connected at port 3, as discussed in the foregoing for other amplifier configurations. Nevertheless, we can compare the gain-bandwidth limitation of the reciprocal amplifier with that of the nonreciprocal one for the ideal brick-wall type of low-pass response when port 3 is terminated in a tunnel diode, again represented by its simplified $-RC$ model of Fig. 5.16. To this end, we rewrite (5.185) as

$$\frac{1}{|S_{33}(j\omega)|} = 2G^{1/2}(\omega^2) - 1. \quad (5.187)$$

An application of (4.282), as we did in (5.85), gives

$$\int_0^\infty \ln [2G^{1/2}(\omega^2) - 1] d\omega \leq \frac{\pi}{RC}, \quad (5.188)$$

which for the ideal brick-wall type of response with a preassigned bandwidth ω_c becomes

$$G(0) = \frac{1}{4}\left[1 + \exp\left(\frac{1}{2RCf_c}\right)\right]^2. \quad (5.189)$$

Again $G(0)$ in (5.189) represents the maximum obtainable constant gain. For a nonreciprocal amplifier, we have from (5.67) or (5.84)

$$G(0) = \exp\left(\frac{1}{RCf_c}\right). \quad (5.190)$$

Since $\exp(1/RCf_c) \geq 1$,

$$\exp\left(\frac{1}{RCf_c}\right) \geq \frac{1}{4}\left[1 + \exp\left(\frac{1}{2RCf_c}\right)\right]^2. \quad (5.191)$$

For large bandwidth f_c, the two sides of (5.191) approach to equality, but for small f_c, it becomes $1 \geq \frac{1}{4}$. In terms of decibels, we conclude that the optimum nonreciprocal amplifier is capable of yielding 6 dB more flat gain than optimum reciprocal amplifier. However, since reciprocal amplifier allows two-way transmission of power, the total gain available for the device is the same.

5.2. Cascade connection

This configuration is depicted in Fig. 5.33 where the three-port N consists of the cascade connection of a frequency-independent three-port network N_α and a lossless coupling two-port network N_β. Let

$$S_\alpha(s) = [S_{ij\alpha}] \quad (5.192a)$$

and

$$S_\beta(s) = [S_{ij\beta}] \quad (5.192b)$$

be the scattering matrices of the three-port N_α and the two-port N_β,

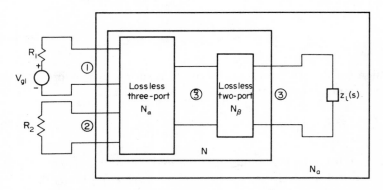

FIG. 5.33. The general configuration of a reciprocal negative-resistance amplifier employing the cascade connection of a frequency-independent three-port network N_α and a lossless coupling two-port network N_β.

THE ACTIVE LOAD

respectively, normalized with respect to the reference impedances R_1, R_2 and $\hat{R}_3$; and $\hat{R}_3$ and $z_3(s) = -z_l(-s)$, where $\hat{R}_3$ is an arbitrary real constant to be determined shortly. Since N_α is assumed to be frequency-independent, the impedance looking into its third port with the other two ports terminating in their reference resistances R_1 and R_2 must be real and positive. This implies that we can choose the reference impedance $\hat{R}_3$ at the port $\hat{3}$ to be its driving-point impedance, so that $S_{33\alpha} = 0$. Applying the interconnection formulas (2.165) derived in Chapter 2 yields the scattering matrix $S(s)$ of the overall three-port network N in terms of those of the component networks N_α and N_β:

$$S(s) = \begin{bmatrix} S_{11\alpha} + S_{11\beta}S_{13\alpha}S_{31\alpha} & S_{12\alpha} + S_{11\beta}S_{13\alpha}S_{32\alpha} & S_{13\alpha}S_{12\beta} \\ S_{21\alpha} + S_{11\beta}S_{23\alpha}S_{31\alpha} & S_{22\alpha} + S_{11\beta}S_{23\alpha}S_{32\alpha} & S_{23\alpha}S_{12\beta} \\ S_{31\alpha}S_{21\beta} & S_{32\alpha}S_{21\beta} & S_{22\beta} \end{bmatrix}.$$

(5.193)

Substituting the elements of $S(s)$ in (5.5) yields the overall scattering matrix of the reciprocal amplifier:

$$\begin{aligned} S_a(s) &= \begin{bmatrix} S_{11\alpha} + \dfrac{(\det S_\beta)S_{13\alpha}S_{31\alpha}}{S_{22\beta}} & S_{12\alpha} + \dfrac{(\det S_\beta)S_{13\alpha}S_{32\alpha}}{S_{22\beta}} \\ S_{21\alpha} + \dfrac{(\det S_\beta)S_{23\alpha}S_{31\alpha}}{S_{22\beta}} & S_{22\alpha} + \dfrac{(\det S_\beta)S_{23\alpha}S_{32\alpha}}{S_{22\beta}} \end{bmatrix} \\ &= \begin{bmatrix} S_{11\alpha} + \dfrac{S_{13\alpha}S_{31\alpha}}{S_{11\beta^*}} & S_{12\alpha} + \dfrac{S_{13\alpha}S_{32\alpha}}{S_{11\beta^*}} \\ S_{21\alpha} + \dfrac{S_{23\alpha}S_{31\alpha}}{S_{11\beta^*}} & S_{22\alpha} + \dfrac{S_{23\alpha}S_{32\alpha}}{S_{11\beta^*}} \end{bmatrix}, \end{aligned}$$

(5.194)

where $S_{11\beta^*} = S_{11\beta}(-s)$,

$$\det S_\beta = \det S_\beta(s) = S_{11\beta}S_{22\beta} - S_{12\beta}S_{21\beta}, \quad (5.195)$$

$$S_{11\beta^*} = S_{22\beta}/(\det S_\beta). \quad (5.196)$$

Equation (5.196) follows from the fact that since N_β is lossless, its normalized scattering matrix $S_\beta(s)$, as indicated in Corollary 2.2 of Chapter 2, is para-unitary, meaning that

$$S'_\beta(-s) = S_\beta^{-1}(s). \qquad (5.197)$$

Let $Z_{11\beta}(s)$ and $\hat{Z}_{11\beta}(s)$ be the impedances looking into the input port of N_β when the output port is terminated in $z_3(s)$ and $z_l(s)$, respectively. Like (5.2), these impedances can be expressed in terms of the immittance parameters $y_{ij}(s)$ and $z_{ij}(s)$ of the lossless two-port N_β (Problem 5.1):

$$Z_{11\beta}(s) = z_{11}(s)\frac{1/y_{22}(s) + z_3(s)}{z_{22}(s) + z_3(s)}, \qquad (5.198a)$$

$$\hat{Z}_{11\beta}(s) = z_{11}(s)\frac{1/y_{22}(s) + z_l(s)}{z_{22}(s) + z_l(s)}. \qquad (5.198b)$$

Invoking the facts that the immittance parameters of a lossless two-port are odd functions and that $z_3(s) = -z_l(-s)$ shows

$$Z_{11\beta}(-s) = -\hat{Z}_{11\beta}(s). \qquad (5.199)$$

Following (5.16), define the current-based active reflection coefficient at the input port of N_β with the output port terminating in $z_l(s)$ as

$$\rho_1^I(s) = \frac{\hat{Z}_{11\beta}(s) - \hat{R}_3}{\hat{Z}_{11\beta}(s) + \hat{R}_3}, \qquad (5.200)$$

resulting in

$$S_{11\beta}(-s) = \frac{Z_{11\beta}(-s) - \hat{R}_3}{Z_{11\beta}(-s) + \hat{R}_3} = \frac{\hat{Z}_{11\beta}(s) + \hat{R}_3}{\hat{Z}_{11\beta}(s) - \hat{R}_3} = \frac{1}{\rho_1^I(s)}. \qquad (5.201)$$

Substituting (5.201) in (5.194) gives

$$S_a(s) = \begin{bmatrix} S_{11\alpha} + \rho_1^I(s)S_{13\alpha}S_{31\alpha} & S_{12\alpha} + \rho_1^I(s)S_{13\alpha}S_{32\alpha} \\ S_{21\alpha} + \rho_1^I(s)S_{23\alpha}S_{31\alpha} & S_{22\alpha} + \rho_1^I(s)S_{23\alpha}S_{32\alpha} \end{bmatrix}. \qquad (5.202)$$

Thus, the transducer power gain is given by

$$G(\omega^2) = |S_{21\alpha}(j\omega)|^2 = \left| S_{21\alpha}(j\omega) + \rho_1^I(j\omega)S_{23\alpha}(j\omega)S_{31\alpha}(j\omega) \right|^2. \quad (5.203)$$

We now proceed to simplify (5.203) with the aid of the symmetric and unitary properties of S_α. Of particular interest are the relations

$$|S_{31\alpha}(j\omega)|^2 + |S_{32\alpha}(j\omega)|^2 = 1, \quad (5.204a)$$

$$|S_{11\alpha}(j\omega)|^2 + |S_{21\alpha}(j\omega)|^2 + |S_{31\alpha}(j\omega)|^2 = 1, \quad (5.204b)$$

$$S_{31\alpha}(j\omega)S_{11\alpha}(j\omega) + S_{32\alpha}(j\omega)S_{12\alpha}(j\omega) = 0, \quad (5.204c)$$

and $S_{ij\alpha} = S_{ji\alpha}$ for $i, j = 1, 2, 3$. From the above relations, we obtain

$$|S_{21\alpha}(j\omega)| = |S_{23\alpha}(j\omega)S_{31\alpha}(j\omega)|, \quad (5.205)$$

which is then substituted in (5.203) to give

$$G(\omega^2) = |S_{23\alpha}(j\omega)S_{31\alpha}(j\omega)|^2 \left| 1 + |\rho_1^I(j\omega)| e^{j\phi(\omega)} \right|^2, \quad (5.206)$$

$\phi(\omega)$ being a phase angle. Clearly, in order to optimize the gain over a frequency band of interest, we must maximize $|S_{23\alpha}(j\omega)S_{31\alpha}(j\omega)|$ subject to the constraint (5.204a), which is found to be (Problem 5.51)

$$|S_{23\alpha}(j\omega)| = |S_{31\alpha}(j\omega)| = 2^{-1/2}. \quad (5.207)$$

Substituting (5.207) in (5.206) yields

$$G(\omega^2) = \frac{1}{4} \left| 1 + |\rho_1^I(j\omega)| e^{j\phi(\omega)} \right|^2. \quad (5.208)$$

In the case where the gain is large, then $|\rho_1^I(j\omega)| \gg 1$ and

$$G(\omega^2) \approx \frac{1}{4} \left| \rho_1^I(j\omega) \right|^2. \quad (5.209)$$

Comparing this with (5.15) together with (5.18) shows that the optimum transducer power gain for a nonreciprocal amplifier is about 6 dB more than that for a reciprocal amplifier, a fact that was pointed out in (5.191). Physically, $|S_{21\alpha}(j\omega)| = |\frac{1}{2} + \frac{1}{2}|\rho_1^I(j\omega)|e^{j\phi}|$ can be viewed to consist of two parts. The part corresponding to the direct transmission from the source to the load is represented by the factor $\frac{1}{2}$, and the part corresponding to reflection due to the active one-port is represented by the factor $\frac{1}{2}|\rho_1^I(j\omega)|e^{j\phi}$. Although the gain for the reciprocal amplifier is 6 dB less than that for the optimum nonreciprocal amplifier, the total available gain for the device, as mentioned in the preceding section, is the same since the amplifier is bilateral.

As for the optimum values of $|S_{ij\alpha}(j\omega)|$, we combine (5.204), (5.205) and (5.207), resulting in $S_\alpha(s)$ which is identical to that given in (5.186). Thus, N_α can be realized by a transformer network as shown in Fig. 5.32. Using these optimum values of $S_{ij\alpha}(j\omega)$ as derived above, the overall scattering matrix for the reciprocal amplifier becomes

$$S_a(s) = \frac{1}{2}\begin{bmatrix} -1+\rho_1^I(s) & 1+\rho_1^I(s) \\ 1+\rho_1^I(s) & +1+\rho_1^I(s) \end{bmatrix}. \quad (5.210)$$

We remark that both the input and the output of the amplifier are mismatched unless $\rho_1^I(s) = 1$, a situation indicative of non-amplification. The complete amplifier together with its load is presented in Fig. 5.34.

As an illustration, consider the amplifier of Fig. 5.34 where the active load $z_l(s)$ is a tunnel diode represented by its simplified model of Fig. 5.16. Suppose that we wish to achieve the nth-order Butterworth transducer power gain that is sufficiently large over the passband so that (5.209) is valid. Obviously, this characteristic cannot be produced by the network of Fig. 5.34, in which from (5.201) and (5.209) we set

$$|S_{11\beta}(j\omega)|^2 = \frac{1}{|\rho_1^I(j\omega)|^2} = \frac{1}{4G(\omega^2)}, \quad (5.211)$$

THE ACTIVE LOAD

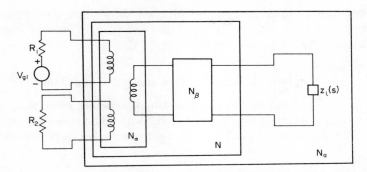

FIG. 5.34. A specific configuration of a reciprocal negative-resistance amplifier.

since $|S_{11\beta}(j\omega)|$ must be bounded by unity for a lossless two-port network. To avoid this difficulty, the following arbitrary choice, as suggested in (5.59), is a reasonable one:

$$G(\omega^2) = \frac{K_n + (\omega/\omega_c)^{2n}}{1 + (\omega/\omega_c)^{2n}}. \quad (5.212)$$

Thus, from (5.209) and (5.201), we obtain

$$S_{11\beta}(s)S_{11\beta}(-s) = (4K_n)^{-1}\frac{1+(-1)^n y^{2n}}{1+(-1)^n x^{2n}}, \quad (5.213)$$

where $y = s/\omega_c$ and

$$x = K_n^{-1/2n} s/\omega_c. \quad (5.214)$$

Proceeding as in §3.4.1, we arrive at (Problem 5.54)

$$K_n \leq \frac{1}{4}\left[1 + \frac{\sin(\pi/2n)}{\pi RCf_c}\right]^{2n}. \quad (5.215)$$

In the limit, as $n \to \infty$,

$$K_\infty \leq \frac{1}{4}\exp\left(\frac{1}{RCf_c}\right), \quad (5.216)$$

which after comparing with (5.190) again shows that the optimum nonreciprocal amplifier is capable of yielding 6 dB more flat gain than optimum reciprocal amplifier.

6. Amplifiers using more than one active impedance

We now turn our attention to the situation where more than one active one-port impedance is employed, as depicted schematically in Fig. 5.35. Again, we assume that the multi-port network N is lossless

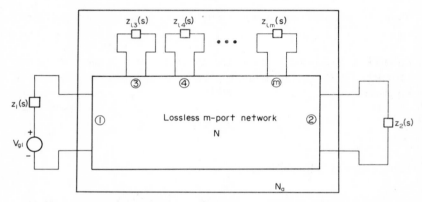

FIG. 5.35. The general configuration of a negative-resistance amplifier using more than one active impedance.

and that all the active impedances belong to the special class defined in § 1. We shall first derive a formula for the transducer power gain of the amplifier in terms of the scattering parameters of the lossless multi-port network N, and then consider the gain for specific configurations of N. One objective is to show that the optimum nonreciprocal matching can never yield more than 6 dB gain over the optimum reciprocal equalization.

Let

$$S(s) = [S_{ij}], \qquad i, j = 1, 2, \ldots, m, \qquad (5.217)$$

be the scattering matrix of the lossless m-port network N normalizing to the strictly passive impedances $z_1(s)$, $z_2(s)$ and

$$z_k(s) = -z_{lk}(-s), \qquad k = 3, 4, \ldots, m, \qquad (5.218)$$

as indicated in Fig. 5.35. Partition the matrix $S(s)$ into the form

$$S(s) = \begin{bmatrix} S_{11}(s) & S_{12}(s) \\ S_{21}(s) & S_{22}(s) \end{bmatrix}, \qquad (5.219)$$

in which the square submatrix $S_{11}(s)$ corresponds to the input and output ports 1 and 2. Using the interconnection formula (2.167) derived in Chapter 2, the scattering matrix $S_a(s)$ of the resulting two-port network N_a normalizing to the impedances $z_1(s)$ and $z_2(s)$ is given by (Problem 5.55)

$$S_a(s) = S_{11}(s) - S_{12}(s)S_{22}^{-1}(s)S_{21}(s). \qquad (5.220)$$

Since the m-port N is lossless, its scattering matrix $S(s)$, as indicated in Corollary 2.2, is para-unitary, giving

$$S_*(s) = S^{-1}(s). \qquad (5.221)$$

The inverse on the right-hand side of (5.221) can be expressed in terms of the submatrices of (5.219), as follows (Problem 5.56):

$$S^{-1}(s) = \begin{bmatrix} S_{11}^{-1} + S_{11}^{-1}S_{12}B_{22}S_{21}S_{11}^{-1} & -S_{11}^{-1}S_{12}B_{22} \\ -B_{22}S_{21}S_{11}^{-1} & B_{22} \end{bmatrix}, \qquad (5.222a)$$

in which

$$B_{22} = (S_{22} - S_{21}S_{11}^{-1}S_{12})^{-1}, \qquad (5.222b)$$

the variable s being omitted in the expressions, for simplicity. Equating the corresponding submatrices in (5.221) results in

$$S_{11*} = S_{11}^{-1} + S_{11}^{-1}S_{12}S_{22*}S_{21}S_{11}^{-1}, \qquad (5.223a)$$

$$S_{21*} = -S_{11}^{-1}S_{12}S_{22*}, \qquad (5.223b)$$

$$S_{12*} = -S_{22*}S_{21}S_{11}^{-1}, \qquad (5.223c)$$

$$B_{22} = S_{22*}. \qquad (5.223d)$$

Combining (5.223b) with (5.223c) and then substituting it in (5.223a) yield

$$S_{11*}(s) = S_{11}^{-1}(s) + S_{21*}(s)S_{22*}^{-1}(s)S_{12*}(s). \qquad (5.224)$$

Performing the para-unitary operation on both sides of (5.224) and then combining it with (5.220) give

$$S_a(s) = S_{11*}^{-1}(s) = \frac{1}{\det S_{11*}(s)} \begin{bmatrix} S_{22*}(s) & -S_{21*}(s) \\ -S_{12*}(s) & S_{11*}(s) \end{bmatrix}. \qquad (5.225)$$

Thus, the transducer power gain of the amplifier is obtained as

$$G(\omega^2) = |S_{21a}(j\omega)|^2 = \left| \frac{S_{12}(j\omega)}{\det S_{11}(j\omega)} \right|^2. \qquad (5.226)$$

Alternatively, $G(\omega^2)$ can be expressed in terms of the determinant of $S_{22}(j\omega)$ by means of the Jacobi's theorem in matrix algebra, which relates the minors of $S^{-1}(s)$ to those of $S(s)$. Applying the theorem to $S_{22}(s)$ in (5.219) in conjunction with (5.221) yields

$$[\det S(s)][\det S_{11*}(s)] = \det S_{22}(s). \qquad (5.227)$$

Since $S(j\omega)$ is unitary, $\det S(j\omega) = \pm 1$, which shows that

$$|\det S_{11}(j\omega)| = |\det S_{22}(j\omega)|. \qquad (5.228)$$

Thus, we have

$$G(\omega^2) = \left| \frac{S_{12}(j\omega)}{\det S_{22}(j\omega)} \right|^2, \qquad (5.229)$$

indicating that the transducer power gain of the amplifier is bounded by

$$G(\omega^2) \leq \frac{1}{|\det S_{22}(j\omega)|^2}, \qquad (5.230)$$

since $|S_{12}(j\omega)| \leq 1$. In particular, if only one active impedance is used, $m = 3$ and (5.230) reduces to (5.9) with $\det S_{22}(j\omega) = S_{33}(j\omega)$.

THE ACTIVE LOAD

We now consider two specific configurations for the lossless m-port network N and discuss their gain-bandwidth limitations. Like the situation where only one active impedance is used, we shall assume that the impedances $z_1(s)$ and $z_2(s)$ at the input and output ports are both real and positive. More specifically, let

$$z_1(s) = R_1 \quad \text{and} \quad z_2(s) = R_2. \tag{5.231}$$

6.1. Nonreciprocal amplifiers

The lossless m-port network N is composed of a cascade connection of the basic units as shown in Fig. 5.36. The complete

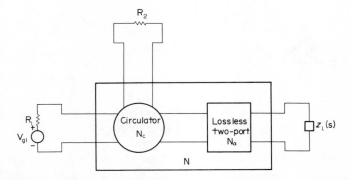

FIG. 5.36. A basic unit for the nonreciprocal negative-resistance amplifiers.

amplifier together with its load is presented in Fig. 5.37. Referring to the network of Fig. 5.37, let $S_k(s)$ ($k = 1, 2, \ldots, m-2$) be the scattering matrices of the two-port networks N_k normalized with respect to the reference impedances $\hat{R}_k$ and $\hat{R}_{k+1}$ where $\hat{R}_k$ are real constants with $\hat{R}_1 = R_1$. Following (5.10)–(5.13) and using (5.5) together with the fact that since N_k are lossless, their scattering matrices must be para-unitary, we obtain (Problem 5.57)

$$S_k(s) = \begin{bmatrix} 0 & 1 \\ S_{11\alpha_k}^{-1}(-s) & 0 \end{bmatrix}, \tag{5.232a}$$

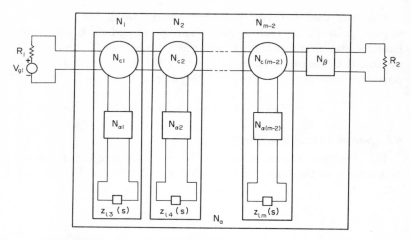

FIG. 5.37. A nonreciprocal amplifier composed of the cascade connection of the basic units of Fig. 5.36.

where

$$S_{11\alpha_k}(s) = \frac{Z_{11\alpha_k}(s) - R_{\alpha k}}{Z_{11\alpha_k}(s) + R_{\alpha k}} \quad (5.232b)$$

is the reflection coefficient at the input port of $N_{\alpha k}$ with its output port terminating in $z_{k+2}(s) = -z_{l(k+2)}(-s)$, as indicated in Fig. 5.38,

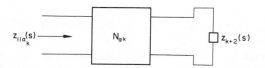

FIG. 5.38. The coupling network for the kth basic unit of Fig. 5.37.

where $R_{\alpha k}$ is an arbitrary real normalizing constant and $Z_{11\alpha_k}(s)$ is the impedance looking into the input port. Appealing to the interconnection formulas (2.165) yields the scattering matrix of the composite two-port network composed of the cascade connection of the two-port networks N_k ($k = 1, 2, \ldots, m - 2$):

$$\begin{bmatrix} 0 & 1 \\ S_{11\alpha_1}^{-1}(-s)S_{11\alpha_2}^{-1}(-s) \cdots S_{11\alpha_{m-2}}^{-1}(-s) & 0 \end{bmatrix} \quad (5.233)$$

normalizing to the impedances R_1 and $\hat{R}_{m-1}$. Appealing once more to the formulas (2.165), and scattering matrix $S_a(s)$ of the amplifier is given by

$$S_a(s) = \begin{bmatrix} S_{11\beta}(s)S_{11\alpha_1}^{-1}(-s)S_{11\alpha_2}^{-1}(-s) \cdots S_{11\alpha_{m-2}}^{-1}(-s) & S_{12\beta}(s) \\ S_{21\beta}(s)S_{11\alpha_1}^{-1}(-s)S_{11\alpha_2}^{-1}(-s) \cdots S_{11\alpha_{m-2}}^{-1}(-s) & S_{22\beta}(s) \end{bmatrix},$$

(5.234)

where

$$S_\beta(s) = [S_{ij\beta}] \quad (5.235)$$

is the scattering matrix of the lossless equalizer N_β, as indicated in Fig. 5.37, normalizing to the impedances $\hat{R}_{m-1}$ and R_2. Thus, the transducer power gain of the amplifier is given by

$$G(\omega^2) = \frac{|S_{21\beta}(j\omega)|^2}{|S_{11\alpha_1}(j\omega)|^2 |S_{11\alpha_2}(j\omega)|^2 \cdots |S_{11\alpha_{m-2}}(j\omega)|^2}. \quad (5.236)$$

In particular, for the cascade connection of n identical units of the type as shown in Fig. 5.36, (5.236) reduces to

$$G(\omega^2) = \frac{|S_{21\beta}(j\omega)|^2}{|S_{11\alpha}(j\omega)|^{2n}} \quad (5.237)$$

with

$$S_{11\alpha}(s) = S_{11\alpha_k}(s), \quad k = 1, 2, \ldots, m - 2; \quad (5.238)$$

indicating that the optimum nonreciprocal amplifier should have a maximum $|S_{21\beta}(j\omega)|$ and a minimum $|S_{11\alpha}(j\omega)|$ over a frequency band of interest. This is essentially the same problem that we discussed in §3 for amplifiers using a single active one-port impedance.

As an example, consider the ideal case where $|S_{21\beta}(j\omega)| = 1$. If, as in (5.67) or (5.84), $S_{11\alpha}(j\omega)$ is chosen so that

$$|S_{11\alpha}(j\omega)|^2 = \exp\left(\frac{-1}{RCf_c}\right) \tag{5.239}$$

for the active impedance of a tunnel diode. Then for n identical tunnel diodes, the corresponding gain in the passband is given by

$$G(\omega^2) = \exp\left(\frac{n}{RCf_c}\right), \tag{5.240}$$

a maximal attainable constant gain for a preassigned bandwidth f_c.

6.2. Reciprocal amplifiers

We now turn our attention to the more delicate question of reciprocal matching. We shall show that irrespective of the number of active impedances employed, the optimum nonreciprocal matching can never yield more than 6 dB gain over the optimum reciprocal equalization for the same number of active impedances used.

In the general configuration of Fig. 5.35, assume that the m-port network N, in addition to being lossless, is reciprocal. Then its scattering matrix $S(s)$ is both symmetric and para-unitary. Partition the matrix $S(s)$ of (5.219) in blocks corresponding to the input port, the output port, and the ports terminating in active impedances, as follows:

$$S(s) = \begin{bmatrix} S_{11}(s) & S_{12}(s) & S'_{31}(s) \\ S_{21}(s) & S_{22}(s) & S'_{32}(s) \\ \hline S_{31}(s) & S_{32}(s) & \mathbf{S_{22}}(s) \end{bmatrix}. \tag{5.241}$$

Note that $S_{22}(s)$ is the reflection coefficient at the output port while $\mathbf{S_{22}}(s)$ is the submatrix of order $m - 2$ corresponding to the ports 3, 4, ..., m. This should not create any confusion since one is in bold face and the other is not. This is similarly valid for other symbols defined in (5.219) and (5.241). Of course, we could have used other

THE ACTIVE LOAD

symbols, but then it may create more difficulties than the present situation, and, furthermore, the notation will not be consistent throughout the chapter. Substituting (5.241) and (5.219) in

$$S^*(j\omega)S(j\omega) = U_m \tag{5.242}$$

yields many equalities, three of which are given by

$$|S_{11}(j\omega)|^2 + |S_{12}(j\omega)|^2 + S_{31}^*(j\omega)S_{31}(j\omega) = 1, \tag{5.243a}$$

$$|S_{12}(j\omega)|^2 + |S_{22}(j\omega)|^2 + S_{32}^*(j\omega)S_{32}(j\omega) = 1, \tag{5.243b}$$

$$\bar{S}_{11}(j\omega)S_{11}(j\omega) + S_{21}^*(j\omega)S_{21}(j\omega) = U_2. \tag{5.243c}$$

From (5.243c) we have

$$|\det S_{11}(j\omega)|^2 = \det [U_2 - S_{21}^*(j\omega)S_{21}(j\omega)]$$
$$= 1 - S_{31}^*(j\omega)S_{31}(j\omega) - S_{32}^*(j\omega)S_{32}(j\omega)$$
$$+ S_{31}^*(j\omega)S_{31}(j\omega)S_{32}^*(j\omega)S_{32}(j\omega) - |S_{31}^*(j\omega)S_{32}(j\omega)|^2$$
$$\geq 1 - S_{31}^*(j\omega)S_{31}(j\omega) - S_{32}^*(j\omega)S_{32}(j\omega), \tag{5.244}$$

since according to the famous Cauchy–Schwarz inequality in matrix algebra

$$S_{31}^*(j\omega)S_{31}(j\omega)S_{32}^*(j\omega)S_{32}(j\omega) \geq |S_{31}^*(j\omega)S_{32}(j\omega)|^2. \tag{5.245}$$

By the addition of (5.243a) and (5.243b), we obtain

$$|S_{11}(j\omega)|^2 + |S_{22}(j\omega)|^2 + 2|S_{12}(j\omega)|^2$$
$$= 2 - S_{31}^*(j\omega)S_{31}(j\omega) - S_{32}^*(j\omega)S_{32}(j\omega). \tag{5.246}$$

With the aid of (5.244), (5.246) can be written as

$$|\det S_{11}(j\omega)|^2 + 1 \geq |S_{11}(j\omega)|^2 + |S_{22}(j\omega)|^2 + 2|S_{12}(j\omega)|^2. \tag{5.247}$$

Since

$$|\det S_{11}(j\omega)| = |S_{12}^2(j\omega) - S_{11}(j\omega)S_{22}(j\omega)|$$
$$\geq |S_{12}(j\omega)|^2 - |S_{11}(j\omega)S_{22}(j\omega)| \tag{5.248}$$

and since
$$|S_{11}(j\omega)|^2 + |S_{22}(j\omega)|^2 \geq 2|S_{11}(j\omega)S_{22}(j\omega)|, \quad (5.249)$$
we have
$$|S_{11}(j\omega)|^2 + |S_{22}(j\omega)|^2 \geq 2|S_{12}(j\omega)|^2 - 2|\det S_{11}(j\omega)|. \quad (5.250)$$

Combining (5.247) with (5.250) and using (5.228) yield
$$1 + |\det S_{11}(j\omega)| = 1 + |\det S_{22}(j\omega)| \geq 2|S_{12}(j\omega)|. \quad (5.251)$$

Applying the inequality (5.251) to (5.229) results in the bona fide gain-bandwidth limitation for the reciprocal amplifier:
$$G(\omega^2) \leq \frac{1}{4}\left[1 + \frac{1}{|\det S_{22}(j\omega)|}\right]^2. \quad (5.252)$$

In particular, if only one active impedance is employed, (5.252) reduces to (5.185) with $\det S_{22}(j\omega) = S_{33}(j\omega)$.

We now compare the optimum transducer power gain for the nonreciprocal amplifier with that for the reciprocal amplifier. To this end, we first compute $|\det S_{22}(j\omega)|$ for the nonreciprocal amplifier. Referring again to Fig. 5.37, the scattering matrix $S(s)$ of the m-port network N can be expressed in terms of those of the component networks $N_{\alpha k}$ and N_β by appealing to the formulas (2.165). For $m = 4$, $S(s)$ was derived in Problem 5.58, which together with Problem 5.60 shows that

$$|\det S_{22}(j\omega)| = |S_{11\alpha_1}(j\omega)||S_{11\alpha_2}(j\omega)| \cdots |S_{11\alpha_{m-2}}(j\omega)|. \quad (5.253)$$

Thus, the transducer power gain for the nonreciprocal amplifier is bounded by
$$G(\omega^2) \leq \frac{1}{|\det S_{22}(j\omega)|^2}. \quad (5.254)$$

Comparing (5.254) with (5.252) shows that the optimum nonreciprocal matching can never yield more than 6 dB gain over the optimum reciprocal equalization.

7. Conclusions

We began the chapter by defining a special class of active load impedances $z_l(s)$ whose associated functions $-z_l(-s)$ are strictly passive. The reasons for doing this are that most practical devices such as the tunnel diode can be approximated by impedances belonging to this special class and that the theory for the special class is much the simplest. We then considered the general configuration of a negative-resistance amplifier in which an active one-port impedance is embedded in a two-port connected between a source and a load. The amplifier can also be represented by a three-port network terminating in a voltage source in series with a strictly passive impedance, a strictly passive load impedance and an active impedance. The three-port network is chosen to be lossless since, intuitively, one would expect that the choice of a lossy three-port would not only lessen the transducer power gain but also severely hamper our ability to manipulate in that the scattering matrix of a lossy multi-port is not necessarily unitary. General formula for the transducer power gain in terms of the scattering parameters of the lossless three-port was derived. This was followed by the consideration of three specific configurations for the three-port network.

The first configuration is composed of an interconnection of an ideal three-port circulator and two lossless two-ports. The circulator provides the needed isolation among the desired parameters, while the frequency shaping is achieved with the lossless two-ports. The resulting amplifier is nonreciprocal and permits only one-way transmission of power. The second arrangement is one in which the active impedance is connected across the input or the output port, which can be considered as the degenerate type of the general three-port configuration. The amplifier is attractive in that it facilitates its design, requiring the use of only one or two transformers. In each of these two configurations, we discussed in detail its gain-bandwidth limitations, and derived formulas for amplifiers having Butterworth and Chebyshev transducer power-gain characteristics of arbitrary order for the active impedance represented by the simplified model of a tunnel diode. A detailed investigation of the stability behavior of the amplifiers reveals that they are stable at each of their ports. Explicit formulas for the sensitivity of the transmission-power amplifiers, operating with different gain characteristics, were derived. Since sensitivity is of the order of the square-root of the gain, in order to minimize drifts in the negative resistance, well-regulated bias supply is required. The third configuration considered is that the lossless three-port network is reciprocal, which allows the amplifier to have power transmission in both directions with exactly the same gain characteristic. We showed that for the use of a tunnel diode, the optimum non-reciprocal amplifier is capable of yielding 6 dB more flat gain than the optimum reciprocal amplifier.

Finally, we considered the situation where more than one active impedance is employed. We showed that the optimum nonreciprocal matching can never yield more than 6 dB gain over the optimum reciprocal equalization, a fact that was pointed out earlier for the case where a single tunnel diode is used.

Problems

5.1. Let the open-circuit and short-circuit immittance parameters of a two-port network be $z_{ij}(s)$ and $y_{ij}(s)$, respectively. Show that the input impedance $Z(s)$ of the two-port with the output port terminating in $z_l(s)$ is given by

$$Z(s) = z_{11}(s) \frac{1/y_{22}(s) + z_l(s)}{z_{22}(s) + z_l(s)}. \tag{5.255}$$

5.2. Applying the interconnection formulas (2.165) to (5.11) and (5.12), derive the normalized scattering matrix $S(s)$ as given in (5.13).

5.3. Show that if the two-port N_β is removed and if the load R_2 is connected directly across the port $\hat{2}$ of the circulator, as shown in Fig. 5.4, then for $\hat{R}_2 = R_2$ the scattering matrix (5.13) becomes

$$S(s) = \begin{bmatrix} 0 & 1 & 0 \\ S_{11\alpha}(s) & 0 & S_{12\alpha}(s) \\ S_{21\alpha}(s) & 0 & S_{22\alpha}(s) \end{bmatrix}. \tag{5.256}$$

What is the transducer power gain for the amplifier?

THE ACTIVE LOAD

5.4. In (5.18), show that for a lossless two-port N_α terminating in $z_t(s)$ we have

$$|\rho_1^t(j\omega)| = |\rho_2^t(j\omega)|. \quad (5.257)$$

[*Hint.* $z_t(s) = -z_3(-s)$.]

5.5. Show that the relations as indicated in (5.49) are valid in general.

5.6. Consider the same problem as in Example 5.1 except now that we wish to achieve the fifth-order Butterworth transducer power gain with a passband tolerance of 3 dB. Realize the amplifier together with its schematic diagram.

5.7. Show that (5.66) approaches to (5.67) as n approaches to infinity. [*Hint.* See (4.31).]

5.8. Consider the scattering matrix (5.11) of the ideal circulator N_c normalizing to the impedances R_1, $\hat{R}_2$ and $\hat{R}_3$. Show that if ports $\hat{1}$ and $\hat{2}$ are terminated in their reference impedances R_1 and $\hat{R}_2$, the impedance looking into port $\hat{3}$ is $\hat{R}_3$. Repeat the problem for port $\hat{2}$ when the other two ports are terminated in their reference impedances.

5.9. Suppose that we choose the scattering matrix (5.22c) instead of (5.11). Are the statements made in Problem 5.8 still valid? [*Hint.* Use (5.22d).]

5.10. In (5.62), once the back-end reflection coefficient $\hat{S}_{22\alpha}(s)$ is determined, the back-end impedance $Z_{22\alpha}(s)$ of the lossless equalizer N_α can be computed from (4.138). Using this $Z_{22\alpha}(s)$, show that the terminating resistance for N_α is given by

$$R_3' = R\left(\frac{K_n^{1/2} - 1}{K_n^{1/2} + 1}\right). \quad (5.258)$$

5.11. Derive a formula, similar to (5.258), for the terminating resistance for the lossless equalizer N_β with $Z_{22\beta}(s)$ being given in (5.70).

5.12. Repeat the problem considered in Example 5.2 with the additional constraint that the variation inside the passband cannot exceed 1 dB.

5.13. Design a low-pass nonreciprocal amplifier for the following specifications:

$$-R = -100\,\Omega, \quad C = 10\,\text{pF},$$
$$R_1 = 100\,\Omega, \quad R_2 = 200\,\Omega,$$
$$f_c = 100\,\text{MHz}.$$

The amplifier is required to have a maximally-flat transducer power-gain characteristic with at least 40 dB within the passband and the passband tolerance is 2 dB.

5.14. For the nonreciprocal amplifier discussed in §3.4.2, show that the terminating resistance for the lossless equalizer N_α is given by

$$R_3' = R\frac{\{[K_n + \epsilon^2 C_n^2(0)]^{1/2} - [1 + \epsilon^2 C_n^2(0)]^{1/2}\}^2}{K_n - 1}. \quad (5.259)$$

[*Hint.* Let $s = 0$.]

400 BROADBAND MATCHING NETWORKS

5.15. Derive a formula similar to (5.259) for the lossless equalizer N_β considered in §3.4.2.

5.16. In (5.258) and (5.259), show that

$$R'_3 \to R'_\infty = R \tanh\left(\frac{1}{4RCf_c}\right) \tag{5.260}$$

as $n \to \infty$ in (5.258) or $n \to \infty$ and $\epsilon \to 0$ in (5.259).

5.17. Compute the driving-point impedance $Z_d(j\omega)$ of the network of Fig. 5.1(a). Show that the real part of $Z_d(j\omega)$ ceases to be negative when the resistive cutoff frequency

$$f_r = (R/R_d - 1)^{1/2}/(2\pi RC) \tag{5.261}$$

is reached or surpassed. Also show that the self-resonant frequency of the network is determined by

$$f_s = (R^2C/L_d - 1)^{1/2}/(2\pi RC). \tag{5.262}$$

5.18. Derive (5.100) and (5.102).

5.19. Consider the same problem specified in Problem 5.13 except now that we wish to design an amplifier having an equiripple transducer power-gain characteristic, everything else being the same. Realize the amplifier together with its schematic diagram.

5.20. Show that in the limit as n approaches to infinity, (5.120) reduces to (5.121).

5.21. Is it true that (5.121) also represents the maximum attainable gain for the ideal brick-wall type of response? If so, justify your conclusion. [*Hint.* See (5.85).]

5.22. In Fig. 5.37, let each active impedance $z_{lk}(s)$, $k = 3, 4, \ldots, m$, be replaced by the parallel connection of a negative resistor with resistance $-R_k$ and a capacitor with capacitance C_k. Using (5.229), show that

$$\int_0^\infty \ln G(\omega^2) d\omega \leq 2\pi \sum_{k=3}^m \frac{1}{R_k C_k}. \tag{5.263}$$

5.23. Following the procedure used to obtain (5.83), derive the inequality (5.128).

5.24. Show that in the limit as $n \to \infty$ and $\epsilon \to 0$, (5.129) reduces to (5.130).

5.25. Derive the gain formula (5.142). What are the normalizing impedances for $S(s)$?

5.26. Let

$$\theta = \sin^{-1}(R/R_2)^{1/2}. \tag{5.264}$$

Prove that (5.120) can be expressed equivalently as

$$K_n \leq \tan^2\theta\left\{\left[1 + \frac{\cos^2\theta \sin(\pi/2n)}{\pi RCf_c}\right]^{2n} - 1\right\}. \tag{5.265}$$

THE ACTIVE LOAD

5.27. Confirm the statement that the terminating resistance for $Z_{22}(s)$, as given in (5.124), for the Butterworth response can be computed directly by the formula

$$R_2' = \alpha R_2 \left(\frac{\hat{K}_n^{1/2} - 1}{\hat{K}_n^{1/2} + 1} \right), \qquad (5.266a)$$

where

$$\hat{K}_n^{1/2} = \left[1 + \frac{\sin(\pi/2n)}{\pi \alpha R_2 C f_c} \right]^n \qquad (5.266b)$$

for $R < R_2$.

5.28. Confirm the statement that the terminating resistance for $Z_{22}(s)$, as given in (5.134), for the Chebyshev response can be computed directly by the formulas

$$R_2' = \alpha R_2 \left(\frac{\epsilon \sinh \gamma - 1}{\epsilon \sinh \gamma + 1} \right) \qquad (5.267a)$$

for n odd, and

$$r_2' = \alpha R_2 \left[\frac{\epsilon \cosh \gamma - (1 + \epsilon^2)^{1/2}}{\epsilon \cosh \gamma + (1 + \epsilon^2)^{1/2}} \right] \qquad (5.267b)$$

for n even, where

$$\gamma = n \sinh^{-1} \left[\sinh \left(\frac{1}{n} \sinh^{-1} \frac{1}{\epsilon} \right) + \frac{\sin(\pi/2n)}{\pi \alpha R_2 C f_c} \right]. \qquad (5.267c)$$

[*Hint.* From (4.93), we have $R_2' = \alpha R_2 (d_0 - \hat{d}_0)/(d_0 + \hat{d}_0)$.]

5.29. Show that the inequality (5.129) can be expressed equivalently as

$$K_n \leq \tan^2 \theta \left\{ \epsilon^2 \sinh^2 \left[n \sinh^{-1} \left(\frac{\cos^2 \theta \sin(\pi/2n)}{\pi R C f_c} \right. \right. \right.$$
$$\left. \left. \left. + \sinh \left(\frac{1}{n} \sinh^{-1} \frac{1}{\epsilon} \right) \right) \right] - 1 \right\}, \qquad (5.268)$$

θ being defined in (5.264).

5.30. Another useful configuration for the negative-resistance amplifier is shown in Fig. 5.39, in which the load resistance R_2 is connected directly across the source having an internal resistance R_1. The tunnel diode is connected at the output

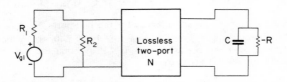

FIG. 5.39. A useful configuration for the negative-resistance amplifiers.

port of the lossless equalizer N. Show that the transducer power gain of the amplifier is given by

$$G(\omega^2) = \frac{R_1 R_2}{(R_1 + R_2)^2}\left|1 + \frac{1}{S_{11}(j\omega)}\right|^2, \quad (5.269)$$

where $S_{11}(s)$ is the reflection coefficient at the input port of N normalizing to a constant resistance $\hat{R}_1$, which is the parallel combination of R_1 and R_2.

5.31. Consider the amplifier configuration of Fig. 5.39. Show that for an ideal brick-wall type of gain response, the maximum attainable constant gain is given by

$$\frac{R_1 R_2}{(R_1 + R_2)^2}\left[1 + \exp\left(\frac{1}{2RCB}\right)\right]^2, \quad (5.270)$$

B being the bandwidth. [*Hint.* See (5.85).]

5.32. Another amplifier configuration is shown in Fig. 5.40, in which the tunnel diode is separated from the source and the load by two matching networks N_α and

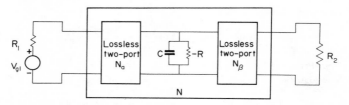

Fig. 5.40. An amplifier configuration in which the tunnel diode is separated from the source and the load by two matching networks.

N_β. Show that the transmission coefficient $S_{21}(s)$ of the composite two-port N can be expressed in terms of the normalized scattering parameters $S_{ij\alpha}(s)$ and $S_{ij\beta}(s)$ of the component two-ports N_α and N_β, respectively, as follows:

$$S_{21}(s) = \frac{2R S_{12\alpha}(s) S_{12\beta}(s)}{(2R - 1) - S_{22\alpha}(s) - S_{11\beta}(s) - (2R + 1) S_{22\alpha}(s) S_{11\beta}(s)}. \quad (5.271)$$

What are the normalizing impedances for the two-ports N_α, N_β and N?

5.33. Consider the amplifier configuration of Fig. 5.40, in which $R = \tfrac{1}{2}\Omega$, $S_{12\alpha}(s) = -S_{12\beta}(s)$ and $S_{22\alpha}(s) = -S_{11\beta}(s)$. Prove that the transducer power gain of the amplifier is given by

$$G(\omega^2) = \frac{[1 - |S_{22\alpha}(j\omega)|^2]^2}{4|S_{22\alpha}(j\omega)|^4}. \quad (5.272)$$

THE ACTIVE LOAD

5.34. In (5.272), assume that $G(\omega^2)$ is a constant over a given band B of frequencies. Show that this constant is bounded above by the quantity

$$\frac{1}{4}\left[\exp\left(\frac{1}{2RCB}\right) - 1\right]^2. \tag{5.273}$$

5.35. Derive the transducer power-gain formula (5.149).

5.36. Using the gain formula (5.149), show that the dc value of gain for a low-pass Butterworth gain response is given by (5.151a), and the constant term in the corresponding Chebyshev response is given by (5.151b).

5.37. Show that (5.151) approaches to (5.152) as n approaches to infinity in (5.151a) or as n approaches to infinity and ϵ to zero in (5.151b).

5.38. Derive the transducer power-gain formula (5.153).

5.39. In all the formulas derived for the transmission-power amplifiers, we assume that $R \neq R_1$ or R_2. Suppose that this is not the case. Can we still design an amplifier using these parameters? If so, give a detailed account of the necessary changes.

5.40. The stability of the transmission power amplifier has been tested for the situation where the tunnel diode is connected across the load. Justify the statement that the same conclusion can be made for the arrangement where the tunnel diode is connected across the source.

5.41. Repeat the problem given in Example 5.4 for the sixth-order Butterworth transducer power-gain characteristic.

5.42. Consider the same problem as in Example 5.4 except now that we wish to achieve an equiripple characteristic for its transducer power gain. What is the smallest n that will satisfy all the requirements? If $n \neq 5$, realize the amplifier.

5.43. Derive the identity (5.158).

5.44. Show that in the limit as $n \to \infty$ in (5.165) or as $n \to \infty$ and $\epsilon \to 0$ in (5.167), (5.165) and (5.167) approach to (5.168).

5.45. The simplest amplifier configuration is shown in Fig. 5.41, in which $Z(s)$ is an arbitrary passive impedance. Show that for a given bandwidth B, the maximum constant gain of the amplifier is given by

$$G = 4(\pi^2 R_1 R_2 B^2 C^2)^{-1}[k + (1 + k^2)^{1/2}]^{-2}, \tag{5.274a}$$

where

$$k = \frac{1}{2\pi BC}\left(\frac{1}{R_1} + \frac{1}{R_2} - \frac{1}{R}\right). \tag{5.274b}$$

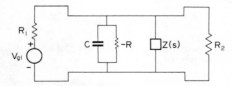

FIG. 5.41. The simplest amplifier configuration employing a tunnel diode.

404   BROADBAND MATCHING NETWORKS

5.46. In the network of Fig. 5.30, show that the dc value of gain of the amplifier, operating in high-gain mode with $R_1 > R$ and having a low-pass power-gain characteristic, is given by the expression (5.172).

5.47. Consider the same amplifier in Problem 5.46. Show that the sensitivity of the dc gain due to a change of the negative resistance R is given by the expression (5.173).

5.48. Let

$$\theta = \operatorname{cosec}^{-1}(R_1/R)^{1/2}. \qquad (5.275)$$

Show that (5.151a), (5.151b) and (5.152) can be expressed equivalently as (5.265), (5.268) and

$$K_\infty \leq \tan^2\theta \left[\exp\left(\frac{\cos^2\theta}{RCf_c}\right) - 1\right], \qquad (5.276)$$

respectively.

5.49. Let

$$\phi = \operatorname{cosec}^{-1}(R_1''/R)^{1/2}. \qquad (5.277)$$

Show that K_n as given in (5.156) can be expressed equivalently as

$$K_n \leq \frac{R_1}{R}\sin^2\phi\,\tan^2\phi \left\{\left[1 + \frac{\cos^2\phi\,\sin(\pi/2n)}{\pi RCf_c}\right]^{2n} - 1\right\} \qquad (5.278)$$

for a Butterworth response, and

$$K_n \leq \frac{R_1}{R}\sin^2\phi\,\tan^2\phi \left\{\epsilon^2 \sinh^2\left[n\,\sinh^{-1}\left(\frac{\cos^2\phi\,\sin(\pi/2n)}{\pi RCf_c}\right.\right.\right.$$
$$\left.\left.\left.+ \sinh\left(\frac{1}{n}\sinh^{-1}\frac{1}{\epsilon}\right)\right)\right] - 1\right\} \qquad (5.279)$$

for a Chebyshev response with the quantities as previously defined.

5.50. Derive the identities (5.193) and (5.194).

5.51. Suppose that we wish to maximize $|S_{23\alpha}(j\omega)S_{31\alpha}(j\omega)|$ subject to the constraint

$$|S_{31\alpha}(j\omega)|^2 + |S_{23\alpha}(j\omega)|^2 = 1. \qquad (5.280a)$$

Show that this will result in

$$|S_{31\alpha}(j\omega)| = |S_{23\alpha}(j\omega)| = 2^{-1/2}. \qquad (5.280b)$$

5.52. Using the angle θ defined in (5.264), show that (5.165), (5.167) and (5.168) can be expressed equivalently as

$$S_R^G(0) = \sec^2\theta \left\{\left[1 + \frac{\cos^2\theta\,\sin(\pi/2n)}{\pi RCf_c}\right]^n - 1\right\} \qquad (5.281)$$

for the Butterworth response;

$$S_R^G(0) = (\epsilon \sinh \gamma - 1) \sec^2 \theta \qquad (5.282a)$$

for n odd, and

$$S_R^G(0) = (1 + \epsilon^2)^{-1/2}[\epsilon \cosh \gamma - (1 + \epsilon^2)^{1/2}] \sec^2 \theta \qquad (5.282b)$$

for n even, γ being defined in (5.166c); and

$$S_R^G(0) = \sec^2 \theta \left[\exp\left(\frac{\cos^2 \theta}{2RCf_c}\right) - 1 \right]; \qquad (5.283)$$

respectively.

5.53. Obtain expressions similar to (5.281), (5.282) and (5.283) for the amplifier of Fig. 5.30, operating in high-gain mode with $R_1 > R$ and having a low-pass power-gain characteristic.

5.54. Derive the gain-bandwidth limitation (5.215).

5.55. Using the interconnection formula (2.167), show that the scattering matrix $S_a(s)$ of the amplifier of Fig. 5.35 can be expressed in terms of the submatrices of the scattering matrix of the m-port network N, as indicated in (5.220).

5.56. Show that the matrix (5.222a) is the inverse of that of (5.219). [*Hint.* $S(s)S^{-1}(s) = U_m$.]

5.57. In Fig. 5.37, show that the scattering matrices $S_k(s)$ of the two-port networks N_k are of the form as shown in (5.232a). [*Hint.* See (5.256).]

5.58. Referring to the network of Fig. 5.37, suppose that we have two sections of the type as shown in Fig. 5.36 in cascade. Show that the scattering matrix of the resulting four-port network can be expressed in terms of the scattering parameters $S_{ij\alpha_1}(s)$ and $S_{ij\alpha_2}(s)$ of the component two-port networks $N_{\alpha 1}$ and $N_{\alpha 2}$, as follows:

$$\begin{bmatrix} 0 & 1 & 0 & 0 \\ S_{11\alpha_1}(s)S_{11\alpha_2}(s) & 0 & S_{12\alpha_1}(s)S_{11\alpha_2}(s) & S_{12\alpha_2}(s) \\ S_{21\alpha_1}(s) & 0 & S_{22\alpha_1}(s) & 0 \\ S_{11\alpha_1}(s)S_{21\alpha_2}(s) & 0 & S_{12\alpha_1}(s)S_{21\alpha_2}(s) & S_{22\alpha_2}(s) \end{bmatrix}. \qquad (5.284)$$

[*Hint.* See Problem 5.3.]

5.59. Repeat Problem 5.58 for $m - 2$ sections in cascade.

5.60. Using the result obtained in Problem 5.59, show that

$$|\det S_{22}(j\omega)| = |S_{11\alpha_1}(j\omega)||S_{11\alpha_2}(j\omega)| \cdots |S_{11\alpha_{m-2}}(j\omega)| \qquad (5.285)$$

with the quantities as defined in §6.1 and §6.2.

5.61. The amplifier shown in Fig. 5.42 is a variant of the general configuration given in Fig. 5.41, in which the tunnel diode is shunted by an RF tank which determines

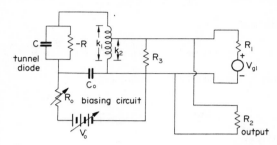

FIG. 5.42. A variant of the general configuration of Fig. 5.41.

the amplifier resonant frequency. Also included are the biasing circuit for the tunnel diode and a by-pass capacitor C_0, which should be made as large as possible to prevent parasitic oscillations in the biasing circuit. Show that the transducer power gain of the amplifier is given by

$$G = \frac{4}{R_1 R_2} \left[\frac{1}{R_1} + \frac{1}{R_2} + \frac{1}{R_3} - \left(\frac{k_1}{k_2}\right)^2 \frac{1}{R} \right]^{-2}. \qquad (5.286)$$

Based on Fig. 5.42, an experimental amplifier was built by Chang (1959) using the following parameters:

$$-R = -375\,\Omega, \qquad C = 40\,\text{pF},$$
$$R_1 = 50\,\Omega, \qquad R_2 = 1\,\text{k}\Omega,$$
$$(k_1/k_2)^2 = 7.65.$$

He observed a gain of 20 dB at the operating frequency of 30 MHz with a bandwidth of 200 kHz. Using the gain formula (5.286), compute the theoretic gain, assuming that R_3 is sufficiently large to be negligible.

5.62. Design an optimum transmission-power amplifier of Fig. 5.24, operating in high-gain mode with $R_2 > R$ and having the second-order low-pass Butterworth gain response. Show that the back-end impedance $Z_{22}(s)$ of the lossless equalizer N with the input port terminating in a resistor is given by

$$Z_{22}(s) = \frac{2s}{\omega_c^2 C(\hat{K}_2^{1/2}+1)} + \frac{\sqrt{2}(\hat{K}^{1/4}+1)}{\omega_c C(\hat{K}_2^{1/2}+1)}, \qquad (5.287)$$

where $\hat{K}_2$ is defined in (5.266b). Compare the constant term with (5.266a).

5.63. Repeat Problem 5.62 for the amplifier configuration of Fig. 5.30, again operating in the high-gain mode with $R_1 > R$.

THE ACTIVE LOAD

5.64. Suppose that we connect a voltage generator of internal resistance R_g at the input port of Fig. 5.1(a). Show that for the network to be stable, we must have

$$\frac{R_d + R_g}{L_d} - \frac{1}{RC} > 0, \qquad (5.288a)$$

$$1 - \frac{R_d + R_g}{R} > 0. \qquad (5.288b)$$

5.65. Applying (4.93), confirm the impedance function $Z_{22\alpha}(s)$ computed in (5.74).

5.66. Repeat Problems 5.13 and 5.19 for the tunnel diode which has the following specifications:

$$-R = -50\,\Omega, \qquad C = 10\,\text{pF},$$
$$R_d = 1\,\Omega, \qquad L_d = 10\,\text{nH}.$$

Also determine the resistive cutoff frequency f_r and the self-resonant frequency f_s of the tunnel diode.

5.67. Using the specifications given in Problem 5.13, design a transmission-power amplifier.

5.68. Consider the amplifier configuration of Fig. 5.43, in which the lossless four-port network N called a *hybrid* has a scattering matrix

$$S(s) = 2^{-1/2} \begin{bmatrix} 0 & 0 & 1 & 1 \\ 0 & 0 & 1 & -1 \\ 1 & 1 & 0 & 0 \\ 1 & -1 & 0 & 0 \end{bmatrix}, \qquad (5.289)$$

normalizing to the impedances $R_1, R_2, \hat{R}_3$ and $\hat{R}_4$; $\hat{R}_3$ and $\hat{R}_4$ being arbitrary real constants. Let

$$S_\alpha(s) = [S_{ij\alpha}] \qquad (5.290a)$$

and

$$S_\beta(s) = [S_{ij\beta}] \qquad (5.290b)$$

be the scattering matrices of the lossless two-port networks N_α and N_β normalizing to the strictly passive impedances $\hat{R}_3$ and $z_3(s) = -z_{13}(-s)$, and $\hat{R}_4$ and $z_4(s) = -z_{l4}(-s)$, respectively. Show that the resulting scattering matrix of the amplifier N_a normalizing to R_1 and R_2 is given by

$$S_a(s) = \frac{1}{2}\begin{bmatrix} S_{11\alpha}^{-1}(-s) + S_{11\beta}^{-1}(-s) & S_{11\alpha}^{-1}(-s) - S_{11\beta}^{-1}(-s) \\ S_{11\alpha}^{-1}(-s) - S_{11\beta}^{-1}(-s) & S_{11\alpha}^{-1}(-s) + S_{11\beta}^{-1}(-s) \end{bmatrix}$$

$$= \frac{1}{2}\begin{bmatrix} \rho_{11\alpha}(s) + \rho_{11\beta}(s) & \rho_{11\alpha}(s) - \rho_{11\beta}(s) \\ \rho_{11\alpha}(s) - \rho_{11\beta}(s) & \rho_{11\alpha}(s) + \rho_{11\beta}(s) \end{bmatrix}, \qquad (5.291)$$

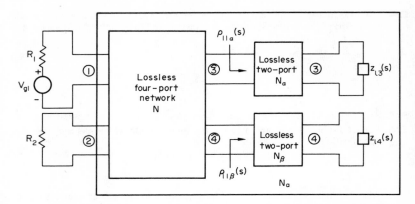

FIG. 5.43. An amplifier configuration employing two active impedances with the lossless four-port network N being a hybrid.

where $\rho_{11\alpha}(s)$ and $\rho_{11\beta}(s)$ are the active reflection coefficients at the input ports of N_α and N_β normalizing to $\hat{R}_3$ and $\hat{R}_4$, respectively, as shown in Fig. 5.43. [*Hint.* Apply (2.165), (2.167) and (5.255).]

5.69. Referring to (5.57), we can write the nth-order low-pass Butterworth transducer power-gain characteristic as

$$G(\omega^2) = \frac{1}{\delta_n + (\omega/\omega_b)^{2n}}, \quad 0 \leq \delta_n \leq 1. \tag{5.292}$$

Following (5.59), choose appropriate $S_{12\beta}(s)$ and $S_{22\alpha}(s)$, and show that the 3-dB bandwidth $\omega_{3\text{-dB}}$ is bounded by

$$\omega_{3\text{-dB}} \leq \frac{2 \sin(\pi/2n)}{RC(\delta_n^{-1/2n} - 1)}. \tag{5.293}$$

5.70. Referring to (5.57), suppose that the transducer power-gain characteristic is specified as

$$G(\omega^2) = \frac{1 + (\omega/\omega_b)^{2n}}{\delta_n + (\omega/\omega_b)^{2n}}, \quad 0 \leq \delta_n \leq 1. \tag{5.294}$$

We can choose $S_{12\beta}(j\omega) = 1$ and $|S_{22\alpha}(j\omega)| = 1/G(\omega^2)$. Show that the 3-dB bandwidth of the resulting amplifier is bounded by

$$\omega_{3\text{-dB}} \leq 2(\delta_n^{-1/2n} - 1)^{-1}(1 - 2\delta_n)^{-1/2n} \cdot \frac{\sin(\pi/2n)}{RC}. \tag{5.295}$$

5.71. In §4.1.1A, let

$$|\hat{S}_{22\alpha}(j\omega)|^2 = \frac{\delta_n + (\omega/\omega_b)^{2n}}{1 + (\omega/\omega_b)^{2n}}, \qquad 0 \leq \delta_n \leq 1. \tag{5.296}$$

Show that the 3-dB bandwidth of the transducer power gain of the resulting amplifier is bounded by

$$\omega_{3\text{-dB}} \leq \left(\frac{1}{R} - \frac{1}{R_2}\right) \cdot \frac{2\sin(\pi/2n)}{C(\delta_n^{-1/2n} - 1)}. \tag{5.297}$$

5.72. Consider the amplifier configuration discussed in Problem 5.33. Let $|S_{22\alpha}(j\omega)|^2$ be the function specified on the right-hand side of (5.296). Determine the gain-bandwidth limitation of the resulting amplifier.

5.73. In §3.4.2, let

$$G(\omega^2) = \frac{1 + \epsilon^2 C_n^2(\omega/\omega_c)}{\delta_n + \epsilon^2 C_n^2(\omega/\omega_c)}, \qquad 0 \leq \delta_n \leq 1. \tag{5.298}$$

Choose $|S_{12\beta}(j\omega)|^2 = 1$ and $|S_{22\alpha}(j\omega)|^2 = 1/G(\omega^2)$. Show that the half-power bandwidth with respect to the gain $1/\delta_n$ is bounded by

$$\frac{2\sin(\pi/2n)\cosh\left(\dfrac{1}{n}\cosh^{-1}[\epsilon^{-1}(1/\delta_n - 2)^{-1/2}]\right)}{RC\left[\sinh\left(\dfrac{1}{n}\sinh^{-1}\dfrac{1}{\epsilon}\right) - \sinh\left(\dfrac{1}{n}\sinh^{-1}\dfrac{\delta_n^{1/2}}{\epsilon}\right)\right]} \geq \omega_{\text{half-power}}. \tag{5.299}$$

[*Hint.* Apply (5.83).]

5.74. By choosing the turns ratios properly, show that the scattering matrix of the three-port network of Fig. 5.32, normalizing to the resistances R_1, R_2, and R_3, can be expressed in the form as shown in (5.186).

References

1. Aron, R. (1961) Gain bandwidth relations in negative resistance amplifiers. *Proc. IRE*, vol. 49, no. 1, pp. 355–356.
2. Boyet, H., Fleri, D. and Renton, C. A. (1961) Stability criteria for a tunnel diode amplifier. *Proc. IRE*, vol. 49, no. 12, p. 1937.
3. Carlin, H. J. and Giordano, A. B. (1964) *Network Theory: An Introduction to Reciprocal and Nonreciprocal Circuits*, Englewood Cliffs, N.J.: Prentice-Hall.
4. Chan, Y. T. and Kuh, E. S. (1966) A general matching theory and its application to tunnel diode amplifiers. *IEEE Trans. Circuit Theory*, vol. CT-13, no. 1, pp. 6–18.
5. Chang, K. K. N. (1959) Low-noise tunnel-diode amplifier. *Proc. IRE*, vol. 47, no. 7, pp. 1268–1269.
6. Chen, W. K. (1975) Design formulas for optimum transmission-power amplifiers having Butterworth and Chebyshev gain responses. *Proc. 9th Asilomar Conf. on Circuits, Systems, and Computers*, Monterey, Calif., pp. 217–221, Nov. 3–5.

7. Chen, W. K. (1976 Explicit formulas for the design of transmission-power amplifiers. *J. Franklin Inst.*, vol. 301, no. 5.
8. Davidson, L. A. (1963) Optimum stability criterion for tunnel diodes shunted by resistance and capacitance. *Proc. IEEE*, vol. 51, no. 9, p. 1233.
9. Esaki, L. (1958) New phenomenon in narrow Ge $p-n$ junctions. *Phys. Rev.*, vol. 109, pp. 603–604.
10. Frisch, I. T. (1964) A stability criterion for tunnel diodes. *Proc. IEEE*, vol. 52, no. 8, pp. 922–923.
11. Hines, M. E. (1960) High-frequency negative-resistance circuit principles for Esaki diode applications. *Bell System Tech. J.*, vol. 39, no. 5, pp. 477–513.
12. Kuh, E. S. and Patterson, J. D. (1961) Design theory of optimum negative-resistance amplifiers. *Proc. IRE*, vol. 49, no. 6, pp. 1043–1050.
13. Kuh, E. S. and Rohrer, R. A. (1967) *Theory of Linear Active Networks*, San Francisco, Calif.: Holden-Day.
14. Sard, E. W. (1959) Analysis of a negative conductance amplifier operated with a nonideal circulator. *IRE Trans. Microwave Theory and Techniques*, vol. MTT-7, no. 2, pp. 288–293.
15. Sard, E. W. (1960) Tunnel (Esaki) diode amplifiers with unusually large bandwidths. *Proc. IRE*, vol. 48, no. 3, pp. 357–358.
16. Sard, E. W. (1960) Gain-bandwidth performance of maximally flat negative-conductance amplifiers. *Proc. Symp. Active Networks and Feedback Systems*, Polytechnic Inst. of Brooklyn, New York, vol. 10, pp. 319–344.
17. Sard, E. W. (1961) Gain bandwidth relations in negative resistance amplifiers. *Proc. IRE*, vol. 49, no. 1, pp. 355–356.
18. Smilen, L. I. and Youla, D. C. (1960) Exact theory and synthesis of a class of tunnel diode amplifiers. *Proc. Natl. Electronics Conf.*, vol. 16, pp. 376–404.
19. Smilen, L. I. and Youla, D. C. (1961) Stability criteria for tunnel diodes. *Proc. IRE*, vol. 49, no. 7, pp. 1206–1207.
20. Sommers, H. S., Jr. (1959) Tunnel diodes as high-frequency devices. *Proc. IRE*, vol. 47, no. 7, pp. 1201–1206.
21. Su, K. L. (1965) *Active Network Synthesis*, New York: McGraw-Hill.
22. Whitson, R. B. (1963) Impedance mapping in tunnel-diode stability analysis. *IEEE Trans. Circuit Theory*, vol. CT-10, no. 1, pp. 111–113.
23. Youla, D. C. and Smilen, L. I. (1960) Optimum negative-resistance amplifiers. *Proc. Symp. Active Networks and Feedback Systems*, Polytechnic Inst. of Brooklyn, New York, vol. 10, pp. 241–318.

APPENDICES

APPENDIX A
The Butterworth Response

TABLE A. COEFFICIENTS OF BUTTERWORTH POLYNOMIALS (3.12).

$$q(s) = 1 + a_1 s + a_2 s^2 + \cdots + a_{n-1} s^{n-1} + s^n$$

n	a_1	a_2	a_3	a_4	a_5	a_6	a_7	a_8	a_9
2	1.41421								
3	2.00000	2.00000							
4	2.61313	3.41421	2.61313						
5	3.23607	5.23607	5.23607	3.23607					
6	3.86370	7.46410	9.14162	7.46410	3.86370				
7	4.49396	10.09784	14.59179	14.59179	10.09784	4.49396			
8	5.12583	13.13707	21.84615	25.68836	21.84615	13.13707	5.12583		
9	5.75877	16.58172	31.16344	41.98639	41.98639	31.16344	16.58172	5.75877	
10	6.39245	20.43173	42.80206	64.88240	74.23343	64.88240	42.80206	20.43173	6.39245

APPENDIX B
The Chebyshev Response

TABLE B.1. CHEBYSHEV POLYNOMIALS $C_n(\omega)$.

$C_0(\omega) = 1$,
$C_1(\omega) = \omega$,
$C_2(\omega) = 2\omega^2 - 1$,
$C_3(\omega) = 4\omega^3 - 3\omega$,
$C_4(\omega) = 8\omega^4 - 8\omega^2 + 1$,
$C_5(\omega) = 16\omega^5 - 20\omega^3 + 5\omega$,
$C_6(\omega) = 32\omega^6 - 48\omega^4 + 18\omega^2 - 1$,
$C_7(\omega) = 64\omega^7 - 112\omega^5 + 56\omega^3 - 7\omega$,
$C_8(\omega) = 128\omega^8 - 256\omega^6 + 160\omega^4 - 32\omega^2 + 1$,
$C_9(\omega) = 256\omega^9 - 576\omega^7 + 432\omega^5 - 120\omega^3 + 9\omega$,
$C_{10}(\omega) = 512\omega^{10} - 1280\omega^8 + 1120\omega^6 - 400\omega^4 + 50\omega^2 - 1$.

TABLES B.2. COEFFICIENTS OF THE POLYNOMIALS $p(s)$ (3.78) ASSOCIATED WITH THE CHEBYSHEV RESPONSE.

$$p(s) = b_0 + b_1 s + b_2 s^2 + \cdots + b_{n-1} s^{n-1} + s^n$$

B.2.1. $\tfrac{1}{2}$-dB ripple ($\epsilon = 0.34931$)

n	b_0	b_1	b_2	b_3	b_4	b_5	b_6	b_7	b_8	b_9
1	2.86278									
2	1.51620	1.42563								
3	0.71569	1.53490	1.25291							
4	0.37905	1.02546	1.71687	1.19739						
5	0.17892	0.75252	1.30958	1.93737	1.17249					
6	0.09476	0.43237	1.17186	1.58976	2.17185	1.15918				
7	0.04473	0.28207	0.75565	1.64790	1.86941	2.41265	1.15122			
8	0.02369	0.15254	0.57356	1.14859	2.18402	2.14922	2.65675	1.14608		
9	0.01118	0.09412	0.34082	0.98362	1.61139	2.78150	2.42933	2.90273	1.14257	
10	0.00592	0.04929	0.23727	0.62697	1.52743	2.14424	3.44093	2.70974	3.14988	1.14007

B.2.2. 1-dB ripple ($\epsilon = 0.50885$)

n	b_0	b_1	b_2	b_3	b_4	b_5	b_6	b_7	b_8	b_9
1	1.96523									
2	1.10251	1.09773								
3	0.49131	1.23841	0.98834							
4	0.27563	0.74262	1.45393	0.95281						
5	0.12283	0.58053	0.97440	1.68882	0.93682					
6	0.06891	0.30708	0.93935	1.20214	1.93083	0.92825				
7	0.03071	0.21367	0.54862	1.35754	1.42879	2.17608	0.92312			
8	0.01723	0.10735	0.44783	0.84682	1.83690	1.65516	2.42303	0.91981		
9	0.00768	0.07061	0.24419	0.78631	1.20161	2.37812	1.88148	2.67095	0.91755	
10	0.00431	0.03450	0.18245	0.45539	1.24449	1.61299	2.98151	2.10785	2.91947	0.91593

B.2.3. 2-dB ripple ($\epsilon = 0.76478$)

n	b_0	b_1	b_2	b_3	b_4	b_5	b_6	b_7	b_8	b_9
1	1.30756									
2	0.82306	0.80382								
3	0.32689	1.02219	0.73782							
4	0.20577	0.51680	1.25648	0.71622						
5	0.08172	0.45935	0.69348	1.49954	0.70646					
6	0.05144	0.21027	0.77146	0.86702	1.74586	0.70123				
7	0.02042	0.16609	0.38251	1.14444	1.03922	1.99353	0.69789			
8	0.01286	0.07294	0.35870	0.59822	1.57958	1.21171	2.24225	0.69607		
9	0.00511	0.05438	0.16845	0.64447	0.85687	2.07675	1.38375	2.49129	0.69468	
10	0.00322	0.02334	0.14401	0.31776	1.03891	1.15853	2.63625	1.55574	2.74060	0.69369

TABLES B.2. (contd.).

B.2.4. 3-dB ripple ($\epsilon = 0.99763$)

n	b_0	b_1	b_2	b_3	b_4	b_5	b_6	b_7	b_8	b_9
1	1.00238									
2	0.70795	0.64490								
3	0.25059	0.92835	0.59724							
4	0.17699	0.40477	1.16912	0.58158						
5	0.06264	0.40794	0.54886	1.41499	0.57443					
6	0.04425	0.16343	0.69910	0.69061	1.66285	0.57070				
7	0.01566	0.14615	0.30002	1.05185	0.83144	1.91155	0.56842			
8	0.01106	0.05648	0.32077	0.47190	1.46670	0.97195	2.16072	0.56695		
9	0.00392	0.04759	0.13139	0.58350	0.67891	1.94384	1.11229	2.41014	0.56592	
10	0.00277	0.01803	0.12776	0.24920	0.94992	0.92107	2.48342	1.25265	2.65974	0.56522

APPENDIX C
The Elliptic Response

TABLES C. COEFFICIENTS OF THE POLYNOMIALS $r(s)$ (3.231) ASSOCIATED WITH THE ELLIPTIC RESPONSE.

$$r(s) = c_0 + c_1 s + c_2 s^2 + \cdots + c_{n-1} s^{n-1} + s^n$$

C.1. Steepness: 1.05 ($k = 0.9523810$)   Passband ripple: $\frac{1}{2}$ dB ($\epsilon = 0.3493114$)

n	c_0	c_1	c_2	c_3	c_4	c_5	c_6	c_7	c_8	c_9
1	2.8627752									
2	1.2743413	0.2751653								
3	1.4342459	1.2583916	1.4502589							
4	0.9382667	1.0876038	2.0210069	1.0599756						
5	0.5755864	1.1892888	1.7434526	2.2156188	1.1509280					
6	0.3849186	0.8787979	1.9447353	2.0199498	2.5676585	1.1335645				
7	0.2296693	0.7657059	1.4938582	2.6411737	2.4021563	2.8777246	1.1349000			
8	0.1536626	0.5248517	1.4457610	2.1472921	3.4889092	2.7569616	3.1974335	1.1333480		
9	0.0916312	0.4220386	1.0470950	2.3312783	2.9404873	4.4246702	3.1180976	3.5155929	1.1326403	
10	0.0613074	0.2789207	0.9325873	1.7746850	3.5027496	3.8415035	5.4657854	3.4779464	3.8343544	1.1320503

C.2. Steepness: 1.05 ($k = 0.9523810$) Passband ripple: 1 dB ($\epsilon = 0.5088471$)

n	c_0	c_1	c_2	c_3	c_4	c_5	c_6	c_7	c_8	c_9
1	1.9652267									
2	1.1672218	0.3141664								
3	0.9845755	1.1629654	1.0788121							
4	0.6921654	0.8610367	1.7548054	0.8757772						
5	0.3951263	0.9765996	1.3156878	1.9993536	0.9212833					
6	0.2799823	0.6469184	1.6088246	1.5601135	2.3367141	0.9121807				
7	0.1576625	0.6017573	1.1064552	2.2500967	1.8620959	2.6510125	0.9125805			
8	0.1117368	0.3777362	1.1632875	1.6150706	3.0212152	2.1493230	2.9705605	0.9116280		
9	0.0629026	0.3248769	0.7644661	1.9224738	2.2301505	3.8865837	2.4398958	3.2892844	0.9111509	
10	0.0445798	0.1984461	0.7365633	1.3133951	2.9395319	2.9338895	4.8557784	2.7297718	3.6083282	0.9107669

C.3. Steepness: 1.10 ($k = 0.9090909$) Passband ripple: $\frac{1}{2}$ dB ($\epsilon = 0.3493114$)

n	c_0	c_1	c_2	c_3	c_4	c_5	c_6	c_7	c_8	c_9
1	2.8627752									
2	1.3846530	0.4298725								
3	1.2393329	1.3454329	1.3201569							
4	0.7837391	1.1337351	1.9190199	1.1195652						
5	0.4486397	1.1071526	1.6137359	2.1572389	1.1470527					
6	0.2855505	0.7745664	1.7349336	1.9243965	2.4670532	1.1391520				
7	0.1620121	0.6309158	1.2904969	2.3898599	2.2709106	2.7650045	1.1371904			
8	0.1031254	0.4110314	1.1897451	1.8838408	3.1506143	2.6105750	3.0660214	1.1354116		
9	0.0585039	0.3115092	0.8287238	1.9445146	2.585819	3.9996324	2.9515848	3.3672873	1.1342152	
10	0.0372395	0.1959180	0.7011888	1.4283170	2.9358923	3.3911759	4.9407073	3.2925319	3.6689730	1.1333368

TABLES C. (contd).

C.4. Steepness: 1.10 ($k = 0.9090909$) Passband ripple: 1 dB ($\epsilon = 0.5088471$)

n	c_0	c_1	c_2	c_3	c_4	c_5	c_6	c_7	c_8	c_9
1	1.9652267									
2	1.2099342	0.4582576								
3	0.8507724	1.2018049	1.0114630							
4	0.5725317	0.8679778	1.6637303	0.9074267						
5	0.3079804	0.8917965	1.2198991	1.9296353	0.9200924					
6	0.2076500	0.5637872	1.4288585	1.4784959	2.2362015	0.9154935				
7	0.1112174	0.4903824	0.9529513	2.0209395	1.7563342	2.5363599	0.9141089			
8	0.0749879	0.2938214	0.9515661	1.4108347	2.7125274	2.0304234	2.8381611	0.9129533		
9	0.0401615	0.2379754	0.6027271	1.5925974	1.9553030	3.4938833	2.3051746	3.1400958	0.9121705	
10	0.0270788	0.1387189	0.5504932	1.0527660	2.4485571	2.5821804	4.3670822	2.5798651	3.4422520	0.9115982

C.5. Steepness: 1.20 ($k = 0.8333333$) Passband ripple: $\tfrac{1}{2}$ dB ($\epsilon = 0.3493114$)

n	c_0	c_1	c_2	c_3	c_4	c_5	c_6	c_7	c_8	c_9
1	2.8627752									
2	1.5067570	0.6609881								
3	1.0623317	1.4259842	1.2442453							
4	0.6412257	1.1344472	1.8363497	1.1522015						
5	0.3460426	1.0086291	1.5092841	2.0909583	1.1507504					
6	0.2091172	0.6668400	1.5491044	1.8230556	2.3731641	1.1440986				
7	0.1126308	0.5085661	1.1105747	2.1505642	2.1458183	2.6549964	1.1404245			
8	0.0680644	0.3142936	0.9706912	1.6400063	2.8362218	2.4675808	2.9385809	1.1379281		
9	0.0366592	0.2243915	0.6477743	1.6063085	2.2626803	3.6030352	2.7896913	3.2229960	1.1361924	
10	0.0221537	0.1340159	0.5189658	1.1353892	2.4412627	2.9775016	4.4517182	3.1119728	3.5079622	1.1349363

C.6. Steepness: 1.30 ($k = 0.7692308$)   Passband ripple: $\frac{1}{2}$ dB ($\epsilon = 0.3493114$)

n	c_0	c_1	c_2	c_3	c_4	c_5	c_6	c_7	c_8	c_9
1	2.8627752									
2	1.5650010	0.8250785								
3	0.9734109	1.4631437	1.2248820							
4	0.5710925	1.1204272	1.8001546	1.1639919						
5	0.2989046	0.9507846	1.4590823	2.0539626	1.1545306					
6	0.1754184	0.6090779	1.4553064	1.7676751	2.3241677	1.1470387				
7	0.0917528	0.4481824	1.0204608	2.0263404	2.0796089	2.5966120	1.1425209			
8	0.0538471	0.2688744	0.8648301	1.5158933	2.6743561	2.3914250	2.8709058	1.1395245		
9	0.0281647	0.1857334	0.5628855	1.4413577	2.0982031	3.3988464	2.7035100	3.1462455	1.1374465	
10	0.0165291	0.1077379	0.4371621	0.9963747	2.1997848	2.7672043	4.2002050	3.0158084	3.4222482	1.1359482

C.7. Steepness: 1.40 ($k = 0.7142857$)   Passband ripple: $\frac{1}{2}$ dB ($\epsilon = 0.3493114$)

n	c_0	c_1	c_2	c_3	c_4	c_5	c_6	c_7	c_8	c_9
1	2.8627752									
2	1.5920793	0.9438940								
3	0.9186494	1.4837636	1.2199499							
4	0.5287467	1.1070699	1.7798206	1.1705639						
5	0.2713306	0.9124353	1.4282914	2.0301998	1.1574354					
6	0.1561835	0.5725279	1.3968781	1.7321205	2.2931935	1.1490984				
7	0.0801257	0.4116078	0.9647832	1.9483859	2.0374310	2.5594819	1.1440033			
8	0.0461220	0.2421399	0.8009761	1.4388655	2.5730490	2.3428490	2.8278098	1.1406482		
9	0.0236616	0.1636924	0.5126440	1.3414975	1.9961461	3.2711237	2.6485336	3.0973246	1.1383279	
10	0.0136201	0.0931385	0.3900262	0.9135984	2.0534335	2.6366660	4.0429661	2.9544521	3.3675842	1.1366583

Symbol Index

The symbols which occur most often are listed here, separated into three categories: Roman letters, vectors and matrices, and Greek letters.

Roman letters

a, a_m	normalized incident wave, 55, 82
A, B, C, D	transmission or chain parameters, 15
$\text{am}(u, k)$	amplitude of u of modulus k, 153
B	bandwidth, 200, 204
b, b_m	normalized reflected wave, 55, 82
$C_n(\omega)$	nth-order Chebyshev polynomial of the first kind, 132
$\text{cn}(u, k)$	Jacobian elliptic cosine function, 153
$\text{dn}(u, k)$	Jacobian elliptic function, 153
$F(k, \phi)$	Legendre standard form of the elliptic integral of the first kind of modulus k, 152
$F_n(\omega)$	characteristic function, 167
f_r	resistive cutoff frequency, 354
f_s	self-resonant frequency, 354
$G, G_{jm}(\omega^2)$	transducer power gain from port j to port m, 22, 84
G_A	available power gain, 21
G_P	power gain, 21
$g(s)$	para-hermitian part of the reference admittance $y(s)$, 62
h_{ij}	hybrid parameters, 14
$I, I(s)$	Laplace transform of $i(t)$, 12
$I_i, I_i(s)$	incident current, 49, 52
$I_r, I_r(s)$	reflected current, 49, 52
$i(t), i_k(t)$	port current, 2
k	modulus of an elliptic function or the selectivity factor of the elliptic response, 153, 179
k'	complementary modulus of k, 155, 166
$K, K(k)$	complete elliptic integral of modulus k, 156, 157, 166
$K', K'(k)$	complete elliptic integral of modulus k', 159, 166
K_1	complete elliptic integral of modulus k_1, 166
K'_1	complete elliptic integral of modulus k'_1, 166
LHS	left-half of the complex-frequency s-plane, 28
N	n-port network, 2
$[n/2]$	largest integer not greater than $n/2$, 126
Re	real part of, 3, 39
RHS	right-half of the complex-frequency s-plane, 28
$r(s)$	para-hermitian part of the reference impedance $z(s)$, 51
$s = \sigma + j\omega$	complex frequency, 3
$S, S(s)$	normalized reflection coefficient, 55
$S_{ij}, S_{ij}(s)$	normalized scattering parameters, 82
$S^I, S^I(s)$	current-based reflection coefficient, 53
$S^I_{ij}, S^I_{ij}(s)$	current-based scattering parameters, 71, 82

SYMBOL INDEX

$S^V, S^V(s)$	voltage-based reflection coefficient, 53	$I_r, I_r(s)$	reflected-current vector, 71
$S_{ij}^V, S_{ij}^V(s)$	voltage-based scattering parameters, 71	$i(t)$	port current vector, 2
sn(u,k)	Jacobian elliptic sine function, 153	$k(s)$	factorization of $g(s)$, 80
		$r, r(s)$	para-hermitian part of the impedance matrix $z(s)$, 68
sn$^{-1}(u,k)$	inverse Jacobian elliptic sine function, 166	$S, S(s)$	normalized scattering matrix, 74, 82
$S_x^I(s)$	sensitivity function, 375	$S^I, S^I(s)$	current-based scattering matrix, 71, 82
t	time, 2		
tn(u,k)	Jacobian elliptic function, 155	$S^V, S^V(s)$	voltage-based scattering matrix, 71
$V, V(s)$	Laplace transform of $v(t)$, 12	$T, T(s)$	transmission or chain matrix, 15
$V_i, V_i(s)$	incident voltage, 49, 52	U_n	identity matrix of order n, 69
$V_r, V_r(s)$	reflected voltage, 49, 52		
$v(t), v_k(t)$	port voltage, 2	$V, V(s)$	Laplace transform of $v(t)$, 12
$w(s)$	$= r_i(s)/z_i(s)$, 221	$V_i, V_i(s)$	incident-voltage vector, 70
x	complex conjugate of x, 23, 51	$V_r, V_r(s)$	reflected-voltage vector, 71
$y_{ij}, y_{ij}(s)$	short-circuit admittance parameters, 13	$v(t)$	port-voltage vector, 2
		$[v(t), i(t)]$	admissible signal pair, 2
$y(s)$	reference admittance, 62	$Y, Y(s)$	short-circuit admittance matrix, 13
$z_{ij}, z_{ij}(s)$	open-circuit impedance parameters, 12	$\hat{Y}, \hat{Y}(s)$	indefinite-admittance matrix, 16
$z(s)$	reference impedance, 53		
$z_*(s)$	$= z(-s)$, 51	$y(s)$	reference admittance matrix, 80

Vectors and matrices

		$Z, Z(s)$	open-circuit impedance matrix, 12
$a, a(s)$	normalized incident-wave vector, 74	$z(s)$	reference impedance matrix, 68
A'	matrix transpose, 2, 68		
A^*	matrix transpose and conjugate, 23	**Greek letters**	
A_h	hermitian part of A, 28	α	attenuation, 137, 177, 178
A_s	symmetric part of A, 28	γ_m	$= m\pi/2n$, 126
$A_*(s)$	$= A'(-s)$, 68	δ_{ik}	Kronecker delta, 89
$b, b(s)$	normalized reflected-wave vector, 74	ϵ	ripple factor, 136, 177
		$\eta, \eta(s)$	real all-pass function, 59, 69
$g, g(s)$	para-hermitian part of the admittance matrix $y(s)$, 80	λ	constant or eigenvalue, 191, 215, 305
$H, H(s)$	transmission or chain matrix, 14	τ	$= 1/RC$, 225, 340
		$\rho, \rho(s)$	normalized or bounded-real reflection coefficient, 220
$h(s)$	factorization of $r(s)$, 68	$\rho^I, \rho^I(s)$	current-based reflection coefficient, 50, 321, 384
$I, I(s)$	Laplace transform of $i(t)$, 12		
$I_i, I_i(s)$	incident-current vector, 70	$\rho^V, \rho^V(s)$	voltage-based reflection coefficient, 50

SYMBOL INDEX

σ neper or imaginary frequency, 3

ω radian or real frequency, 3

ω_c radian cutoff frequency, 116, 132, 165

ω_s edge of the stopband frequency, 178

$\eta(s)$ diagonal matrix whose elements are real all-pass functions, 69

Subject Index

ABCD parameters 15
Active impedance 314
Active reflection coefficient, current-based 321, 384
Activity of an n-port network 7
Admissible signal pair 2
Admittance matrix 13
All-pass function 59
Amplifier
  equiripple 346, 364, 369, 372
  negative-resistance 316
  nonreciprocal 319, 391
    stability 355
  reciprocal 378, 394
    cascade connection 382
    transmission-power 357
      equiripple 364, 369, 372
      maximally-flat 361, 369, 372
    sensitivity 375
    stability 373
  tunnel diode 338, 346
    equiripple 346
    maximally-flat 338
Amplitude of the elliptic integral 153
Approximation problem 116
Available power gain 21

Basic coefficient constraints 222
Bode–Fano–Youla broadband matching problem 218
Bode's parallel RC load 224
  Butterworth transducer power-gain characteristic 225
  Chebyshev transducer power-gain characteristic 235
  elliptic transducer power-gain characteristic 246

Bounded-real function 91
Bounded-real matrix 91, 92, 97
Bounded-real reflection coefficient 65, 220
Butterworth function 118
  poles of 118
Butterworth LC ladder network 125
Butterworth network 122
Butterworth polynomial 120
  coefficients of 120, 412
Butterworth response 116
  LC ladder network of 125
  network of 122
  poles of 118
Butterworth transducer power-gain characteristic 225, 274

Cascade connection 382
Cauer-parameter filter 167
Causality of an n-port network 9
Chain matrix 15
Chain parameters 15
Characteristic function 167
Chebyshev function 138
Chebyshev LC ladder network 146
Chebyshev network 144
Chebyshev polynomial 132, 413
  generalized 139
  of the first kind 132
  of the second kind 209
Chebyshev response 132
  LC ladder network of 146
  network of 144
  poles of 138
Chebyshev transducer power-gain characteristic 235, 281

428   SUBJECT INDEX

Classification of zeros of transmission 222
Complementary modulus of the elliptic integral 155
Complete elliptic integral of the first kind 156
Complex frequency 3
Conjugately matched 51
Constant transducer power gain 292
Current
  incident 52
  port 2
  reflected 52
Current-based active reflection coefficient 321, 384
Current-based reflection coefficient 50, 53
Current-based scattering matrix 71
Current-based scattering parameters 53, 71

Darlington type-C load 274
  Butterworth transducer power-gain characteristic 274
  Chebyshev transducer power-gain characteristic 281
  elliptic transducer power-gain characteristic 287
Definite hermitian form 24
Definite hermitian matrix 24

Electrical network 1
Elliptic filter 152
Elliptic function 152
  period of 155
    imaginary 158
    real 157
Elliptic integral of the first kind 152
  complete 156
Elliptic network 190
Elliptic response 165
  network of 190
  poles of 183
  steepness in 179
  zeros of 183
Elliptic transducer power-gain characteristic 246, 287

Equalization 217
Equalizer 217
Equalizer back-end impedance 257, 269, 290
Equicofactor matrix 17, 18, 19
Equiripple low-pass amplifier 346, 364, 369, 372
Equiripple response 132, 136 (*see also* Chebyshev response)
Equiripple transducer power-gain characteristic 346 (*see also* Chebyshev transducer power-gain characteristic)

Filter
  Butterworth 122
  Cauer-parameter 167
  Chebyshev 144
  elliptic 152
Frequency
  complex 3
  imaginary 3
  mid-band 201
  neper 3
  radian 3
  real 3
  resistive cutoff 354
  self-resonant 354
Frequency transformation 196
  band-elimination 204
  band-pass 200
  high-pass 197
Function
  all-pass 59
  bounded-real 91
  Butterworth 118
  characteristic 167
  Chebyshev 138
  elliptic 152
    periods 155
  Jacobian elliptic 153
    sine 153
  positive-real 29, 36, 37
  reactance 205
  sensitivity 375
  time-delay 207

Gain (*see* Power gain)

SUBJECT INDEX 429

Hermitian form 23, 24
  definite 24
Hermitian part of a matrix 28
Hurwitz polynomial 58, 62
Hurwitz test 37
Hybrid 407
Hybrid matrix 14
Hybrid parameters 14

Imaginary frequency 3
Impedance matching 217
Impedance matrix 12
Incident current 52
Incident-current vector 70
Incident voltage 52
Incident-voltage vector 70
Incident wave, normalized 55
Incident-wave vector, normalized 74
Indefinite-admittance matrix 16, 19
Inverse Chebyshev characteristic 209
Inverse hybrid matrix 15
Inverse transmission matrix 15

Jacobian elliptic functions 152
  addition theorems for 162
  complex argument 162
  imaginary period of 158
  pole of 159
  real period of 157
  zero of 159
Jacobian elliptic sine function 153
Jacobi's imaginary transformation 154

Kronecker delta 89

Leading principal minor 26
Legendre standard form of the elliptic integral 152
  of the first kind 152
Linear n-port network 5
Lossless n-port network 8

Maclaurin series expansion 117

Matching network 217
Matrix, admittance 13
  bounded-real 91, 92, 97
  chain 15
  current-based scattering 71
  equicofactor 17, 18, 19
  hermitian 24
  hermitian part of 28
  hybrid 14
  impedance 12
  indefinite-admittance 16, 19
  inverse chain 15
  inverse hybrid 15
  inverse transmission 15
  negative-definite 45
  negative-semidefinite 45
  nonnegative-definite 24
  non-positive definite 45
  normalized scattering 74, 88
  of the hermitian form 24
  open-circuit impedance 12
  para-hermitian 68
  para-unitary 97
  positive-definite 24
  positive-real 28, 31, 35, 42
  positive-semidefinite 24
  reference impedance 68
  scattering 66, 97
    current-based 71
    normalized 74, 88
    voltage-based 71
  short-circuit admittance 13
  symmetric part of 28
  transmission 15
  unitary 89
  zero-row-sum and zero-column-sum 17
Maximally-flat low-pass amplifier 338, 361, 369, 372
Maximally-flat response 116 (*see also* Butterworth response)
Maximally-flat transducer power-gain characteristic 338
Mid-band frequency 201
Minimum-phase reflection coefficient 124
Minor
  leading principal 26
  principal 26

SUBJECT INDEX

Modulus
  of elliptic integral 152
  of Jacobian elliptic sine function 153

Negative-definite matrix 45
Negative-resistance amplifier 316
Negative-semidefinite matrix 45
Neper frequency 3
Network
  Butterworth 122
  Butterworth LC ladder 125
  Chebyshev 144
  Chebyshev LC ladder 146
  elliptic 190
  n-port network 1, 2
    active 7
    causal 9
    linear 5
    lossless 8
    nonlinear 5
    nonreciprocal 10
    passive 1, 6, 7
    real 3
    reciprocal 10
    strictly passive 8
    time-invariant 4
    time-varying 4
  n-port network parameters 11
    $ABCD$ 15
    admittance 13
    chain 15
    hybrid 14
    impedance 12
    inverse chain 15
    inverse hybrid 15
    inverse transmission 15
    open-circuit impedance 12
    scattering 48, 50, 71
      current-based 53, 71
      normalized 75, 82, 90
      voltage 53, 71
    short-circuit admittance 13
    transmission 15
Nonlinear n-port network 5
Nonnegative-definite matrix 24
Non-positive definite matrix 45
Nonreciprocal amplifier 319, 391
  stability 355
Nonreciprocal n-port network 10

Normalized incident wave 55
Normalized incident-wave vector 74
Normalized reflected wave 55
Normalized reflected-wave vector 74
Normalized reflection coefficient 54, 55, 62, 64, 83, 84
Normalized scattering matrix 74, 88
Normalized scattering parameters 55, 75, 82, 90
Normalized transmission coefficient 84

Open-circuit impedance matrix 12
Open-circuit impedance parameters 12

Para-hermitian matrix 68
Para-hermitian part of a matrix 51, 68
  factorization 57, 59
Para-unitary matrix 97
Passband 115
Passivity 1, 6, 7
Polynomial
  Butterworth 120
  Chebyshev 132, 413
    generalized 139
    of the first kind 132
    of the second kind 209
  Hurwitz 58, 62
Port 1
Port current 2
Port-current vector 2, 12
Port voltage 2
Port-voltage vector 2, 12
Positive-definite matrix 24
Positive-real function 29, 36, 37
Positive-real matrix 28, 31, 35, 42
Positive-semidefinite matrix 24
Power gain 21
  available 21
  Butterworth transducer 225, 274
  Chebyshev transducer 235, 281
  constant transducer 292
  elliptic transducer 246, 287
  equiripple transducer 346
  inverse-Chebyshev transducer 209
  maximally-flat transducer 338
  transducer 22
  transitional Butterworth–Chebyshev transducer 213

SUBJECT INDEX

Principal minor 26
  leading 26

Radian frequency 3
Reactance function 205
Real frequency 3
Reciprocal amplifier 378, 394
  cascade connection 382
Reciprocity of an n-port network 10
Reference impedance 53
Reference impedance matrix 68
Reflected current 52
Reflected-current vector 71
Reflected voltage 52
Reflected-voltage vector 71
Reflected wave, normalized 55
Reflected-wave vector, normalized 74
Reflection coefficient 52
  bounded-real 65, 220
  current-based 50, 53
  current-based active 321, 384
  minimum-phase 124
  normalized 54, 55, 62, 64, 83, 84
  voltage-based 50, 53
Resistance cutoff frequency 354
Ripple factor 136, 177

Scattering matrix 66, 97
  current-based 71
  normalized 74, 88
  voltage-based 71
Scattering parameters 48, 50, 71
  current-based 53, 71
  normalized 55, 75, 82, 90
  voltage-based 53, 71
Selectivity factor 179
Self-resonant frequency 354
Sensitivity 375
Sensitivity function 375
Short-circuit admittance matrix 13
Short-circuit admittance parameters 13
Stability 355, 373
Steepness 179
Stopband 115
  edge of 178
Strictly passive n-port network 8
Symmetric part of a matrix 28

Time-delay function 207
Time-invariant n-port network 4
Time-varying n-port network 4
Transducer power gain 22, 359, 368, 372
  Butterworth 225, 274
  Chebyshev 235, 281
  constant 292
  elliptic 246, 287
  equiripple 346
  inverse Chebyshev 209
  maximally-flat 338
  transitional Butterworth–Chebyshev 213
Transitional Butterworth–Chebyshev response 213
Transmission coefficient, normalized 84
Transmission matrix 15
Transmission parameters 15
Transmission-power amplifier 357
  equiripple 364, 369, 372
  maximally-flat 361, 369, 372
  sensitivity 375
  stability 373
Tunnel diode 314
  equivalent circuit 315
  resistive cutoff frequency 354
  self-resonant frequency 354
Tunnel diode amplifier 338, 346
  equiripple 346
  maximally-flat 338

Unitary matrix 89

Vector
  incident-current 70
  incident-voltage 70
  normalized incident-wave 74
  normalized reflected-wave 74
  port-current 2
  port-voltage 2
  reflected-current 71
  reflected-voltage 71
Voltage
  incident 52
  port 2
  reflected 52

Voltage-based reflection coefficient 50, 53
Voltage-based scattering matrix 71
Voltage-based scattering parameters 53, 71

Wave
  normalized incident 55
  normalized reflected 55

Youla's theory of broadband matching 219

Zero of transmission 221
  classification of 222
Zero-row-sum and zero-column-sum matrix 17